"A blistering critique of the disconnect between what transportation engineers do and what the research tells us about traffic safety. This eminently readable book is a wake-up call for all of us."

—Susan Handy, Professor, University of California, Davis; author of
  Shifting Gears: Toward a New Way of Thinking about Transportation

"In *Killed by a Traffic Engineer*, Wes Marshall lays bare the dogmas that give traffic engineering an appearance of objective scientific authority. A century of lavish investment in the most costly and extensive road network in the world has given Americans the worst traffic safety record among high-income countries.

As a trained and experienced expert, Wes is an insider who proves his loyalty to the profession by exposing its departure from its own avowed purposes. He reveals how engineering standards that are supposed to make streets safer or more efficient actually do neither. They invite speeding, make walking more dangerous, and often don't even save drivers' time. Wes is a constructive critic who offers remedies for the afflictions he diagnoses.

Killed by a Traffic Engineer is serious but readable, disturbing but witty. Wes exposes the exclusive traffic engineering establishment to the people who actually need transportation, equipping us as citizens to demand the reforms we need. In writing it, he has done us all a uniquely valuable service."

—Peter Norton, Associate Professor of History, Department
  of Engineering and Society, University of Virginia; author of
  Fighting Traffic and *Autonorama*

T0309096

About Island Press

Since 1984, the nonprofit organization Island Press has been stimulating, shaping, and communicating ideas that are essential for solving environmental problems worldwide. With more than 1,000 titles in print and some 30 new releases each year, we are the nation's leading publisher on environmental issues. We identify innovative thinkers and emerging trends in the environmental field. We work with world-renowned experts and authors to develop cross-disciplinary solutions to environmental challenges.

Island Press designs and executes educational campaigns, in conjunction with our authors, to communicate their critical messages in print, in person, and online using the latest technologies, innovative programs, and the media. Our goal is to reach targeted audiences—scientists, policy makers, environmental advocates, urban planners, the media, and concerned citizens—with information that can be used to create the framework for long-term ecological health and human well-being.

Island Press gratefully acknowledges major support from The Bobolink Foundation, Caldera Foundation, The Curtis and Edith Munson Foundation, The Forrest C. and Frances H. Lattner Foundation, The JPB Foundation, The Kresge Foundation, The Summit Charitable Foundation, Inc., and many other generous organizations and individuals.

The opinions expressed in this book are those of the author(s) and do not necessarily reflect the views of our supporters.

KILLED BY A
TRAFFIC ENGINEER

KILLED BY A
TRAFFIC ENGINEER

Shattering the Delusion that Science Underlies
Our Transportation System

WES MARSHALL

ISLANDPRESS | Washington | Covelo

Library of Congress Control Number: 2023947670

All Island Press books are printed on environmentally responsible materials.

Manufactured in the United States of America
10 9 8 7 6 5

Keywords: auto industry, autonomous vehicles, civil engineering, congestion, crosswalk, design speed, driver education, education, enforcement, engineering, engineer speak, factor of safety, functional classification system, human error, induced demand, level of service (LOS), *Manual of Uniform Traffic Control Devices* (MUTCD), pedestrian, road safety, road user, safety data, scofflaw bicyclist, speed, speed limit, traffic crash, traffic safety, urban freeway

For my amazing wife, Debbie, and our awesome kids,
Rowan, Luke, and Mae.
This book is for you (whether you read it or not).
Love you lots.

And yes, 88 chapters about road safety does seem like a lot.

Contents

—

Part 1

What Are We Doing Here?

———

01 Bad Medicine

—

Let me tell you a story about the time I ended up in a Cabo San Lucas emergency room where a doctor slowly and methodically dug around each of more than a hundred sea urchin needles and pulled them out of my feet like vertical splinters.

The localized anesthesia shots hurt just as much as his digging, so I figured I might as well skip the shots. My approach to pain management was staring intently at some Spanish version of the Hippocratic oath hanging in a rusty frame on the wall. I tried to translate it in my head. I thought about the life of Hippocrates. I felt happy to live in a time where we must know a heck of a lot more about medicine than he did.[a]

In the years since, I've learned that it's been a long and painful journey for modern medicine to get to this point.

Don't get me wrong; we need to give doctors their due. Anesthesia. Germ theory. X-rays. Antibiotics. Vaccines. DNA. Organ transplants. Chemotherapy. Magnetic imaging. The human genome. The list of medical advances that we now consider commonplace is *loooong*. Everyday medical treatments of today would have been considered a medical miracle not that long ago. Before the discovery of insulin in 1921, diabetes was a death sentence.

But for at least the first 1,000 years after Hippocrates, it wouldn't be hard to make a case that doctors killed more people than they saved. On his *Against the Rules* podcast, author Michael Lewis says, "For roughly 5,000 years, people called themselves doctors and pretended to know all sorts of things that they didn't know and were as likely to kill you as to cure you. These doctors existed because sick people desperately wanted to believe them."[1]

---

[a] I was also happy that it happened on the last day of spring break (instead of the first) and that I had learned how to surf—at least a little bit—before the sea urchin–covered rocks ended the fun. Why my friend's Cabo-based brother brought us to a surfing location where you can't safely walk from the waves to the shore (and didn't tell us) is beyond me. Also, two-plus hours of "surgery" cost me about $20 in cash.

Doctors in the early 1900s treated coughs and sore throats—in kids—with heroin.

During this time, doctors used X-rays to remove unwanted hair ... and give people cancer.

Through the 1940s, doctors used the radioactive substance radium in an attempt to cure anything from asthma to blindness.[b]

Doctors continued to recommend smoking to relieve throat irritation well past the US surgeon general's report on smoking and health in 1964.

As late as 1967, Dr. Walter Freeman performed lobotomies (in his Lobotomobile) in an attempt to cure mental illness, hyperactivity in kids, and minor neck pain. Freeman's 14 percent fatality rate on more than 3,500 lobotomies doesn't do justice to those left in a permanent vegetative state.

Want a transportation-related example? Doctors treated everything with bed rest and scoffed at the benefits of exercise—unless "*taken in Olympian doses*"[2]—until the mid-1970s, when Laurence Morehouse and Leonard Gross's book, *Total Fitness in 30 Minutes a Week*, redefined "exercise" as "increased physical activity" where even walking counts.[3] It took until 1986 for the research[4] to catch up and show that increased physical activity helps with mortality.[c]

Even now, many doctors won't advise walking or biking for transportation as a health intervention. Why? Because they assume that the road safety and air pollution risks outweigh the physical activity benefits.[d] That isn't the case everywhere. But given that much of the transportation system was built by civil engineers, these doctors may have a point.

This is something we need to fix.

---

[b] Right up through the 1940s, we put radium in toothpaste, lipstick, and even butter. Victims of radiation poisoning included the Radium Girls, whose job at the Radium Luminous Material Corporation in Newark, New Jersey, was to paint watch dials with radium-infused glow-in-the-dark paint. Doctors were adamant that the radium was harmless.

[c] In the 1897 paper "The Hidden Dangers of Cycling" published in the *National Review*, a doctor named Arthur Shadwell wrote that bicycling causes goiters, appendicitis, headaches, insomnia, nervous depression, dementia, and worst of all for women, "bicycle face."

[d] Multiple doctors told me this at an American Public Health Association conference.

Civil engineers have also been around for about 5,000 years. Like doctors, they've learned a lot in this time.

When people think about civil engineering, most think about structural engineering and public works projects such as roads, bridges, and dams. But civil engineering is a huge umbrella that encompasses most anything related to the design, construction, and maintenance of the built or natural environment. Civil engineers developed the first sewer systems, and modern sanitation alone apparently added 20 years of life expectancy.[5]

Traffic engineering falls under the umbrella of civil engineering, but traffic engineering has been around for less than 100 years.

A 1965 British report explains the origin story of traffic engineers: "In order to deal with traffic problems in a satisfactory way, it has been necessary to build up a scientific discipline" and, as a result, "a relatively new branch of engineering has therefore been evolved, that of traffic engineering."[6] I also found a 1959 paper published in *Traffic Quarterly*, the preeminent traffic-engineering journal at the time, stating that "only the last decade has witnessed the adulthood of traffic engineering" and that "there are few things today that are more important to the welfare of the modern community than good traffic engineering. It is right up there with medicine and education as a factor in the everyday life of every citizen."[7]

What do we mean by traffic engineering?ᵉ The British report simply describes it as "that part of engineering which deals with the traffic planning and design of roads, of frontage development and of parking facilities and with the control of traffic to provide safe, convenient and economic movement of vehicles and pedestrians."[8]

Safe is the key word there, but like doctors did for a thousand-plus years, traffic engineers spent the last hundred years killing as many people as they've saved.

At first, traffic engineers tried to rationalize road deaths, akin to doctors in an experimental field:

"It's a new discipline," they'd say.

"We're on a learning curve."

But errors by engineers suggest incompetence—even when due

ᵉ Transportation engineering and transportation planning are broader and more comprehensive than traffic engineering, but for the sake of this book, I'm going to use them somewhat interchangeably. When I want to differentiate, I'll make it clear.

to the newness of the science—and engineers don't want to portray anything other than extreme competence. So, it wasn't long before traffic engineers presented a nascent scientific field as if it was a finished product.

While the early traffic engineers knew what they didn't know, subsequent generations of traffic engineers started to believe that they knew more than they did. These traffic engineers then built out the transportation system without ever taking a good, hard look at the science (or lack thereof) behind it all. It's time to take that good, hard look.

In this book, I'll show how traffic engineers designed and built a system that incites bad behaviors and invites crashes. The bad behaviors let traffic engineers off the hook, but most deaths and injuries are the predictable, systemic outcomes that traffic engineers inadvertently caused—thus, the title of this book—but that can be fixed.

Just knowing this fact isn't enough to fix the problem. Traffic engineers hold too much power, as well as the superpower to shut down most arguments. They do so by saying they are following the methods and standards. So if you want to win an argument with a traffic engineer, you need some ammunition. This book provides that by looking at what traffic engineers do wrong today and exposing the pseudoscience that led traffic engineering to create these methods and standards in the first place.

I felt the need to write this book when I realized that we aren't going to come close to fixing our road safety problems based on what I was taught in engineering school. I began digging into all the systemic things that traffic engineers do wrong. The more I considered the problems we face, the more I realized I needed to figure out *why* traffic engineers do what they do.

Traffic engineers don't want you to peek behind this curtain, because most traffic engineers have no idea what's behind it. But if we want to fix the carnage on our streets, we need to shine a spotlight on how little there is behind the *why* question. Following the science means first shattering the pseudoscience. Only then can we start making the world a safer place on a larger scale.

The sad reality is that we need a dramatic transformation of transportation, and we need the right kind of mind-set to make it happen. This goes for traffic engineers in particular because they are still the ones in charge of our streets. They won't change ***what they do*** without first changing *__how they think__*. If your friendly, neighborhood traffic engineer reads this book, great. If not, this book will teach you how to effectively argue with your friendly, neighborhood traffic engineer.

When you finish reading this book, you're going to look at your city and streets much differently. Things that never bothered you before are going to drive you crazy.

You don't need to be a traffic engineer or transportation planner—or any kind of expert—to read this book or join the cause. That is one of the great things about the transportation discipline. Everyone has opinions (lots of them), and (almost) anyone can speak intelligently on their own experiences.

So if you want to learn the truth about traffic engineering and road safety,

If you want to fix our streets and communities,

If you want to make how we get around safer,

If you want to help start us on a path to saving millions of lives,

I'm happy to have you come down this rabbit hole with me.

If not, stop reading now and drive to work tomorrow.

02 Deal or No Deal

Guido Calabresi is a former federal judge and the former dean of the Yale Law School. For decades, he used to present his class with this offer from God:

God is going to give us a wondrous invention that will allow us to see new places, spend more time with family and friends, and do heretofore-unthinkable jobs.

In exchange for this invention, God is going to select 1,000 people at random each year, including kids, and strike them dead.

Would you take the deal?

Most of his students say no, of course not. That's a terrible deal.

Calabresi then says that cars and trucks kill around 40,000 Americans each year. Around the world, the number is closer to 1.35 million deaths annually. These numbers don't include all the serious and debilitating injuries, but still, we accept this deal without even thinking about it. Why? Well, that's the way it is and has been our whole lives. We can't imagine a world without cars.

This book isn't meant to imagine a world without them. I'm mostly looking for a world where a car isn't your only viable option.

I grew up just outside of Boston, in Watertown, Massachusetts. As a kid, my friends and I walked and biked all over town. We walked to school and biked to each other's houses. As teenagers, we hopped on the 57 bus from Watertown Square[a] to Fenway Park. If we wanted to get to Harvard Square, it was easy to walk to the 71 or 74 buses. Back then, Harvard Square was an experience in itself, but from there, we could take the MBTA subway Red Line to what felt like anywhere.

I spent my college years in Charlottesville, Virginia. While not as walkable or bikeable as the Boston area, it wasn't bad in and around the Grounds.[b] Other than the awesome sandwich shop at the Exxon

---

[a] For those who may be interested, Watertown Square is shaped more like a rounded pork chop than a square.

[b] At the University of Virginia, campus is referred to as the Grounds for whatever reason.

7

gas station or the Charlottesville Airport, I could get most everywhere I needed to go without a car.

For the next few years, I worked as a civil engineer at Sasaki Associates, a design firm located along the Charles River back in my hometown. As a young civil engineer, I was learning to design projects, applying whatever **standards** I found in the manuals without a second thought. Whoever wrote these 1,000-page books must know more than I did, I thought. The only time I ever felt a slight hesitation came toward the end of these projects. What I saw in person was never as good as the existing places I knew and loved.

I then took a civil engineering job at Clough, Harbour & Associates—more of a traditional engineering firm than Sasaki—and moved to Connecticut. It was the first time I lived in a place where I couldn't reasonably leave my street without being in a car. Sala's Corner Store was less than a half mile away, but as they say up in Maine, you can't get there from here. When I asked my friends and neighbors what they thought about this, they didn't have an opinion. It is what it is.

They had no idea what they were missing.

There is an episode of *The Simpsons* where the Simpson family visits Boston. Homer hates everything about Boston until he discovers candlepin bowling. Having grown up on candlepin,[c] I had the opposite experience when I first came across 10-pin bowling. I was in middle school when I went up to the second floor of the bowling alley and saw a whole new type of bowling. While not nearly as dramatic as Harry Potter seeing Diagon Alley for the first time, it was still a shock to not only find out that 10-pin bowling existed, but that 10-pin bowling was what most people considered bowling.

The same can be said of living on a street that I couldn't leave without a car. It was a bit of a shock to realize that such places existed but even more of one to find out that this was the norm for most people. What I experienced growing up around Boston, where a kid could independently get around town, was unquestionably atypical.

It started to drive me crazy.

Why were these streets like this? Why wasn't there a back road

---

[c] My mother won multiple candlepin bowling trophies when I was a toddler. She left me in what was apparently the bowling alley daycare.

that could get me there? Why was every building an island in a sea of parking?

I started digging a little deeper whenever I designed a project at work. I wanted to understand why, for example, the books told me that driving lanes should be so wide or that my project needed so much parking.

At the time, I had no idea how deep this rabbit hole was, but I only had to scratch the surface to see that there wasn't nearly as much science behind the numbers as the 1,000-page manuals make it seem.

Most of my colleagues didn't have the time or patience for such nonsense. Their focus was on getting the project done and moving on to the next. The results were never as good as the empirical examples I grew up with, but fixing this problem would mean digging a lot deeper. So the next thing you know, I was applying to graduate school.

I had just gotten my Professional Engineering (PE) license—the profession's highest standard of competence—when I told our regional manager that I was quitting to go to graduate school. He asked me why and told me that a PE is all I would ever need and that a PhD isn't going to be useful at a company like his. I didn't say it out loud, but I remember thinking, "That's because I don't want to work here anymore." What I was really feeling was a disconnect between reaching a so-called pinnacle of professional competence with this licensure while at the same time only starting to realize just how much I didn't know.

03   Murder Incorporated

In the 1964 James Bond movie *Goldfinger*, James Bond (Sean Connery) tries to reason with Auric Goldfinger about his plan to release nerve gas into the atmosphere over Fort Knox:

> BOND: You'll kill 60,000 people uselessly.
> GOLDFINGER: Ha! American motorists kill that many people every two years!

Goldfinger underestimated the number ... by a lot.

In 1964 alone, 45,645 people died on US roads. Combine that with 1965, and Goldfinger could've said that the number of people killed by American motorists was more than 90,000 people.[a]

The death toll on US streets continued to climb and climb, reaching 54,589 deaths in 1972 before it started to fall with the 1973 oil crisis.

The Vietnam War lasted from 1954 to 1975, and 58,209 Americans died in that conflict. I remember watching TV shows like *The Wonder Years* and thinking about how horrible that time period must've been. And it was, but over that same time span, drivers killed 974,567 people on US roads *by accident*.[b] These aren't accidents; these are results.

In truth, more Americans died on US roads than in all US conflicts combined, including the American Revolution.

There are countless other ways to try to put these numbers in perspective.

---

[a] US motorists killed a total of 92,734 people in 1964 and 1965 combined.

[b] You may already know this, but *accident* isn't the right word. *Accident* implies bad luck and implies that nobody—including the traffic engineer—is to blame. It further implies that there isn't much we can do to keep it from happening again. Some want to use the word *collision*, but that also feels like it downplays the problem. Instead, I'll use the word *crash* whenever I'm not quoting somebody who used *accident* or *collision*.

In the United States alone, we see millions of injuries and tens of thousands of deaths every year. In most years, road safety saps more productive years of life than any other disease, including cancer and heart disease combined. A 1968 *Traffic Quarterly* paper, written by a lawyer who used to prosecute organized crime, says that "neither Al Capone nor all his contemporaries, their predecessors and successors, aided and abetted by Murder Incorporated[c] could account for the fifty thousand persons killed on the nation's highways in 1966, nor for the eight billion dollars out for the personal injury and property damage in that same year." He goes on to say that traffic safety "takes a far greater toll in our society in one year than organized crime could account for in one hundred years."[1]

Less than six months after the terrorist attacks of September 11, 2001, U2 played the halftime show at Superbowl XXXVI in New Orleans.[d] While Bono sang "Where the Streets Have No Name," the band paid tribute to the 2,977 victims of the terrorist attacks via a vertical scrim that scrolled through every name. With three names side by side, it took about 5 minutes. As we cross four million total traffic deaths in the United States alone since the advent of the automobile, we'd be looking at 110 hours of scrolling through names to pay tribute to our road safety victims today.

The Centers for Disease Control and Prevention (CDC) regularly points to traffic deaths[e] as the number one cause of unintentional injury death for all Americans aged 4 through 34.[2] This isn't new. I found a 2002 paper saying that "in the US, injury, both intentional and unintentional, is the leading cause of death from age 1 to age 45. Because it so disproportionately strikes the young, it is also the leading cause of lost years prior to age 75, more than either cancer or heart disease. Motor vehicle injury is the largest single component of these losses."[3] I also found a 1983 paper lamenting that "automobile crashes are the leading cause of death for all persons between the ages of six months and age 47."[4] And a 1965 paper warned us that "traffic accidents are the leading cause of death among all persons from age 5 to 31!"[5]

---

[c] Murder, Inc. was the unofficial name of the murder-for-hire group used by organized crime in the 1930s.

[d] The Patriots won.

[e] For kids 19 and younger, gun violence overtook car violence for the first time in 2020, according to the CDC database.

The World Health Organization (WHO) estimates that each traffic death equates to an average of 30 lost years of life expectancy.[6] With 1.35 million road deaths worldwide each year, more than 2 percent of *all* deaths from *all* causes, this adds up to 40 million years of life lost annually on our streets.[f] WHO also estimates that for every road death, there are at least 10 to 15 people hospitalized due to road crashes.[7] This means 13 million to 20 million hospitalizations and millions more that probably should've gone to the hospital.

But the United States must be doing better than our international counterparts, right?

Right?!?

The Organisation for Economic Co-operation and Development (OECD) ranks countries by road fatalities per 100,000 population. When I first looked at these numbers for 2004, no OECD country was worse that the United States.[8]

It's not much better now. The last time I checked, Argentina, Russia, and Georgia were the only OECD countries with a worse road safety fatality rate than the United States. It's also gotten relatively worse in the United States over time. Back in the 1970s, there were at least a dozen major countries with worse road safety records.

So what the hell have traffic engineers been doing for the last hundred years?

---

[f] Black, Brown, and Indigenous communities bear a disproportionate brunt of all this traffic violence. So do pedestrians and bicyclists.

04 Hand-Me-Downs

—

Just like infrastructure gets handed down from generation to generation, so have many of our traffic engineering standards and guidelines (along with an enduring traffic engineering mind-set). Current traffic engineers look at these books and assume that the people who came before us did enough research to know what works best. But when you dig in—even just a little—you begin to realize that most transportation engineering decisions and designs do not put safety first.

There is a fundamental disconnect between what we want our places to look and function like versus what engineers give us. To understand *why* traffic engineers, including me, did what we did, I went deep down dozens of rabbit holes.

In some cases, I would wonder why we did something one way and think to myself, "Man, were we dumb back then." Then I'd find an original source and realize, yup. For instance, Edward Fisher and Robert Reeder write in their 1974 book, *Vehicle Traffic Law*, "Most of the early pieces of legislation, studied from the vantage point of experience 50 years later, demonstrate utter lack of comprehension of the real problems."[1] In other cases, I'd come across a 100-year-old book and realize that we weren't dumb whatsoever. What we did was intentional. We knew exactly what we were doing.

It's fun to imagine an old-timey traffic engineer who looks like Dastardly Whiplash coming up with our so-called standards. On rare occasion, that wasn't far off. Instead, it felt more like we put a bunch of foxes in charge of the henhouse. In other words, some of early decision makers were more concerned with selling cars than with, for instance, accurately assessing safety.

William P. Eno (1858–1945) was not one of the foxes. Considered the father of the early traffic engineering movement, he lived through the emergence of the automobile age.[a]

---

[a] William Eno never learned to drive, but he was given an honorary driver's license by France in 1912.

In January 1900, Eno published his *Rules of the Road*, printing and distributing thousands of copies over the years. He proposed that drivers stay to the right and signal with their arms when turning or stopping. Eno recommended that we regulate the speed of cars, and in a follow-up piece, he suggested a 10 mile per hour maximum (and 3 mph maximum when turning or crossing a busy street).[2] He wrote New York City's first traffic regulations in 1909 as well as six books about traffic. Eno is credited with innovations such as the stop sign, pedestrian island, one-way street, and traffic circle.[b] In 1921, nine years before the Institute of Traffic Engineers[c] was founded, he created the Eno Foundation for Highway Traffic Control.[d]

One of the reasons I bring Eno up now is that, beginning in 1947, the Eno Foundation first published *Traffic Quarterly*, widely considered the preeminent, independent traffic engineering journal of the time.[e] For years, the Eno Foundation printed annual editions embossed with the name of the recipient on the cover. I have a few copies from the early 1970s, and each includes a handwritten note of appreciation from the editor.

Traffic Quarterly changed its name to *Transportation Quarterly* in 1982 and was published through 2003. As longtime editor Colonel Robert Goetz said in his initial 1947 editorial, *Traffic Quarterly* was made for the "many experts and professional men of long experience in the field of traffic" that "have sound and logical ideas and methods for the improvement of the present chaotic conditions in traffic."[3] The other big dogs on the traffic engineering journal block didn't come around until much later. *Transportation Research Record* (originally known as the *Highway Research Record*) kicked off in 1963, while the *Transportation Research* series started in 1967.[f] For

---

[b] Eno called them "rotary or gyratory traffic systems" and was able to get New York City to try the idea with Columbus Circle in 1905.

[c] Now known as the Institute of Transportation Engineers.

[d] Now known as the Eno Foundation for Transportation.

[e] For years, the journal's slogan was, "An Independent Journal for Better Traffic." It later became "An Independent Journal for Better Transportation."

[f] The *Transportation Research* journal series split into Parts A and B in 1979 but now has six parts (A through F). On a related note, the *Journal of the American Planning Association (JAPA)* (originally known as the *Journal of the American Institute of Planners*) dates to 1935. While not as useful for trying to get into the minds of traffic engineers, several *JAPA* papers helped my cause.

my purposes, *Traffic Quarterly*'s 2,427 papers and editorials provided a longitudinal record of what traffic engineers were thinking over the course of 57 critical years.

It was interesting that it took *Traffic Quarterly* six years to publish a paper written by a woman.[g] It was a bit wild to see a 1962 *Traffic Quarterly* paper blame the traffic congestion that men have to deal with on all the "housewives returning from shopping expeditions." This same paper goes on to argue ***against sprawl*** because it would harm the "young girls temporarily employed while seeking a tenure position in marriage."[4]

Sexist papers notwithstanding—and we'll see some racist ones as well—this collection of papers turned out to be a deep and useful rabbit hole. These papers, combined with hundreds of other books, papers, and reports from the last 100-plus years of traffic engineering, helped me wrap my head around ***why*** traffic engineers do what we do. What you are about to read, however, is not a history book. I'm using all these resources as a tool to tell the bigger story about why traffic engineering needs a tectonic shift.

After more than 50 years of publishing papers, *Transportation Quarterly* admitted in a 1999 paper that traffic engineers don't know as much as they think they do, noting that "most of the conventional responses to our contemporary transportation problems are doomed to be ineffective. We have a body of both theoretical and empirical research on transportation dating back more than 40 years that tells us so."[5]

---

[g] The first *Traffic Quarterly* paper authored by a woman was written by Dorothy Lee, the mayor of Portland, Oregon. She bragged about arresting pedestrians for all infractions. The second woman-authored paper came a year later when Clara Lagace, a planning commissioner from Gloucester, Massachusetts, wrote about building a seawall so as to add off-street parking along the waterfront.

05   Passing the Buck

___

Conference presentations about transportation safety often start with a statistic about how human error accounts for at least 90-something percent of traffic crashes or fatalities. The speaker eventually tells the audience that autonomous vehicles will eliminate **all** these human error crashes from the system.

Sounds pretty good, right?

This line of thinking is fundamentally flawed. It's also one of the most insidious impediments to building safer streets.

Why?

Because traffic engineers wholeheartedly believe that crashes and deaths are rarely, if ever, the fault of the traffic engineer. We did our job. The traveling public didn't do theirs. Back to business as usual.

Human error isn't blameless, but it's not the root cause of most safety problems. As is, human error makes it easy for traffic engineers to pass the buck. Instead, traffic engineers should consider human error a sign of deeper problems in our transportation system. Those deeper problems often directly relate to what the traffic engineer is responsible for: design.

Conventional traffic engineering methods have us designing major streets for traffic volumes 20 or 30 years down the line. Given all the human error, we want these streets to be forgiving, so traffic engineers design them to be wide and straight.

At the same time, we know that people tend to drive whatever speed they feel most comfortable driving. While speed limits play a role, street design plays a bigger one.

We also know that higher speeds have a direct connection with more serious crash outcomes, particularly with pedestrians and bicyclists.

So when somebody speeds on this wide and straight street, crashes, and kills someone, it'll get chalked up as one of the 90-plus percent of human error crashes. The driver made the mistake of speeding.

Who cares if the engineer designed a street built for speed, just begging you to go fast, as long as the guidebook was followed? The courts don't. Few others do.

The engineer will say that people need to behave better or that we need more enforcement. Case closed. Engineers don't learn anything about how to do their job better. There is no feedback loop.

This is a simple example, but as Sidney Dekker says in *The Field Guide to Understanding "Human Error,"* "Human error as explanation for accidents has become increasingly unsatisfying.... 'Human error' does not come out of the blue, and is not separate from the system in which it occurs. Error has its roots in the system surrounding it."[1]

What are the roots of error in traffic crashes? Street design is one, but it goes deeper. It includes the network around that street. And the viability of other modes. And how we design cars. And where we put land uses. And what those land uses are. And how we fund transportation. And how much the different options cost. Dekker goes on to say that "the deeper you dig, the more you will understand why people did what they did, based on the tools and tasks and environment that surrounded them. The further you push on into the territory where the errors came from, the more you will discover that the distinction between human and system failure does not hold up."[2]

Much of the human error rabbit hole that I fell into stems from the work of British psychologist James Reason, whose first book on this topic came out in 1990. In that book, Reason alludes to the overconfidence of traffic engineers: "Problem solvers and planners are likely to be overconfident in evaluating the correctness of their knowledge.... They will justify their chosen course of action by focusing on evidence that favours it and by disregarding contradictory signs."[3]

In a research project I led,[a] we surveyed 18,000 people and found that every single road user admitted to some form of law breaking.[4] If everyone is breaking the law, it's easy to place blame on everyone ... except the traffic engineer.

We'll talk more about human error later in the book. For now, suffice it to say that our current approach lets engineers off the hook in almost every situation.

---

[a] Called the scofflaw survey.

06 Won't Someone Please Think of the Children?

———

We now kill well over 7,000 pedestrians each year in the United States.[a] This number would hit 8,000 per year in the late 1970s and early 1980s. It dropped into the low 4,000s in the early 2010s following the Great Recession.[1] Traffic engineers like to point out that these fluctuations are beyond our control. A 2020 pedestrian fatality report states that "many factors outside the control of traffic safety officials contribute to annual changes in the number of pedestrian fatalities, including economic conditions, demographics, weather, fuel prices, vehicle miles traveled, and the amount of time people spend walking."[2]

If walking were actually safe, most of these contributing factors wouldn't matter much. Fuel prices might not help, but fuel prices don't kill pedestrians.

Let's look at child pedestrians. One recent study says that "a safe pedestrian must manage to perceive, attend to, and process multiple sources of stimuli within a very short time period. Young children struggle to do so safely."[3]

So we know that kids have difficulties assessing things like the speed and distance to a moving car, which in turn leads to difficulties in figuring out when a gap in traffic is enough. Kids also have limited peripheral vision and trouble locating sounds. Yet traffic engineers don't design to help kids out. We'll design everything as we normally do and then put up some signs that try to limit speeds around schools at certain times of the day and certain times of the year. That's it.

Kids under age 18 represent nearly a fourth of the US population, but other than signage, traffic engineers make little effort to design for them. Denver won't even lower the school zone speed limit to under 35 mph if it considers the adjacent street important for cars. Denver isn't alone.

Parents ask for a crosswalk and pedestrian signal. Traffic engineers say there aren't enough pedestrians crossing the street to warrant them.[b]

———

[a] We climbed to 7,508 pedestrian deaths in 2022.

Let's say we do put in a pedestrian signal. Traffic engineers will time it for a healthy, determined adult rather than an inattentive, meandering kid.

Whatever the case, let's say a kid gets hurt crossing the street. Who's at fault? It might be the driver. It could be the kid. In the court of public opinion, it could be the parents for letting their kid walk on a dangerous street or the school for letting the parents let their kid walk on a dangerous street. It could even be an act of God, like an icy street.

It's never the traffic engineer.

It's easy for traffic engineers to make sure we don't get blamed. We've been doing it forever. In 1960, *Traffic Quarterly* published an editorial by its longtime editor, Colonel Robert Goetz, about road safety. What is the biggest road safety problem? Obviously, it's the "deliberately hazardous driving."[4] In his editorial the next year, the Colonel reminds us that traffic engineers have done everything they possibly can: "Obviously no effort has been spared to build safety into our highways."[5] So, why is our nation's safety record so bad? The Colonel says that maybe you should look in the mirror, American people: "The highway engineer cannot build causation or safety into the minds of unqualified drivers or willful speeders.... Some improvement in the national character of our driving is urgently needed."[6] That same year, *Sports Illustrated* put out an issue dedicated to **Safe Driving: An Important New Series**, with a typical highway filled with cars as the cover model.[c]

A special effects artist who goes by Xiaojie created a video showing two kids about to cross the street. As the kids enter the crosswalk, a dozen spiked bollards, or posts, come flying out of the ground to keep drivers from entering the crosswalk. At the same moment, a giant medieval morning star swings back and forth in front of the waiting cars.[d]

This medieval-protected crosswalk scenario is fiction—and

---

[b] Did we count how many people swam to San Francisco before building the Golden Gate Bridge in the 1930s? Of course not.

[c] Yes, *Sports Illustrated*.

[d] A "morning star" is a spiked ball on a shaft that you might see in a *Mortal Kombat* video game or wielded like a club in *Game of Thrones*. For those who care, one of Joffrey's Kingsguards in *Game of Thrones* uses a morning star in the Battle of the Blackwater.

maybe a bit extreme—but there is no real reason traffic engineers can't build a system that better protects people, particularly kids who are walking or biking.

We choose not to. We have priorities other than safety.

Plus, it's not like it's our fault if somebody gets hurt.[e]

---

[e] As you can probably tell, I will sometimes refer to all the traffic engineers out there—and those that came before—by using "we" or "our."

07 Little Lies

—

In a 1971 *Traffic Quarterly* paper titled "The First Consideration," FBI director J. Edgar Hoover says too many drivers "tend to make speed, not safety, their first consideration."[1] But for engineers, Hoover assures us that it's safety first: "The safety of the public is to be the first consideration of the staff."

In other words, Safety Is Our #1 Priority!

This slogan is embraced by generations of traffic engineers.[a]

But let me put it to you straight: safety first is a lie. Safety has never been the top priority.

Transportation exists for a reason, and that reason isn't safety. We want to move people and goods from here to there and quickly back again. People make (and spend) a lot of money on this.

In one of the older road safety books I was able to track down, *Seven Roads to Safety* (1939), Paul Hoffman states, "First, a qualification. Safety is not necessarily the most important thing on earth. 'Safety first' is a misnomer. Sometimes we want other things first. But all will agree, I believe, that if we can have safety along with other things, the result is more pleasant."[2]

We'll talk a lot more about Paul Hoffman later.

Jack Leisch is another guy who had a long and influential career in traffic engineering. In response to a paper lauding roundabouts, Leisch put our priorities in order for us: "Let us first consider the course that the United States pursued in its highway development and what goals were to be achieved in traffic operations: increase in capacity; increase in speed, followed by improvement in safety."[3] And Leisch is a guy who was inducted into the exclusive Institute of Transportation Engineers (ITE) Honorary Member club. At last count, only 96 people have been inducted since

---

[a] I recently made a presentation at the Colorado Department of Transportation and learned that their Wi-Fi password was S@fetyF1rst. If they need to change it now, sorry about that.

1933.[b] Leisch's ITE profile says that one of his most significant contributions "was his belief that the design of a roadway must be viewed from the standpoint of the motorist—the real user of the facility."[4]

When testifying before Congress in 1973, Highway Safety Foundation executive director and traffic engineer William Sacks said that "we are dealing with accidents which are byproducts of the system whose purpose is really not safety, it is mobility."[5] He then suggests that the main problem with road crashes is that they reduce system productivity.

So, why do we want to reduce crashes? We do so because crashes increase traffic congestion and reduce mobility. We also do so because crashes cost us money. Misidentified priorities lead right into how we mismeasure transportation, which we'll talk about later. At best, we can say that we have **conflicting objectives** in transportation. Sure, traffic engineers want to improve safety, but before doing so, we want to minimize congestion, maximize mobility, minimize costs, and so on.

Safety is never first on the list.

Otto, the defunct autonomous vehicle company bought by Uber,[c] was a bit more direct. It used the motto **Safety Third** when talking about its work. Engineers pasted "Safety Third" stickers all over Otto's San Francisco headquarters because they knew their founder would love them.[6]

Soon after being acquired by Uber—and prior to getting approval to test Otto's self-driving cars in San Francisco—a driver's dashcam video caught one of Uber's test cars blowing through a red light and straight through an enormous crosswalk in front of the Museum of Modern Art.[7]

Uber used the incident to try to drum up support for autonomous cars. It said that at the time of the incident, the driver, not the software, was in control of the car. Human error is the obvious culprit, which is "why we believe so much in making the roads safer by building self-driving Ubers," said a company spokesperson.[8]

---

[b] The ITE Honorary Member club inductions date back to 1933, and as far as I could tell, the first woman inducted (Willa Wilcox) wasn't until 1992. The first—and seemingly only at the time of this writing—Black engineer in this ITE Honorary Member club (Garrett A. Morgan) died in 1963 but wasn't inducted until 1998.

[c] Otto was founded in January 2016. Uber bought it for $680 million in August of that same year.

It turned out that wasn't the case. According to a *New York Times* article, "The self-driving car was, in fact, driving itself when it barreled through the red light, according to two Uber employees, who spoke on the condition of anonymity." The autonomous test car went through six red lights that day.[9]

While traffic engineers may not be near Uber on the deceit spectrum, the reality is that traffic engineers never start the design process with safety. Truth be told, traffic engineers rarely ever think about safety because we assume that the system will be safe if we follow our guidebooks.

I've heard folks like Chris Argyris, a former Harvard Business School professor, call this the difference between espoused theory versus theory-in-use. Traffic engineers say that safety is our first priority, but such organizations, as Sidney Dekker suggests, "lack a conherent strategy on continous improvement ... [and] if you recommend design or organizational changes, you are likely to invite more opposition. This will cost money, after all, or suggest that some people higher up were not doing their jobs."[10]

Even though Dekker wasn't talking about traffic engineering in this quote, anybody who has ever proposed a potential solution to a traffic engineer in a public meeting can attest that it hits close to the mark.

Personally, I have a hard time watching public meetings about road safety unfold. My main irritant has to do with what I call ***engineer speak***.

That's what engineers use when they want the public to sit down and shut up. It's the overly technical language meant to appease or confuse the audience. It's the catchphrases meant to deflect blame. It's the sneaky terms said to make things sound better or worse than they are. When we like something, we describe it as an improvement or investment in a public good. When we don't, we use words like *degrade* or say it requires a public subsidy.

Here are some of my favorite least favorite examples of engineer speak.

"Safety is a shared responsibility."

When the public tells the engineer that they don't feel safe crossing the street here or bicycling there, traffic engineers will remind everyone that ***safety is a shared responsibility*** or that ***personal responsibility*** needs to play a bigger role. In other words, they don't plan to fix the problems with design changes; they just want everyone to follow the rules we've laid out for them. I've found versions of this almost as far back as the traffic engineering literature goes. A 1955 paper tells us that "safety is everybody's problem" and that "if each accepts his individual share of the responsibility, the task [of improving road safety] will be easy."[11] Some of today's more progressive traffic safety movements use this phrase as well. They want "safety is a shared responsibility" to mean that engineers and policy makers need to share in this responsibility of creating a safe transportation system. Unfortunately, traffic engineers and policy makers continue to co-opt it in a way that shifts blame onto road users.

"We need to take a balanced approach."

In many cases, the public wants more space for pedestrians and bicyclists. Or they want the engineer to help slow cars to make streets safer for other modes. The engineer reminds the audience that *we need to balance everyone's needs*, which in theory means that we need to find an *appropriate balance* between things like design consistency, safety, aesthetics, environmental issues, and community goals (espoused theory). What it really means is that driving and cars are still more important (theory-in-use). This can flip when a vocal minority is complaining about giving up space or priority to pedestrians or bicyclists at what seems like the expense of cars. The engineer might remind those folks that we need to take a balanced approach to the transportation system and shut them up as well.

"Efficiency."

When a traffic engineer talks about prioritizing *efficiency*, this means increasing traffic flow. And when traffic engineers talk about an efficient traffic flow, this means more cars moving faster. The better traffic engineers now focus on moving people instead of cars. This gives more weight to high-occupancy vehicles or transit (and sometimes even pedestrians and bicyclists).[d] What's always left unsaid is that focusing on efficiency means limiting necessities like the amount of time we give to pedestrians trying to cross the street. In other words, *efficiency* means that we've got to give priority to vehicle speeds and vehicle movements.

---

[d] Although when engineers measure people flow, no engineers account for car lanes being wider than bike lanes.

"Improvement."

Traffic engineers will refer to pretty much everything they do as an *improvement*.[e] But what's left unsaid is that *improvement* usually means an improvement for those in cars. In other words, anything we do to improve vehicular traffic flow gets classified under the umbrella of an improvement. Does that improvement make things worse for pedestrians, bicyclists, or transit users? Or how about the people who live or work along this "improved" street? The answer to these questions is beside the point.

"Not my job."

More often than not, the phrase "Not my job" gets used when traffic engineers try to defend themselves for things not turning out as people imagined they would. For example, traffic engineers from the heyday of the interstate era liked to point out—after the fact—that they were told to build highways but to do so as inexpensively as possible. So when they tore up the poorest neighborhoods to build these new highways, they remind us how they fulfilled the task they were given and what a bang-up job they did. Worrying about the people who live where these highways end up is *not my job*. We'll later see how untrue this is—because many of the old traffic engineering books highlight slum clearance as a bonus of highway building—but it's a convenient retcon.[f]

"We need to study the problem further."

Traffic engineers offer *to study the problem further* simply to blunt and divert public attention by kicking the can down the road. In the old days, we'd appoint a blue-ribbon commission. Whatever the term, this approach lets traffic engineers avoid dealing with complaints until a later unspecified date. Whether the traffic engineer plans to actually study the problem further is irrelevant, but it shuts the discussion down and buys some time. How much time does it buy? Once the furor dies down, few come knocking on your door

---

[e] Traffic engineers sometimes use the words *upgrade* or *enhancement* instead.

[f] A retcon—or *ret*roactive *con*tinuity—is a plot device used to change a previously established narrative. One of the early examples, well before the term existed, happened when Sir Arthur Conan Doyle brought Sherlock Holmes back from the dead. Bored with writing Sherlock stories, Doyle killed him off. Doyle faced a lot of backlash. In the next book, it turned out that Sherlock faked his own death by secretly climbing onto a ledge and hiding until he could escape. Convenient.

to ask for that study. If they do, then the study perpetually remains almost done. "We need to study the problem further" is often a nice way to put the problem off indefinitely.

"The standards say …."

This is a big one. Traffic engineers like to tell the public that whatever the public is proposing won't work because **the standards say** this or **the standards say** that. The truth is that they are referring to guidelines, not standards. Other than the *Manual on Uniform Traffic Control Devices* (MUTCD) and the ADA Standards for Accessible Design, none of the guidance encased in our massive national manuals should be considered "standards." Even the MUTCD—often used to tell the public that a crosswalk or pedestrian signal isn't warranted—gives the engineer discretion to overrule the infamous warrants with **engineering judgment**.[g] If you want to give traffic engineers the benefit of the doubt, you could argue that it's just a slip of the tongue and that *guidelines* and *standards* are synonymous anyway. If you want to know the truth, many traffic engineers know exactly what they are doing.[h] The use of the word *standard* is intentional. First, it shuts people up because it's hard for people to argue with a standard. Second, it keeps us from getting our hands dirty. It keeps us above the fray. As a 1973 *Traffic Quarterly* paper says, traffic engineers "frequently cloak themselves in the mantle of neutrality."[12] In other words, traffic engineers don't take sides; we only do what our traffic engineering books tell us to do. Casually referring to something as a *standard* helps us do that.

"Public participation."

One common misconception is that traffic engineers go to public meetings because they want to hear what the public has to say and adjust their designs accordingly. In most cases, this couldn't be further from the truth. By the time most traffic engineers step in front of the public, they've already decided what they want to do. Their job that night is to announce what they've decided and defend it by showing some carefully selected, inferior alternatives. The DAD acronym (decide, announce, defend) was something my old boss joked about prior to my first public meeting when I was a young engineer.

---

[g] *Engineering judgment* is another loaded term, which we'll talk about later.

[h] On the other hand, some traffic engineers probably do believe them all to be standards.

My job was to use all the above strategies to get the meeting over as quickly and painlessly as possible. Any changes to our design would be the opposite of quick and painless. Historically, as the same 1973 *Traffic Quarterly* paper points out, "most of these hearings receive little notice, and are attended primarily by" people "who congratulate each other on the high degree of consensus reflected at the hearing."[13] These days, ***public participation*** is a required part of most projects, but traffic engineers still think about this obligation merely as a box that needs to be checked off. The amount of opposition a traffic engineer sees at a public meeting is quite inconsequential.[i] We write them off as squeaky wheels and the usual vocal minority that always come out to complain. As a 1959 paper in *Traffic Quarterly* says, "The 'squeaking wheel' minority cannot be the only determining factor. Certainly, the economic well-being of our great city is closely tied in with its traffic circulation."[14] Plus, we all know that people would never bother coming to public meetings to say you're doing a good job. It's going to be the squeaky wheels no matter what we do.

Traffic engineers also like to argue that public participation processes will lead to worse outcomes. A 1977 *Traffic Quarterly* paper lays out this thinking. Because of public participation, "transportation facilities may cost more, be less safe, and have a longer trip time." Why? Well, all these pesky people and their opinions mean that we can no longer focus on "classic engineering." Moreover, "participants may not be representative of the entire community," so why should we listen to them? If we do, "the results of participatory planning will not be as optimal in the classical sense as traditional planning."[15]

Yup.

---

[i] I wouldn't hate hearing this sentence said by Dr. Evil.

08   Science versus Faith

—

One of the best things about science is that when it's done right, it's self-correcting. Science is in a constant state of acquiring knowledge, and if new knowledge contradicts old knowledge, we try to figure things out and change courses if necessary.

The discipline of traffic engineering often acts more on faith than science. Dan Brown's book *Origin*, from the Robert Langdon series, discusses the difference. Faith, by "its very definition, requires placing your trust in something that is unseeable and indefinable, accepting as fact something for which there exists no empirical evidence."[1]

As for science, Dan Brown suggests that "science is the antithesis of faith…. Science, by definition, is the attempt to find physical proof for that which is unknown but not yet defined, and to reject superstition and misperception in favor of observable facts.[a] When science offers an answer, that answer is universal. Humans do not go to war over it; they rally around it."[2]

Traffic engineering always wanted to be a science. We came up with theories, put them into practice, and waited to see what happened. We made adjustments, sometimes to improve safety, but usually to improve traffic flow. Still, we learned from our mistakes. We had a feedback loop based, at least in part, on the scientific method.

But much of what traffic engineers do today isn't that different than what traffic engineers did generations ago. That isn't a problem in itself. The problem is that we continue to do so despite evidence to the contrary. Take roundabouts, for example. Why did they fall out of favor in the 1940s and 1950s? The reason, according to one 1988 paper, is that "in those days, the loudest voice and best oratory generally prevailed in establishing highway standards when supporting data were lacking."[3]

---

[a] Given the antiscience trends we've seen recently, Amy Poehler and Seth Meyers might say "Really!?!" to this definition.

And data was indeed lacking. A 1968 *Traffic Quarterly* paper tells us that "the quality of the available data is so poor that it is impossible to establish benchmarks for any kind of measurement: we are today largely unable to determine whether the accident rate is increasing or decreasing or to evaluate the efficacy of any specific countermeasure." As a result, "accident research in general is not characterized by a high degree of sophistication, and mere awareness of the problem does not provide a solution."[4]

Heck, we traffic engineers back then readily admitted that we didn't know that much. The author of a 1971 *Traffic Quarterly* paper[b] specifically about road safety says that "usually the approaches were based more on hope and logic than on practical facts."[5]

Congress also knew this to be a problem. In 1965, a House of Representatives report on road safety bluntly states that "much of what has been done has never been tested, evaluated, or applied."[6] But a Senate report from later that year implores us to take action "without waiting completion of long-term research projects."[7]

The next year, Congress created a new federal agency on traffic safety, and President Lyndon Johnson appointed William Haddon, a doctor by training, to head that up.[c] Not long into the job, Haddon pointed out that crash research is still in what could be considered its prescientific stage. Haddon said that "increasing attention is being paid in accident research to the evaluation of the efficacy of preventative measures. But this is a substantially new trend. Relatively few accident countermeasures—some in use for decades—have yet be subjected to adequate scientific scrutiny."[8]

Traffic engineers didn't even invent the striped zebra crosswalk until London first tried it in 1948. In this case, they treated it like a science experiment by testing the design at 1,000 sites across the United Kingdom. After a statistically significant drop in pedestrian

---

[b] The author of this 1971 paper also warns us that most of the money we get for road safety research comes from "the major insurance, oil, and automobile companies."

[c] In 1966, William Haddon, who is widely considered the father of modern-day injury epidemiology, was an administrator in two newly created agencies: the National Traffic Safety Agency and the National Highway Safety Agency. These agencies were consolidated into the National Highway Safety Bureau (NHSB) in 1967, which, in 1970, became what it is known as today, the National Highway Traffic Safety Administration (NHTSA). NHTSA is pronounced "NITZ-ah" in case you want to talk the talk.

fatalities as compared to the control group was seen, this painted crosswalk design became the official crosswalk treatment in 1951.[d]

Seat belts date back to an 1885 patent by Edward J. Claghorn of New York. It took 99 years for New York to be the first state to require their use. Even in the early 1960s, "it was widely accepted as a 'fact' that in an automobile accident occupants who were thrown from the car were safer than those who remained inside. Indeed, the connotation of 'thrown clear' was that the ejectee would otherwise have been killed."[9] This "widely accepted ... fact" contradicted reports from a decade earlier telling us that being thrown clear usually led to death, not safety.[10] It took us longer than it should have to listen.

We later figured out that seat belts reduce the risk of fatality by 45 percent and injury by 50 percent. This should've been a good lesson for us when airbags were invented in the early 1950s. But it took more than four decades before the federal government required those in new vehicles.

In her last book, *Dark Age Ahead*, Jane Jacobs calls traffic engineering "a 'perfect' example of antiscience masquerading as the science it betrayed."[11] Years earlier, in *The Death and Life of Great American Cities*, she warns us that the traffic planning profession is a "pseudoscience" that "seems almost neurotic in its determination to imitate empiric failure and ignore empiric success."[12]

William Haddon doesn't disagree. In his 1965 *Traffic Quarterly* paper, he makes it clear that traffic engineers base "programs on unsupported presumptions ... often with dogmatic public assurance as to their efficacy."[13]

Traffic engineers still do so today. We have faith in those who came before us. We trust that those who came before us knew what they were doing. Those who came before us, however, didn't know much. How do I know they didn't know much? They said so ... time and time again.

This undeserved faith has stifled the science. In 1993, the prominent road safety researcher Leonard Evans wrote about how little we've accomplished: "It has been more than 50 years since driving behavior and traffic safety were first analyzed in a technical way....

---

[d] The Abbey Road crosswalk wasn't installed until the late 1950s, only a decade before the Beatles crossed.

However, when compared to advances in the traditional sciences, increases in knowledge about traffic safety are less impressive."[14]

When you treat something as a given, you don't have much reason to look for better alternatives. But as a 1973 congressional road safety report made clear, "We cannot remain wedded to the past and embrace uniformity when there is just reason to suspect that certain practices may be uniformly bad. New and better ways must be found to warn, guide, and control the motorist and the pedestrian. The current system simply is not adequate."[15]

It's time to do better. It's time to be better.

09 Killed by a Traffic Engineer?

—

I should start by making it clear that traffic engineers—as far as I know—are **not** out there trying to cause deliberate harm to anyone.[a]

What traffic engineers are guilty of is creating a transportation system whose designs remain largely based on plausible, but unproven, conjecture.

Some traffic engineers are guilty of assuming that "the succession of standards committees that formulated and improved design standards did so on the basis of factual knowledge about how their decisions affect crashes."[1]

The traffic engineers already aware of this disconnect are guilty of implying that safety is built into our so-called standards when, in reality, "knowledge of the crash frequency and severity consequences of design decisions played no discernible role in the formulation and in consecutive revisions of important design standards and procedures."[2]

The above quotes are from a 2000 paper about geometric design by renowned road safety researcher Ezra Hauer. In that same paper, he makes it clear that *Safety Third* in traffic engineering might be overly optimistic: "Many believe that roads designed to standards are safe roads. Roads are neither safe nor unsafe; … their safety is largely unpremeditated."[3]

In other words, there is less of a link between design and actual safety outcomes than you would think possible. In some cases, there is none. In other cases, we focus on measuring things we think *may* impact actual safety outcomes. For instance, when developing shared-lane bike markings, also known as sharrows, the "safety" research only looked at differences in the lateral spacing of cars and bicyclists. Since the driver seemed to be giving the bicyclist a little more space, the bicyclist should theoretically be safer, so we added sharrows to our manuals. Will sharrows improve actual road safety

---

[a] And to answer my daughter, Mae, no, Uncle Rob is not a murderer (as far as I know).

outcomes? When we added them to our manuals, we had no idea.[b] Actual safety outcomes were not part of our process.

Despite manuals lacking the requisite science to back them up, traffic engineers put their faith in these manuals and adhere to them. Traffic engineers also don't take too kindly to those criticizing these holy texts.

It wouldn't be hard to argue that this sounds more like a religious cult than a science-based discipline.

The unfortunate reality? The system isn't broken; we built it this way.

Every undergraduate civil engineering student at my university is forced to memorize the seven Fundamental Canons of Ethics put out by the American Society of Civil Engineers. The very first one kicks off with safety: "Engineers shall hold paramount the safety, health and welfare of the public and shall strive to comply with the principles of sustainable development in the performance of their professional duties."[4] The National Society of Professional Engineers only has six canons, but again, the first one is that engineers shall "hold paramount the safety, health, and welfare of the public."[5] In contrast, the first canon of ethics for the Institute of Transportation Engineers also mentions safety, but it seems a bit less important: "The member will have due regard for the safety, health and welfare of the public in the performance of professional duties."[6] What does *due regard* mean as opposed to *hold paramount*? I would guess that it's the difference between being mindful of safety versus making safety chief in importance.

Yet traffic engineers don't quantify the safety implications of different design alternatives. Even worse, few traffic engineers could do so if they tried. Most of our manuals provide little data regarding the safety consequences of different design choices. Nor is it common for traffic engineers to collect before-and-after crash data. Unless we live locally or there is a lawsuit, traffic engineers are unlikely to ever hear about the crashes that happened on a street they've designed.

Without a strong feedback loop, how do we learn anything? How do we get better? How do we get safer?

Some traffic engineers are guilty of indifference. Most are guilty

---

[b] Research today suggests that sharrows don't help safety and may make things worse by enticing bicyclists to ride on dangerous streets.

of tolerating all the above practices as the status quo. Einstein is quoted as saying, "The world is a dangerous place to live, not because of the people who are evil, but because of the people who don't do anything about it."

Now that we know better, traffic engineers have become the not-so-innocent bystander. If traffic engineers continue to conduct business as usual, more people will die on our streets, indirectly killed by the generations of traffic engineers who came before them.

So when I say "*killed by a traffic engineer*," what I want is for traffic engineers to treat every road death as if we are the only ones who can fix it. No more blaming human error or a lack of enforcement.

It's on us to save the day.

10 The Three E's

—

Most books about road safety—or any kind of safety for that matter—usually focus on what we call the three E's: ***engineering, education,*** and ***enforcement.***

Even in 1939, Paul Hoffman was calling them the "classic three" needed to improve road safety. In theory, these three should be able to handle just about any problem (other than crashes caused by what he calls an "act of God"). In 1947, Colonel Robert Goetz wrote this in his *Traffic Quarterly* editorial: "Engineering, education, and enforcement are generally considered to be the three essentials in the solution of the traffic problem."[1]

In 1957, the American Association of State Highway Officials (AASHO)[a] informs us that "safety depends principally on good engineering practice, proper law enforcement, and effective public education."[2]

So what do we mean by the three E's? A 1965 *Traffic Quarterly* paper does a nice job breaking them down into (1) physical change (engineering); (2) behavior change, motivated either through carrots or sticks (education); and (3) population change, by keeping the bad apples off the road (enforcement).[3]

Traffic engineers remind us that they've already done everything they can on the ***engineering*** front (that is, unless you have some more money for us to spend).[b]

Law enforcement reminds us that they've done everything they can on the ***enforcement*** front and don't have the resources to put more officers on traffic duty[c] (that is, unless you have some more

---

[a] Founded in 1914, AASHO changed their name in 1973 to AASHTO by adding "*and Transportation*" (American Association of State Highway and Transportation Officials).

[b] "Money, pleeeeease."—Mona Lisa Saperstein (*Parks and Recreation*).

[c] A Denver police officer told me at a public meeting in 2011 that Denver has seven officers assigned to traffic duty. That was not seven officers out there at any given time; it was seven officers total.

money for them to spend). We also lack the political will to pursue extensive camera enforcement.

Education is the low-hanging fruit.[d] Traffic engineers have been saying as much for decades. A 1947 *Traffic Quarterly* paper explains why that is the case: "It seems reasonable to assume that safe use of streets and highways depends more upon the skilled movement of the vehicle and pedestrian than upon established signs, pavement markings, police supervision or any other type of supervisory or regulatory action."[4] Thus, traffic engineers have the mind-set that road safety is not up to engineers or police officers. "It is upon the individual himself that dependence must be placed for the prevention of accidental death or injury from hazards" on our streets, that paper states.[5]

As Sidney Dekker's *Field Guide to Understanding "Human Error"* points out, those in charge of safety tell us that "in order to not have safety problems, people should do as they are told."[6] In other words, we put these rules in place for your own good. All people need to do is comply with what traffic engineers have spent more than a hundred years figuring out for them. If they don't, it's their own fault. The best we can do is educate them. More specifically, he tells us that "it is because of people's negative attitudes which adversely affect their behaviors. So more work on their attitudes (with poster campaigns and sanctions, for example) should do the trick."[7]

The funny thing is that traffic engineers have known for generations that the three E's aren't enough. Here is a *Traffic Quarterly* paper from 1955 in which Hallie Myers, a man who served on the National Safety Council board of directors, talks about how far the field has come: "I am amazed at the number of things I knew twenty years ago that aren't true now and probably never were." He then goes on to discuss the three E's, and guess what? They aren't enough. "Had this article been written twenty years ago it would have stated positively that the three E's—engineering, education, and enforcement—constitute the answer to the traffic safety problem. Since that time, observation of their fighting many losing battles … has considerably shaken my faith in the three E's as a cure-all."[8]

A *Traffic Quarterly* paper from 1962 then lets us in on a little secret: "The 'Three E's'—Education, Enforcement, and Engineering—are

---

[d] Although technology, and autonomous vehicles, remains the pipe dream.

effective instruments in achieving better traffic flow and capacities."[9] In terms of safety? Not so much.[e]

We even had evidence that the three E's didn't work.

A 1963 *Traffic Quarterly* paper boasts that "the Tarheel State is on the threshold of the greatest advancement in traffic accident prevention in its history" because starting in 1962, "Project Impact" would put a complete focus on the three E's.[10]

Wow! Sounds promising, right?

Unfortunately, North Carolina had more road fatalities in 1962 than any year up to that point. It proceeded to get worse every single year for nearly another decade. There weren't any follow-up papers to let us know any of that.

In his 1955 paper, Hallie Myers also tries to shed light on how to make our *education* efforts more effective. His answer is that we need religious leaders—rather than engineers—telling people how to behave on our roads. And in a 1958 report, AAA echoes similar sentiments and recommends that clergy start giving sermons about pedestrian safety.[11] Maybe that would help, but as safety researcher Ezra Hauer warns us, most transportation agencies "are more concerned with looking good than doing good."[12]

So, traffic engineers know that the three E's aren't going to solve the problem, but traffic engineers also need it to look like we are doing *something* about the problem. Sidney Dekker tells us what *something* really means: "That 'something' can include telling everybody else to try harder, to watch out more carefully. It may be putting up posters exhorting people to do or not do certain things. It may be closing out an investigation by recommending a new policy or stricter compliance with an existing procedure. Those are not really investments in safety, as much as they are investments in showing that you did something about the problem."[13] So rather than looking at actual safety outcomes, we focus on public relations.[f]

This usually means road safety *education* campaigns.

---

[e] In that same 1962 *Traffic Quarterly* paper, Arthur Henderson goes on to say that if we truly cared about safety, we "would eliminate conflict between pedestrian and vehicular traffic."

[f] Starting in 1958, France's incoming president would pardon most traffic tickets for the 6 to 12 months prior to the presidential inauguration. This wild public relations tradition ended in 2002 after research showed a significantly worse road safety record in the months preceding every new French president.

11 You Could Learn a Lot from a Dummy

—

The need to look like we are doing *something* is why most road safety efforts focus on ***education*** "solutions" such as the "***Click It or Ticket***" campaign that has been around since 2003.[a] You also might remember the 1985 "***You Could Learn a Lot from a Dummy***" campaign that promoted seat belt use with Vince and Larry, the crash test dummies, who were placed in the Smithsonian in 2010.

Scare tactics were the go-to for road safety education over the years. The *Death on the Highway* series (a pamphlet and movie created by a Florida company called The Suicide Club) sold more than two million copies in the 1950s and was shown to tens of thousands of high school kids over the years. It depicted "decapitations, multiple compound fractures, intestines strewn over the road, flesh burned from bones, and arms and legs torn from the human body."[1]

In 1956, the National Safety Council published road safety ads with parents shopping for a small child-sized coffin with captions such as,

"How does it feel to plan for a coffin?"

They'd then flip it and focus on the parents dying, saying,

"This child is scheduled to be an orphan tomorrow."

Or the husband dying:

"Will your wife be a widow this year?"[2]

Some scare-tactic ads seem better than others. One of my favorites is a Volkswagen commercial from Hong Kong that starts off with people finding their seats in a movie theater.[3] The previews begin, and you see the inside of a car speeding down the road. Then, with the help of a location-based emergency broadcast signal, everybody in the movie theater receives a text message at the same moment. When people start pulling out their phones to look, the car on the screen crashes loudly, jolting the entire audience. The commercial

---

[a] The last time I checked, the US Department of Transportation was spending $600 million annually on road safety ***education***.

then reminds people to keep their eyes on the road because mobile phone use has become a leading cause of road fatalities.

But we've long known that scare tactics don't work that well. A 1961 *Traffic Quarterly* paper expands upon this: "In traffic safety campaigns,[b] scare techniques are common. So are claims for their effectiveness. But there is little evidence behind these claims. In fact, psychological studies in related fields show that scare campaigns may do more harm than good."[4]

Australia started to figure this out in 2007 when it transitioned its road safety education from a series of horrifically realistic crashes[c] to those focused on ridicule. The most famous was the "pinkie campaign"[d] in which several women, of all different ages, wave their pinkie fingers at speeding male drivers, implying that the size of the drivers' manhood led to the need to speed.

The good news is that the ridicule ads at least have some scientific basis in social norming theory. That's the same reason your energy or water company sends you a report telling you how much more electricity or water you use as compared to your neighbors.

A *Traffic Quarterly* paper from 1948 tries to make a similar point. In discussing the culture at the General Motors test track,[e] the author says, "We do not believe in luck, good or bad: and since there are people who do not have accidents, there must be something wrong with the driver who has them." So, if you come back "with so much as a dented fender, you are immediately the target for all manner of ribbing" and are subject to comments like "What's the matter, can't you handle an automobile? Why not go out and take lessons in driving? Why not get a wheelbarrow and practice on it awhile?" He even says that serious car emergencies—like blowing out a tire on a high-speed road—would not happen if we had better drivers.[5]

Most education campaigns cross into victim blaming. The Colorado Department of Transportation (CDOT) has put out a few

---

[b] The word *campaigns* is written as *compaigns* in the actual text. I assume that it was a typo.

[c] An example of one of those commercials can be seen at www.youtube.com /watch?v=B_EKVtPh82Q.

[d] The *Speeding: No One Thinks Big of You* commercial can be seen at www.youtube .com/watch?v=jvACGhyC9XM.

[e] It is called the Proving Ground and is located in Michigan. We'll talk about it again later in parts 3 and 5.

masterpieces in the time I've lived here. One series of fake info-mercial ads has a mustachioed man named Hank in a 1970s polyester suit ask us, "Are you tired of having to look up from your phone at crosswalks? Sick of having to walk all the way to crosswalks in the first place? Well now, you don't have to! Introducing Hank's How To Get Hit By A Car series on Beta or VHS."[f]

More recently, CDOT's gigantic eyeball campaign reminds pedestrians (and dogs) to "make eye contact" with drivers before crossing at a crosswalk. CDOT even dressed up a half dozen employees in giant eyeball masks to walk around downtown Denver. The underlying premise of the campaign is that our crosswalks are not safe and that we need pedestrians to make eye contact with drivers. Pedestrians need to make sure the driver acknowledges them before "stepping into the road," as CDOT's traffic safety manager says.[g]

Do we know if road safety education campaigns like this help road safety outcomes? Nope. We don't know, and we don't really care to know. I admit that tying safety outcomes to such campaigns is tricky, but the bigger problem is that transportation agencies care more about looking like they're doing something than actually doing something about road safety.

The other problem becomes clear when we think about the implications of what this means for pedestrians who don't "make eye contact" with a driver before getting run over in the crosswalk. Does it imply that the crash was the pedestrian's fault, even if they were in a crosswalk, because they didn't first make eye contact? What about visually impaired pedestrians who cannot make eye contact before crossing the street?

My question is, why don't we just fix the real problem and better educate the traffic engineers?

That is exactly what I'm trying to do in this book.

---

[f] One example of Hank's "infomercial" that will only cost you five easy payments of $9.99 (plus $3 shipping and handling) can be found at www.youtube.com /watch?v=6JT6wrUMTug.

[g] Two years after this campaign ended, I saw this 2022 tweet by John Riecke (@ JCRiecke): "Just got yelled at by a motorist for not making eye contact while crossing the street in a crosswalk at a stop sign. Thanks, DOT, for giving drivers the idea that I have to bless them for following the law."

12   License to Drive

___

If public safety campaigns don't help, how else might we bestow *education*? In theory, it could come via driver education.

Driver education tends to be tied to driver's licensing programs, which date back to 1903 in Massachusetts. By the 1930s, about half of US states had them. By the 1950s, Wyoming was the lone holdout. Heck, we didn't even share driver records between states until 1961. As a 1949 *Traffic Quarterly* paper points out, "Driver-licensing is regarded primarily as a revenue-producing measure."[1] It isn't much different today.

Here is what I had to do in early 1990s Massachusetts. First, I watched 20 hours of 1950s-era videos in a dark room full of napping teenagers. I was then supposed to "observe" someone driving for a minimum of 20 hours, which the driving school signed off on without question. Last, I needed to drive with an instructor for 20 hours. This requirement was the only one that helped at all.[a] The rest of it was a joke. Think back to your own driver's education, and it probably wasn't that different.

So it's not all that surprising when the research shows driver's education to be ineffective. In the early days, most papers on this topic were "low-quality research" fueled by "overzealous attempts by the supporters of drivers education." The research is better now, but we are still in need of "an increase in the quality of driver education, resulting from high quality research and experimentation," as said in a 1962 *Traffic Quarterly* paper.[2]

Driver's education hasn't changed much either, in part because we've always treated a driver's license as a rite of passage.[b] A 1968

___

[a] As I write this, Massachusetts requires 30 hours of classroom instruction but only 6 hours of on-road observation and 12 hours of behind-the-wheel on-road instruction.

[b] "That thing in your wallet, that's no ordinary piece of paper. That is a driver's license. And it's not only a driver's license …, it is a license to live, a license to be free, to go wherever, whenever, and with whomever you choose."—Dean, played

Traffic Quarterly study surveyed 500 teenagers, and most saw the course only as a way to reduce insurance costs. This paper also found no difference in attitudes toward road safety before and after taking the course. But driving was consistently important to these teenagers, even more so than "music and attire." The authors were shocked to see that this was even the case for "both sexes."[3]

Teenagers have changed and now tend to prioritize their cell phone over driving. The driver's tests haven't changed. Most driving tests remain based on a bygone era, before power steering or automatic transmissions, where we force people to make three-point turns or come to a complete stop on a steep hill. Do we teach them how to handle a car that is skidding on ice? Nope.[c]

Colorado, like many other states, now outsources most of the actual license testing to private companies. So instead of the months-long appointment backlog—which I experienced when I was 16—you can take the test whenever you are willing to pay another 35 bucks. If you don't pass today, you can take the test tomorrow. Or the next day. Or the day after that. Are you a better driver on day 4 than you were on day 1? Probably not, but welcome to our streets!

This could come as early as 14 years, 3 months, in South Dakota. In New Jersey, you can't get a driver's license until you are 17. While most states now have graduated licensing programs with curfews and passenger restrictions for younger drivers, there is little chance that you will ever have to take another driver's test for the rest of your life.[d]

The safety data for young drivers is not good. According to the National Highway Traffic Safety Administration, young drivers in the United States are twice as likely to be in a fatal crash as compared to adult drivers.[4] That's why some countries keep pushing driver's licensure off until later and later. Drivers in some parts of Australia

---

by Corey Feldman, talking to Corey Haim's character in *License to Drive* (1988) while the "Battle Hymn of the Republic" plays in the background.

[c] I paid an extra $249 to have my oldest child, Rowan, take a defensive skills course. While not required for licensure, the course teaches students to control a car during a skid (or not skid in the first place) and forces them to drive while wearing drunk goggles. Here is a video: www.youtube.com/watch?v=j41v3VuOjOE.

[d] When Daniel LaRusso got his driver's license in the first *Karate Kid* (1984) movie, the only advice he got was Mr. Miyagi saying, "Just remember, license never replace eye, ear, and brain."

can't get a full driver's license until they are 22 years old. Before then, they face severe restrictions, including set maximum speeds that are enforceable even on roads with higher speed limits. Other countries, like the Netherlands, force older drivers to come back and take the test again.

This makes sense because the cars of today are not like the cars of yesterday. It takes me a good 5 minutes to figure out how to change the radio volume whenever I rent a car. And when airline pilots switch to a new aircraft, they take a four- to six-month technical course and simulator training. If they go back to their old aircraft, they take a refresher course. Will I ever need to take a refresher course on driving? Maybe if I accumulate enough demerit points, but if not, my Magic 8 Ball tells me that its sources say no.

The demerit points system started in the United States in the 1960s and is probably one reason we don't put much effort into driver's education and licensing. The thinking is that we can wait for the bad drivers to do bad things and sort them out later. Paul Hoffman lays this thinking out in his 1939 book: "Do we assume that people are criminals? No, we wait till they commit crimes. Similarly, we need not assume that people are unsafe drivers. We can wait till they have accidents."[5]

Rather than being on the receiving end of such "accidents," I'd prefer that we teach people how to drive safely and give them a transportation system that helps make sure they do. The other problem with the demerit points system isn't that it doesn't work; it's that we don't know if it works or not. There is not much research out there to link it to actual road safety outcomes.

The same can be said for vehicle registration policies, required nationally starting in 1967. A *Traffic Quarterly* paper from a half dozen years after that says that the "foremost goal is to identify the vehicle and owner rapidly" so as to prevent the second goal, the "registration of stolen vehicles."[6]

Safety seems to be third yet again. In this case, the third goal is to make sure vehicles are safe to drive. Yet many states do no vehicle safety inspections whatsoever. Colorado, for instance, focuses on emissions.[e] For those states that do periodically conduct a motor

---

[e] And does so by forcing dozens of drivers to idle in their cars for longer than anyone wants while waiting to get tested.

vehicle inspection, we still have little research to tell us if they work or not.[f] That didn't stop traffic engineers from suggesting a national motor vehicle inspection program, such as in a 1968 *Traffic Quarterly* paper. At least the author admits that the recommendation isn't based on much: "The author is aware of no published studies which prove any direct relationship between defects and accidents, much less between inspection and accidents."[7]

The fourth goal of vehicle registration is to keep so-called deviants off the road. But depending on the state, upward of 90 percent of driver's license suspensions come from failing to pay a fine as opposed to what a person does on the road.[8]

On a related note, I came across a fun 1921 *Saturday Evening Post* editorial suggesting a "scarlet license plate" that would use "sharply contrasting designs for the plates of those convicted of serious traffic offenses." And if drivers don't like it, the author didn't seem to care: "If the humiliation proved intolerable, the motorist would always have the option of leaving his car in the garage."[9] And this was in response to a fraction of the traffic deaths that we now see.

We don't use license plates as a scarlet letter, but we do use them for at least one safety-related goal: to force people to use retroreflective license plates. This speaks to the fifth goal of vehicle registration: reducing rear-end crashes at night.[g]

It would help the safety cause if we made this data (vehicle registration and anonymous driver licensure, infraction, and crash records) more readily available to safety researchers.

But safety isn't what it's for.

---

[f] In one of the few good studies on the topic, Peter Christensen and Rune Elvik looked at the extensive motor vehicle inspection program started by Norway in 1995 and found that it didn't help improve safety outcomes. Christensen and Elvik, "Effects on Accidents of Periodic Motor Vehicle Inspection in Norway," *Accident Analysis and Prevention* 39, no. 1 (2007).

[g] You may be interested to know that the research didn't find much of a difference in rear-end crashes with or without reflective license plates. However, there was a drop in the number of drivers hitting parked cars, especially in rural areas. For an overview of the half dozen papers looking into crashes and retro-reflective license plates, see Paul Olson, "Review of Research Evidence Bearing on the Desirability of Using Retroreflective License Plates in Michigan," Highway Safety Research Institute (1977).

13 Good Cop, Bad Cop

___

Only a few months before we enacted the national vehicle registration standard, the FBI initiated the National Crime Information Center, also known as the NCIC.

FBI director J. Edgar Hoover, in a 1971 *Traffic Quarterly* paper, describes the NCIC as "an electronic clearinghouse of criminal justice information (mug shots, crime records, etc.) that can be tapped into by police officers in squad cars or by police agencies nationwide." He then brags that the system has already been used to catch murderers, car thieves, and a fugitive with an illegal alligator.[a] Hoover mentions the first-ever NCIC system "hit" happened when "a New York City police officer—suspicious of a parked car—radioed in a request for an NCIC search of the license plate. Within 90 seconds, he was informed that the car had been stolen a month earlier in Boston."[1]

More than 50 years later, White police officers in Aurora, Colorado, used this system to force four Black girls—ages 6, 12, 14, and 17—to the ground, with the 12- and 17-year-olds handcuffed. Why? Because the license plate on their Colorado SUV had the same numbers as that of a stolen Montana motorcycle. Unfortunately, our automated license plate readers are not sophisticated enough to recognize the difference between states or vehicle types. Neither were the officers.

Other than to point out the absurdity of putting both police officers and the general public—particularly minorities—at risk with traffic stops that could be handled by cameras, I'm not looking for this book to be a referendum on police or even cameras.[b] Nor do I

___

[a] In the early days, the FBI also trained local police officers how to conduct traffic enforcement.

[b] Traffic enforcement cameras were first used in 1983 in Victoria, Australia. They weren't used in the United States until 1993 in New York City. On a related note, John Jacob Astor IV published his book, *A Journey in Other Worlds: A Romance of the Future*, in 1894 about the year 2000. In it, he laid out his vision for future thoroughfares, saying that speed limits would be enforced by speed cameras. And yes,

even want to spend much time talking about police *enforcement*. The average citation takes up between 23.4 and 27.2 minutes of an officer's time, not including the time lost in the court system, and hasn't been shown to improve road safety outcomes.[2] It also puts drivers—especially those who are not White—in danger. So, I'm going to refrain from wasting any more time on in-person police enforcement of traffic laws than I have to.

It is worth pointing out, however, that education and enforcement are related. The fifth *Traffic Quarterly* paper ever published calls traffic enforcement by police officers "the greatest educational force."[3] Yet a 1959 *Traffic Quarterly* paper reminds us that in most crashes, "enforcement is almost helpless. Prevention lies in the judgment and know-how of the individual driver."[4]

It is also worth pointing out that enforcement—akin to education—is a crutch that traffic engineers use to keep the focus off of them.[c] The bottom line? If we need to rely entirely on enforcement to police our streets, we've done a bad job of designing our streets.

Traffic engineers never consider enforcement like this, as an admission of failure. Instead, traffic engineers remind people that our streets would be safer if only we had more enforcement.

Better yet, we like to shift the enforcement emphasis away from drivers and onto other road users. A common refrain we hear today is that drunk walking is just as dangerous as driving, if not more. But we never ask for whom.

In other words, who is drunk walking dangerous for? Answer: the drunk walker.

Who is the drunk driver dangerous for? Answer: everybody.

There is a massive ethical difference between the two. But traffic engineers don't differentiate.

In 1958, the AAA Foundation for Traffic Safety released what became an influential report[d] focused on the road safety of pedestrians. One of the first things they say is that "drinking and walking in traffic compares in deadliness to drinking and driving." So what

---

we still have too many camera enforcement programs focused more on increasing revenue than safety.

[c] "Pay no attention to that man behind the curtain."—The Great and Powerful Oz

[d] Influential is relative, but in this case, I only say so because of just how often I have seen this report cited in academic papers.

should we do about it? In the same 1958 report, AAA says that "it is time that we become concerned with pedestrian violations and unwise walking practices" and then highlights all the progress on this issue in cities like Detroit, which arrested 19,765 pedestrians for crossing against the signal but only 8,662 drivers for violating the pedestrian's right-of-way. The report noted that San Francisco arrested 165 pedestrians for crossing between intersections as compared to 7,304 drivers arrested for violating the pedestrian's right-of-way. But don't let the numbers fool you; San Francisco also arrested 32,968 pedestrians for public intoxication.[5]

Yes, they are talking about arrests, not citations. And yes, this was in 1956 alone.

Such thinking wasn't atypical during this era. A 1967 *Traffic Quarterly* paper advocated for the citation and arrest of "all people every time they violate the law, no matter what the circumstances" with the goal of "creating a deterrent to unlawful behavior by all potential violators."[6] The author then argues for more undercover cops doing traffic enforcement akin to what we would assign to a murder investigation. An earlier *Traffic Quarterly* paper suggests even more, because "for every person murdered, three die needlessly in traffic accidents" (and that "for every person inconvenienced by criminal acts, hundreds are hindered and delayed by traffic congestion").[7] In the 1930s, Paul Hoffman urges "better enforcement of legal restrictions upon pedestrians" and for "cities to arrest and convict pedestrians."[8]

Midcentury Portland, Oregon, also decided that the best way to "protect its pedestrians" was to arrest 15,196 of them for crossing against the signal and another 9,923 for crossing between intersections. In a 1952 *Traffic Quarterly* paper, the mayor of Portland says that starting in early 1945, "police officers began to make straight arrests for pedestrian violations." What do they mean by a violation? They mean everything because "there is no such thing as a 'minor' traffic violation." But guess what? The mayor then tells us that "the city of Portland has nearly 100 percent pedestrian observance of traffic signals."[9]

What happened to traffic fatalities in Portland? Between 1944 and 1945, they *increased* by more than 50 percent, from 46 deaths to 70.

As for drivers? Portland arrested 857 drivers. Given the whole "there is no such thing as a 'minor' traffic violation," how is that possible?

On one hand, we can't possibly pull over all the drivers who break the law. Why? Not only don't we have time to do so, but as a 1957 *Traffic Quarterly* paper argues, it would be really bad for the economy because "if licenses were suspended on any large scale, as so many demand, the economy would be severely injured, since that economy is so dependent on the motor transport."[10]

On the other hand, we could just get rid of things drivers don't pay attention to anyway—like crosswalks. Here is that 1952 *Traffic Quarterly* paper again: "Portland traffic officials have taken the view that the painting of crosswalks has disadvantages as well as advantages."[11] Why they would say this? The answer is that one "disadvantage of marking crosswalks is that the pedestrian is inclined to place too much confidence in the painted markings, whereas the motorist is often inclined to ignore them."[12]

This is interesting. After all, they just bragged about getting 100 percent pedestrian compliance by arresting all of them, and as the *99% Invisible* podcast says, "the whole point of a marked intersection is that it shows you: this is where it's safe to cross."[13] Yet for whatever reason—perhaps for the sake of the economy—they didn't enforce driver compliance in the same way, which resulted in unsafe crosswalks that "the pedestrian is inclined to place too much confidence in." Their solution? Remove the crosswalks.

Who cares about midcentury Portland, you ask? While we don't (usually) arrest pedestrians for minor infractions like we used to, we still remove crosswalks because drivers don't stop as often as they should. In Honolulu, for instance, 45 crosswalks were eliminated because they were "deemed unsafe" for pedestrians. As the city traffic engineer told the city council in 2019, "We feel we cannot keep pedestrians safe just by the markings."

And if they aren't removed? "The city could be held legally liable if a pedestrian is seriously injured or killed in the painted stripes at those locations based on the latest Federal Highway Administration standards."

Where should pedestrians cross? Honolulu says they "can still legally cross at those spots," but without a marked crosswalk.[14]

OK, but who is to blame if a pedestrian is seriously injured or killed "at those spots" where there is no crosswalk but they can still legally cross? Back when we had a crosswalk, it might've been the drivers. Now, it won't be the drivers unless they are drunk, texting, or outlandishly speeding. Nor the police officers for failing to enforce the rules. Nor the traffic engineers for failing to engineer a safe place to cross.

Unfortunately, the pedestrian is to blame for trying to cross a street that the three E's couldn't figure out how to make safe.

The overall literature isn't as kind. In the words of a 1999 *Transportation Quarterly* paper, "Public education and enforcement campaigns are often used to promote pedestrian safety but generally have not produced tangible and long-lasting benefits."[15]

Nevertheless, in an effort to reduce the millions of crashes caused by so-called human error, we spend millions of dollars trying to influence people's behavior through better *education* or increased *enforcement*. Putting our eggs in the education and enforcement baskets is much easier than reconsidering the design of our existing streets. It's also easier to continue conducting business as usual—with the same designs we've always used—than to call into question the design of our entire transportation system.

14   Can We Fix It?

___

There is a 1964 *Traffic Quarterly* paper written by the head of the "society of police chiefs," whatever that is. When it comes to traffic *enforcement*, he rails against the use of terms like *speed trap* or *entrapment* because it's not his fault people break the law.[a] He writes that "both these terms imply that by some inducement or action the police have caused or enticed a person to violate traffic laws. The police deplore the use of these or similar terms. The police do not entice people to violate traffic laws."[1] Traffic engineers might be inclined to tell you the same thing. Once we've built the street, how fast people drive is out of our hands. If people drive too fast, they need more *education*. If that doesn't work, we need more *enforcement*.

The truth is that we need better *engineering*.

When I was in college at the University of Virginia, I was a civil engineering major. Within that major, we had to choose a concentration. The three choices were structural engineering, environmental engineering, and transportation engineering.[b] Thinking that I may want to be an architect, I chose structural.

Fundamental to structural engineering is the *factor of safety* concept. In other words, we try to design things like bridges and buildings to handle something on the order of three times the design load. So if we need to size a beam, we do all the calculations and come up with a minimum size for that beam. Anything bigger may

---

[a] In the unanimous 1996 Supreme Court ruling *Whren v. United States* (517 US 806), Justice Anthony Scalia finds that "any traffic offense committed by a driver was a legitimate legal basis for a stop" and that search and seizure with "reasonable suspicion" does not violate Fourth Amendment rights. However, it isn't hard to argue that most in-person traffic law enforcement is arbitrary and could therefore be considered unreasonable.

[b] I'm currently a professor at the University of Colorado Denver, and we have six different concentrations within the Department of Civil Engineering: transportation, structural, hydrologic/environmental/sustainability, geotechnical, construction, and geomatics/GIS. We don't require the undergraduates to declare a concentration, however.

cost more but is considered better. Why? Because it would further increase the factor of safety. Environmental engineers do something similar when designing, for instance, a culvert. They figure out how much water it may need to hold and go bigger so as to give it a factory of safety.

One of the underlying problems in traffic engineering is that we've taken on this factor of safety mentality. We calculate that two lanes will accommodate the expected traffic but assume that four lanes would be **better** because they give us a bigger factor of safety. We think we need 20 parking spots but have room and money for 30, so why not? It'll give us a nice factor of safety.[c]

It's not just in the mind-set; it's also in the words we use. Our guidebooks tell us that the travel lane has a recommended minimum width of 11 feet or that the shoulder has a recommended minimum width of 6 feet. The word *minimum* implies the lowest threshold of adequacy. Who wants to be barely adequate? Anything bigger is considered better.

That's exactly how I interpreted the word *minimum* in our guidelines when I was a young engineer. I specifically remember designing a parking lot, and given the layout, I needed a minimum of 22 feet between the aisles of parked cars. We had the room for 24 feet, and it wasn't going to cost much more, so that is what went into the plans. I figured I had given the world a little bit of a factor of safety.

I also gave the world some unnecessary asphalt, perhaps more of a heat island effect, and less room for other more useful uses.

In terms of safety, I provided more room for drivers to maneuver, and when drivers have more room to maneuver, drivers go faster, and when drivers go faster, you tend to find more severe crashes.

Sorry about that.

This example isn't an egregious one—particularly given how auto manufacturers have turned the Canyonero[d] from a *Simpsons* joke into the typical SUV—but it speaks to how transportation engineers go about our jobs with a structural engineering mentality.

I also wasn't wrong to interpret the word *minimum* like that. The

---

[c] As you can see, many of the transportation-related factors of safety aren't about protecting humans from injury; they protect us from capacity problems.

[d] "Can you name the truck with four wheel drive, smells like a steak and seats thirty-five? … 12 yards long, 2 lanes wide, 65 tons of American Pride!"

guidelines say as much: "Above-minimum design values should be used, where practical."[2]

I'll be talking more about this later in parts 4 and 5, but for now, I want to introduce one of the most insidious offenders of this mentality: ***design speed***.

What is design speed? It represents the theoretical safe limit that we are designing the street to, akin to a design load for a bridge or building. For most of the years since the design speed concept originated in the mid-1930s, we defined it as the "maximum safe speed" for a given section of roadway.

The idea behind design speed is simple enough. Let's say I'm trying to design a major street where I'm expecting the speed limit to be 30 or 35 mph. As a designer, I'm aware that some people will drive faster than that. My guidelines[3] tell me that I need to make sure those people are safe too: "The selected design speed should fit the travel desires and habits of nearly all drivers expected to use a particular facility."[e] That is good to know, but what does it mean for design speed? What the guidelines say is that other than on local streets, "every effort should be made to use as high a design speed as practical."[4]

So after setting the design speed as high as possible, I then design everything to make sure people can safely drive at that speed.

When structural engineers determine their design load, they do so with the expectation that if all goes well, they'll never come close to reaching it, let alone exceeding it. Design speed is a different animal because people are different animals.

People crossing a bridge have no idea what the structural engineers used for their design load. Drivers may not know the exact design speed either, but they adapt to meet it.[f] When the design speed is higher, drivers drive faster. When drivers drive faster,

---

[e] Here's the full quote: "Except for local streets where speed controls are frequently included intentionally, every effort should be made to use as high a design speed as practical to attain a desired degree of safety, mobility, and efficiency with the constraints of environmental quality, economics, aesthetics, and social or political impacts."

[f] Krammes's 2000 paper estimated the design speed for 138 highways across five states and found that most drivers exceed the design speed. Raymond Krammes, "Design Speed and Operating Speed in Rural Highway Alignment Design," *Transportation Research Board* 1701, no. 1.

we lose more lives on our streets. Killed, albeit indirectly, by a traffic engineer.

The problem is that too many of our engineering solutions treat people like inanimate carbon rods. For example, we've long treated traffic flow like water in a pipe. As a 1973 *Traffic Quarterly* paper about urban traffic problems says, "The build-up of traffic flows is analogous to the flash-flooding of water courses—rainfall intensity, duration of downpour, and length, grade, and cross-section of tributaries all have their traffic counterparts."[5]

We don't have much control over how much rainfall nature delivers, but we do have control over the design of the pipe. Building a bigger pipe doesn't entice nature to rain more. But if we expand a street from two lanes to four lanes and the parking lot from 20 to 30 spaces, walking to the store isn't as easy as it used to be. So maybe I'll drive instead of walk. Other people may agree. Traffic engineering impacts human behavior, which in turn impacts road safety.

The more enlightened traffic engineers now realize that traffic is more like a gas. Why? Because people change their behavior based on the transportation system we put in front of them. When you say it out loud, it seems obvious, right? At first, I figured this was a good example of how dumb we were. Then I found an early *Traffic Quarterly* paper complaining about it: "Many attempts to explain the movement of traffic ignore the human element, and treat traffic as something very much like hydraulics." In this 1952 paper on the theory of traffic control, Theodore Caplow says that "those who think of traffic … as a liquid moving through connecting channels are likely to see the whole problem as a need for more expressways and grade elevations."[6] Another *Traffic Quarterly* paper from the early 1970s even alludes to traffic as a gaseous state of matter, as if we've always known so, saying, "Yet it must be realized that … traffic expands to fill the space available for its circulation."[7]

At first, it makes perfect sense to build a factor of safety into our transportation system to keep people safe. But if that factor of safety tempts people to drive faster and more dangerously, a higher design speed isn't the factor of safety we signed up for.

When it comes to road safety, many of our engineering solutions aren't what we signed up for.

15   Fast Times

Given how we engineer our roadways to entice unwanted speeds, it doesn't help the safety cause that cars have speedometers with numbers approaching 140 mph. I've even seen a Corvette with a speedometer that tops out at 220 mph.

In 2018, there was a fatal crash in Fort Lauderdale, Florida, when an 18-year-old drove his Tesla off the road while going 116 mph in a 30 mph zone. He died, along with one of his passengers. I read about this crash in a National Transportation Safety Board (NTSB) brief.[a] Although NTSB makes it very clear that it "do[es] not assume fault or blame," the board lists the **probable cause** as "excessive speed."[1] If you look up this crash in our national fatal crash database, it'll be grouped with all the other human error crashes with speeding listed as the cause.[b]

Was it?

Just two months before the Tesla crash in Fort Lauderdale, the same driver got a speeding ticket for going 112 mph in a 50 mph zone. His father got Tesla to place the vehicle in "loaner" mode, which gives the car a maximum speed of 85 mph. A month later, the 18-year-old talked the dealership into turning that restriction off. At the time of the crash, according to Tesla's data, he had the accelerator pedal pressed down to only 74 percent of the max and was still going well over 100 mph.[c]

---

[a] The National Transportation Safety Board (NTSB) is a federal transportation agency that investigates transportation-related crashes and publishes reports about what it found and the *probable cause*. You've probably heard about the NTSB after a plane or train crash, but it also investigates traffic crashes from time to time. As you'll see throughout this book, these NTSB reports turned out to be another useful resource (from time to time).

[b] FARS (the Fatality Analysis Reporting System) doesn't actually say that speeding *caused* the crash; it will say that the *critical precrash event* is that the vehicle (not the driver) lost control due to traveling too fast for conditions.

[c] In this case, I also blame the traffic engineer for tossing a 30 mph sign on a road (Seabreeze Boulevard, also known as Route A1A) that looks as wide and straight as

Why do we allow this? Doesn't the "e" in **engineering** also refer to vehicle engineering?

When I was a teenager, I paid $800 for a 1983 Chrysler LeBaron that could barely scratch 75 mph on a good day. I was lucky because, like millions of other dumb teenagers, I drove as fast as I could pretty much everywhere I went. It would not have been good for teenage me to have a car that topped 100 mph.

Clearly, we have the ability to limit vehicle speeds. We just choose not to use it.

In reality, we've always had it. *Scientific American* wrote about automatic speed governors as far back 1907.[2] A few years later, New York State—fed up with pedestrian deaths—introduced a bill that would have forced car manufacturers to limit speeds to 20 mph. It failed. Peter Norton's fantastic book *Fighting Traffic* covers a ballot referendum in Cincinnati in 1923.[3] The vote was going to force car manufacturers to equip cars with speed governors, or restrictors, this time setting the max speed to 25 mph. When Peter tells this story in his presentations, he first shows the tiny "vote yes" advertisement that reminds us about safety and warns us of death. He then shows a giant, graphic-filled "vote no" ad that compares speed governors to the Great Wall of China.[d] This campaign, funded by the auto industry, also hired pretty girls to bring men to the polls to vote against the measure. The measure failed.

In 1942, only 30 percent of cars could exceed 40 mph. By 1957, almost 90 percent could do so.[4] Most cars today can triple that and do so with ease.

There have been scattered articles calling for speed governors—such as a 1968 *Atlantic Monthly* piece called "Automobile Selfishness"—but few real attempts. Even though speed governors would save lives, they seem un-American.

But when cities were first inundated with electric scooter rentals in 2018 or so, people treated them like they did cars in the early 1900s. Many cities responded with strict oversight. Some responded with speed governor restrictions. Denver wanted the electric scooters to be equipped with "geofencing" technology, limiting electric scooters to

---

an airplane runway. It looks perfect for driving and driving fast. In fact, the driver didn't lose control until the road curves left and changes to a 25 mph zone.

[d] The ad warns us that "China is the most Backward of All Nations."

3 mph in certain areas. Washington, DC, did something similar on the National Mall. Few complained about scooter speed governors. For whatever reason, this was not considered overly un-American.

We could use geofencing to limit car speeds within cities, or certain parts of cities, or even by time of day or day of week such as when schools or bars let out. Doing so would save lives. We choose not to.

Europe isn't exactly limiting speeds, yet. But as of 2022, Europe requires that all new cars sold use GPS and street sign detection to determine if a driver is speeding and if so, warn the driver. Starting in 2021, Volvo limits top speeds on new cars to 112 mph. It's not revolutionary—nor that helpful to pedestrians and bicyclists—but it's better than nothing.[e]

Limiting how fast a car can go pisses some folks off to the point where they are trying to find ways to finagle around this 112 mph limitation. I'll talk a lot more about how engineers handle speed later, but for whatever reason, most drivers don't make the connection between speed and bad safety outcomes. As a 1959 *Traffic Quarterly* paper tells us, many think the opposite: "Apparently, many drivers think that excessive power is a safety feature."[5]

Texas State Highway 130—running from Austin to Seguin—has an 85 mph speed limit, one of the fastest in the United States. Seat belts and airbags are not designed for such speeds.[f] But when questioned about the highway's poor safety record, a traffic engineer from the Texas Department of Transportation said that ***personal responsibility*** is the real problem.

Another *Traffic Quarterly* paper estimated that only 43 percent of those injured in car crashes are the ones driving.[6] What should we tell the other 57 percent of injured people about personal responsibility?

---

[e] In 2005, Volvo set a "Vision 2020" goal of Volvo cars not killing or seriously injuring anyone—even pedestrians or bicyclists—by 2020. They haven't achieved this goal, but they've added many safety-related features over the years since, including external pedestrian airbags in the 2013 Volvo V40 and Volvo's City Safety system that uses radar and cameras to try to detect pedestrians and bicyclists. There remain many caveats with technology like this—including the inability to detect kids well and the prospect of drivers over-relying on potentially unreliable features—but at least Volvo is trying.

[f] The National Highway Traffic Safety Administration runs its crash tests at 35 mph, and it's hard to figure out a specific speed at which current seat belt and airbag designs may fail.

But rather than fixing our problems and misconceptions, it's easier to chalk speeding up to an issue of personal responsibility. In other words, if you speed and get hurt, it's your fault.[g]

That isn't the whole story.

---

[g] In an episode of *The Ranch* on Netflix, an old Ford with no seat belts or airbags is found in the woods. Luke (Dax Shepard's character) jokes that drivers were much better when a fender bender would kill your whole family. In many situations, he might be right.

16   Safety for Whom?

—

The human body can only withstand so much force. Fundamentally, we can reduce the force that we need to dissipate via lower vehicle speeds ... or with lighter cars.[a]

So if we want to prioritize road safety, the laws of physics say we should make cars slower and lighter. Car manufacturers—even Volvo—don't seem to be heading in that direction.[b]

Why is that?

The question of safety is always trickier than it seems. Heavier cars tend to be larger, which can give them more space for larger crumple zones to absorb the force of the crash over a slightly longer time and greater distance. Heavier cars also tend to force the force absorption onto the smaller cars with which they crash.[c]

The great safety researcher Leonard Evans delved into vehicle weight and safety, finding that—given the same crash[d]—heavier vehicles tend to be safer for only those inside that heavier vehicle.[1]

After the oil embargo in the 1970s, safety researchers fretted that fuel-efficient "minicars" would "capture 40% of the automobile market by the year 2000" and devastate the US road safety record[e] (whereas by 1990, they predicted it would approach 70,000 fatalities

---

[a] Newton's second law of motion tells us that force = mass × acceleration. So speed matters, but when a crash happens, the real problem is the near instantaneous deceleration of a fast-moving car from very fast miles per hour to zero miles per hour.

[b] Electric cars weigh about 20 percent more than the conventional version of the same car.

[c] This tendency suggests that large but light cars could be safer.

[d] We also need to keep in mind that "given the same crash" is not always a fair comparison. Bigger, heavier vehicles tend to have about double the risk of a rollover crash than smaller cars, and other than in a head-on collision, they are one of the most dangerous crash types.

[e] At the time, there wasn't much research to support this. Even as late as 1993, researchers were still telling us that other than "for single vehicle run-off-road crashes, no association between car weight and driver injury was apparent." Errol

each year) while bankrupting US auto manufacturers due to liability claims.[2] Reports like this habitually fail to point out that heavier vehicles are *less safe for everybody else* on the street. That includes pedestrians or bicyclists as well as those in other cars.[f]

So, the question is, safe for whom?

If you are a car manufacturer, the answer to this question will lean toward those buying your cars. Does the auto industry care about the safety of people in other cars or those who are walking or biking? Sure it does, but not as much as the safety of those paying the bills. An auto dealership could mention that this SUV more than doubles their risk of killing a pedestrian as compared to that sedan.[3] Or that if you hit people in a car with this pickup truck, you nearly triple the risk of their death than if you were driving that sedan.[4] But it's tougher to "sell" the safety of others over the safety of those doing the shopping. And even if a pedestrian dies, the car manufacturer can blame the pedestrian or driver for the so-called human error that led to the crash.

The bottom line? Vehicle *engineering* safety focuses on those buying the vehicles rather than on the greater good.

When it comes to killing pedestrians and bicyclists, it isn't just the weight of the vehicle that matters. If I'm walking across the street and get hit, there is a difference between getting hit in the legs by a Honda Civic and rolling onto the hood and into the windshield versus getting hit in the chest by a Chevy Suburban and my head slamming into the pavement before getting run over. It's most likely the difference between a leg injury and a head injury. It could also be the difference between life and death.

We could design SUVs to protect pedestrians. Lower bumpers. Breakaway mirrors. Softer and more flexible materials. Instead, as a 2010 meta-analysis paper suggests, "the bumper systems of light trucks are basically designed to prevent or limit physical damage to

---

Noel, Bala Akundi, and Aladdin Barkawi, "Safety and Operational Impacts of Minicar," *Transportation Quarterly* 47, no. 3 (1993).

[f] A really good 2019 paper by Alena Høye said that "the increasing weight of more recent cars may contribute to improved safety for those who drive heavier cars, but overall the effect of increasing weight is probably small or even negative because heavier vehicles impose greater risk on other car drivers, pedestrians, and cyclists." Alena Høye, "Vehicle Registration Year, Age, and Weight—Untangling the Effects on Crash Risk," *Accident Analysis and Prevention* 123 (2019).

the 'expensive' components of the vehicle, and thereby reduce insurance costs."[5]

Part of the problem has to do with liability (and I'll talk more about liability in part 8). If I kill a pedestrian with a bubble-wrapped car, I'm no more or less liable than if I kill them with a three-ton Ford F-250 outfitted with a steel bull bar.[g] We know the latter is far more dangerous to pedestrians.[h] We just don't advertise or legislate anything close to a pedestrian safety rating. In fact, car owners get very little safety-related information at all. A 1999 *Transportation Quarterly* paper points this out: "For example, if interested in a Chevrolet Camaro, it is unlikely that consumers will be informed that this particular model has a driver death rate higher than any other model."[6] Of course this statistic is influenced by the type of person (and driver) likely to buy a Chevy Camaro. Still, the Camaro's high death rate was something few people knew about (at least it was back in the 1990s, when a Chevy Camaro was the coolest[i]).

I don't want to spend a lot of time on crash testing—or even on car design. But despite a 1971 *Traffic Quarterly* paper bragging about how our crash tests have "duplicated nearly every conceivable type of highway crash situation," the US government didn't get into the crash testing business until 1978, first requiring it on model year 1980 cars.[7] Even then, it was only frontal crash testing, and the Transportation Research Board made it clear that "frontal crash testing is an inadequate representation of real-world crash configurations, as size and weight problems exist and the current tests required for compliance represent only a small fraction of real-world crashes."[8] Yet for the next 18 years, we only conducted front crash tests at 35 mph.[j] Not until 1996 did we start testing side impact protection. At

---

[g] Things have shifted in Europe, where car manufacturers agreed to stop installing bull bars on new cars in 2002. Aftermarket bull bars were then banned in 2005. While existing vehicles with bull bars remain legal in Europe, drivers could be subject to proposed manslaughter rules.

[h] There is even research from the early 1980s (such as the 1980 Australian report from Chiam and Tomas) showing how dangerous bull bars can be for pedestrians.

[i] Although I prefer the T-top version of the Pontiac Firebird Trans Am that *Billy Madison* (1995) pulls up to Knibb High School in.

[j] The Insurance Institute for Highway Safety, a nongovernment agency representing the interests of the insurance industry, also does some crash testing. They use what is called the offset barrier crash test, where each vehicle's front end hits a deformable

that time, "current automobiles provide impact protection of 30 to 35 mph in a barrier collision or 60 to 70 mph in an impact with a stationary vehicle of the same weight. Side-impact protection is even worse, protecting only up to 12 to 15 mph."[9]

In 2010, we started measuring side impact at more reasonable speeds but still do so with less weight than that of a typical sedan.[k] We also don't physically test things like rollovers, but instead calculate a "crash avoidance rating" based on vehicle width and center of gravity. There is one other crash test where we pull a car at an angle into a pole at 20 mph, but that is pretty much the crux of our so-called five-star safety ratings program.

We also don't look at the safety of anybody other than the crash test dummies in the car. Not the occupants of other cars. Not pedestrians. Not bicyclists. And the legal system and government agencies don't give the auto industry much reason to do so.

---

barrier at 40 mph. This crash type represents about 0.04 percent of real-world crashes (although it could increase if we'd use more bollards).

[k] In our current side barrier crash test, a 3,015-pound barrier slams into the side of a stationary vehicle at 38.5 mph. According to recent *Edmunds* data, the average sedan weighs 3,351 pounds. What about SUVs and pickup trucks? Both average more than 4,700 pounds.

17 Full of Hot Air

——

In the 1960s, Ralph Nader told us that the auto industry makes a lot of money when we crash. It also bears few of the costs.

Over the years, car manufacturers have made this fact abundantly clear. Consider their fights against life-saving technologies such as airbags. Neil Arason's safety book mentions a 1972 *Geneva Times* story about the "airbag crisis" confronting the auto industry. In the article, the CEOs joked that they'll next be asked to put popcorn in the dashboard. Henry Ford II called airbags "a lot of baloney," and they all jested that seat belts were the best they could do, short of "building a womb with a view."[1] The funny thing is that the same old boys club had made fun of seat belts in similar ways a decade earlier.

The auto industry characteristically digs its heels in against safety features because they cost too much. For decades, the auto industry told us that airbags were "too costly and only marginally effective," trying to distract us with other so-called safety features.[2] The US government first required airbags in 1970, giving car manufacturers until 1974 to figure it out. Instead of complying, several sued. Others lobbied directly to President Richard Nixon.

They got their way, and instead of airbags, we got the seat belt interlock system. This forced you to buckle up before you could start the car. It was less expensive than airbags. It also didn't work well. The requirement was dropped in 1974.

Airbags were put on the back burner until 1977 when the US government said that either airbags or automatic seat belts needed to be phased in, beginning in 1981, and be on every new car by 1984. Lawsuits followed, along with lobbying to President Ronald Reagan. In a 1981 statement submitted to Congress, General Motors says that when it comes to passive restraints like airbags, we need to "eliminate [the] requirement, or at a minimum, delay and reverse order." Why? Well, GM tells us that airbags cost about $1,000 per car, and, the company says, "We believe that consumer resistance will be so great ... that the safety benefits will not justify the costs of this controversial regulation."[3]

What's worse? GM also tells us that Japan is going to have a competitive advantage, so "there is an immediate need to avoid the sharp economic impediment that these requirements ... would place on the domestic car market."[4]

The car manufacturers got their way, again, and the airbag requirement went away—that is, until a 1983 Supreme Court case. In the opinion of the US Supreme Court, "For nearly a decade, the automobile industry waged the regulatory equivalent of war against the airbag and lost—the inflatable restraint was proved sufficiently effective. Now the automobile industry has decided to employ a seatbelt system which will not meet the safety objectives.... This hardly constitutes cause to revoke the Standard itself. Indeed, the Act was necessary because the industry was not sufficiently responsive to safety concerns."[5]

Sure, this opinion gave the government the wherewithal to reissue the airbag requirement in 1984. But it was another five years before we saw mandatory driver's side airbags in passenger cars and four more years until the passenger's side got them. They weren't mandatory in light trucks until 1998.

In the decades between when the US government first tried to require airbags and when they successfully fought off the auto industry's complaints, we unnecessarily lost tens of thousands of lives on our streets. Since then, the federal government tells us that airbags save close to 2,800 lives each year.[6]

That's pretty good, right?

Well, let's say it still costs about $1,000 for a manufacturer to add airbags. There are about 17 million new cars sold in the United States each year, which adds up to $17 billion spent annually on airbags. At that rate, it costs us about $6 million per life saved. Is it worth it?

These are rough numbers, but that is essentially the calculation car manufacturers make when deciding whether or not to add a safety feature.[a]

---

[a] To be more specific, they probably ask themselves whether they can recoup the money spent on the safety feature by selling more cars for more money. In some countries, only rich people have airbags.

The Ford Pinto represents an infamous example of this.[b] In that case, the engineers explicitly decided that the cost of a better fuel tank wasn't worth a certain number of fiery deaths.

Likewise, we know that side guards on large trucks help protect cars from running up under them as well as pedestrians and bicyclists from being run over. They only cost a few hundred bucks.[c] How many lives do we need to save to make them worth the money? Sadly, the answer is way more than one.

We also know that as SUVs and pickup trucks grow, so does the blind spot in front of them. WTHR Channel 13 in Indianapolis tested this by sitting a dozen kids directly in front of a Cadillac SUV. The driver had no idea the kids were there until the 13th one sat down. Even then, the driver was only able to see the top of that kid's head.[7]

But front cameras aren't free, and we don't run over enough kids to make them a worthwhile investment.

As someone who walks and bikes around town, I'd certainly appreciate brake lights on the *front* of cars and trucks. That way, I'd know they are slowing down before I start crossing the street. Again, those cost money (and annoying people like me isn't enough of an incentive).

A 1999 *Transportation Quarterly* paper lays out the math as to why seat belts on school buses aren't worth the money.[8] The author first tells us that, on average, in the United States only 11 kids die each year on school buses. That isn't so bad, right? He then says that even if we put seat belts on all school buses, it's not like kids will actually use them. So he applies an arbitrary noncompliance assumption—that 80 percent of kids won't use seat belts—to the costs and calculates that it would take $40 million to save one kid. Unfortunately, that isn't a good enough return on investment.[d]

I'm not bringing all this up to implicate the auto industry. Traffic engineers make the same cost-benefit calculations when trying to decide whether a "safer" design is worth it.

In other words, how much is a life worth?

---

[b] "In-famous is when you are more than famous. This man, El Guapo, is not just famous, he's IN-famous."—Ned Nederlander (¡*Three Amigos!*).

[c] Europe now requires them.

[d] As of now, only eight states require seat belts on new school buses.

18 How Much Is Your Life Worth?

When traffic engineers commit to improving safety, what do they mean? A 1981 *Traffic Quarterly* paper—written by a man named James Taylor—explains the basic dilemma: "How much, in public funds, should be spent to obtain a given reduction in highway fatalities, injuries, and property damage accidents?"[1]

In the early 1970s, President Richard Nixon's auto industry task force set the number at what would be $885,000 in today's dollars. By 1975, the US Department of Transportation (USDOT) estimated a life as being worth $1.4 million in today's dollars. When the USDOT value of life reached $2.5 million in 1998, it suddenly made sense to "install bars at the rear of trucks to prevent passenger vehicles from sliding underneath them in a collision."[2]

People today seem worth a lot more because the USDOT now uses $11.8 million.[a] Interestingly, every federal agency seems to value a life differently.[b]

The reason I cringe when seeing numbers like these used in road safety calculations isn't due to a sense of moral outrage. Truth be told, these numbers exist *because* they allow traffic engineers to resolve moral and ethical dilemmas with math.

Instead, my main issue is that these calculations implicitly assume that we can estimate lives saved with any with any sort of accuracy. When it comes to road safety, we know much less than you'd think.

The other issue I have comes in seeing how we've mangled these calculations over the years.

---

[a] A 2021 paper by Justin Tyndall estimated that replacing SUVs with cars would have saved 1,100 pedestrians between 2000 and 2019. That equates to more than $10 billion in lives lost, which, unfortunately, is only a fraction of the size of the American SUV market. Tyndall also found no evidence that SUVs improved car occupant safety. Justin Tyndall, "Pedestrian Deaths and Large Vehicles," *Economics of Transportation* 26 (2021).

[b] The Environmental Protection Agency, for instance, uses $7.4 million. Either way, if this book helps save just one life on our streets, I'm not making nearly enough.

Let me start with a little background. One of the fundamental mistakes we often made in the early road safety research was treating a fender bender no differently than a fatal crash. We did so in part because it made the math easier. If we want to gauge the safety of a treatment at a particular intersection, we fortunately don't get enough fatalities to get a statistically significant answer. It was easier to count all crashes—regardless of severity—and assume that reducing all crashes will concurrently reduce injuries and fatalities.

This assumption turned out to be very wrong. In most cases, more fender benders happen in places with fewer severe injuries and fatalities. Why? Because they tend to happen at slower speeds. Conversely, we see more severe injuries and fatalities in places with fewer fender benders. Why? Because when a crash happens on a high-speed street, it's likely to be serious.

When it comes to prioritizing road safety interventions, we estimate how many fewer crashes we'll see and pick interventions until we ran out of money. But once we realized the lack of harmony between fender benders and injuries/fatalities, these decisions became a value judgment.

And engineers don't like making value judgments.

So instead of deciding whether we valued locations with lots of crashes or those with fewer crashes but more injuries/fatalities, we weighted our crash data. In the 1970s, for instance, Taylor's paper tells us that the Kentucky DOT weighted a fatal crash as being worth 9.5 times that of a crash where nobody got hurt.[3]

Let's say we want to compare two intersections, one with 20 fender benders and the other with two fatal crashes. Given Kentucky's weighting system, the intersection with two fatal crashes is safer than the intersection with 20 fender benders but no injuries whatsoever. Why? Well, each fatal crash is worth 9.5 fender benders, so two fatal crashes only add up to 19 fender benders, and 20 > 19.

Sticking to the math gives traffic engineers the go-ahead to proceed. No value judgments necessary.

But aren't we making a value judgment by saying that one fatal crash is equivalent to 9.5 fender benders? What if we used 20× like the Pennsylvania DOT did back then? Or 175× like the National Safety Council did in 1978? Or 550× like the National Highway Traffic Safety Administration did in 1975? It gets even more complicated when we weigh injury crashes somewhere between fender benders and fatalities.

Where do these numbers even come from?

Although a 1956 *Traffic Quarterly* paper said that "dollar value cannot be placed upon a human life nor on national freedom," that didn't stop us from trying.[4]

That same year, D. J. Reynolds from the Road Research Laboratory in England[c] complains in a paper that "it is beyond the competence of the economist to assign objective values to the losses suffered" due to road crashes. If the economists weren't going to help, Reynolds, lucky for us, was up to the task. His methods focused on economic output. If you have a job and die, economic output is reduced. If you work part time, you're worth only half as much. Things get tricky when he considers what a housewife is worth: "For example, if a housewife is killed or disabled so that her services are lost to her family, the national income is not affected even though the community has suffered a loss which (theoretically at least) is measurable." How does he measure it? He gives these housewives the benefit of the doubt and assumes that "the average output of housewives is equal to that of the average female worker," which he estimates as 53.3 percent of the average male worker.[5] Given this logic, I wouldn't be surprised to hear him argue that somebody on paid disability dying on our streets would be a net benefit.

This same paper may also be the first use of the term *vulnerable road user* when Reynolds said in the appendix that "in general, safety measures will tend to benefit the more vulnerable road user such as the pedestrian, cyclist, and motor-cyclist."[6] He follows with an example of a new pedestrian crossing that could lead to an 8 percent drop in pedestrian fatalities. He asks whether the cost of the pedestrian crossing is more than or less than the value of those lives. If the crossing costs more, stop now. No crosswalk for you. If not, build it. Using his math, we would get very different answers if a businessman died instead of a housewife.[d] And if a kid gets killed crossing the

---

[c] The Road Research Laboratory used to be a government agency. It was privatized in 1996 and is now known as the Transport Research Laboratory.

[d] A 1946 English report on the same subject by John Jones valued women's lives as equal to men's lives.

street, I hope that kid had a job or owned a factory downtown.[e] The kid may not be worth much otherwise.

Today, the estimated cost of all motor vehicle traffic crashes in the United States alone is on the order of $340 billion annually.[7] These tangible costs include lost productivity, workplace losses, legal and court expenses, medical costs, emergency medical services, insurance administration costs, congestion costs due to the crash, and property damage costs. If we include nontangible losses such as those related to loss of quality of life, it jumps to $1.76 trillion.

If we sum the revenue of the top 20 professional sports leagues around the world, we get a total of $76 billion dollars each year.[f] In other words, the estimated societal cost of road crashes in just the United States is more than 23 times the combined revenue of the top 20 professional sports leagues in the world.

Given these numbers, what the hell are we waiting for? Shouldn't we be willing to spend hundreds of billions of dollars each year trying to improve road safety?

To answer that, let's revisit the question of what traffic engineers actually mean when they commit to improving safety. A 1975 *Traffic Quarterly* paper says that "an improvement in safety is defined as a change in vehicle or highway design that, ceteris paribus, reduces accident loss rates. That is, holding volume of traffic, speed, and environmental conditions the same."[8]

In other words, better safety is fine but not if it means fewer cars at slower speeds.

---

[e] "And my son Bart, he owns a factory downtown."—Homer Simpson while introducing his family to Frank Grimes (in probably my favorite *Simpsons* episode).

[f] The highest is the National Football League at approximately $18 billion.

19   The Cost of Doing Business

—

In March 2020, Senator Ron Johnson of Wisconsin was trying to put the initial COVID-19 pandemic deaths in perspective: "We don't shut down our economy because tens of thousands of people die on the highways. It's a risk we accept so we can move about."[1]

Despite the deaths and outrageous dollars we can attribute to them, that's how we think about people dying on our streets: they are the cost of doing business. In other words, we accept these deaths as an inevitable by-product of mobility and our modern lifestyle.

This isn't new.

In 1924, at the first President's National Conference on Street and Highway Safety, then secretary of commerce Herbert Hoover complained that "it is high time that something should be done about … a waste in human lives each year" but observed how interesting it is "to note the attitude of our great daily papers." Even though "street and highway accidents were most common, the accounts of them rarely appeared on the front page." You'd be lucky to get "a paragraph or two somewhere near the classified advertising section."[2]

At that same 1924 conference, President Calvin Coolidge agreed about the timing: "With deplorable and continuing increase in highways and mortality and injury, the time is highly appropriate for a comprehensive study of the causes, that we may have proper understanding of conditions and so may intelligently provide remedies."[3] How did we heed his call to action for a comprehensive road safety research program? As a 1965 *Traffic Quarterly* paper suggests, we didn't do very much: "During the next 25 years only some scattered research on shoestring budgets was undertaken independently."[4]

The results of this "scattered research on shoestring budgets" didn't make much of a difference, so we needed a second President's National Conference on Street and Highway Safety in 1946. President Harry Truman, in his remarks to the conference, laid out the extent of the problem:

The problem before you is urgent…. At the present rate, someone in the United States will die and a score will be injured during the few minutes I am speaking to you here today. During the three days of this Conference, more than one hundred will be killed, and thousands injured. Now when I was in the Senate, I made a study of this problem, and I found at that time that more people had been killed in automobile accidents than had been killed in all the wars we had ever fought, beginning with the French and Indian wars. That is a startling statement. More people have been injured, permanently, than were injured in both the World Wars from the United States. The property damage runs into the billions. Never less than a billion dollars a year … it would pay off half the national debt. Just think of that![5]

Given all that, it makes sense why the overwhelming sentiment was that we need to do whatever it takes. Here is the chairman of that second conference: "If we want traffic safety, we must be willing to pay the price for traffic safety. Otherwise, we may expect to see a repetition of the old 'spasm' method of enforcement, which has never produced permanent results."[6]

But again, nothing much changed, so we needed a third President's National Conference on Street and Highway Safety in 1949. Truman addressed the conference again, this time talking about all the Americans killed on the roads over the recent Memorial Day weekend, saying, "Now, if a town had been wiped out by a tornado or a flood or a fire and killed 429 people, there would be a great hullabaloo about it. We would turn out the Red Cross, and we would have the General declare an emergency, and I don't know what-all. Yet, when we kill them on the road …, we just take it for granted. We mustn't do that."[7]

In 1951, *Traffic Quarterly* published an editorial by its longtime editor, our old friend the Colonel Robert Goetz, about the United States reaching "our millionth victim" on the roads. He calls it "a shocking record, shocking enough to awaken a vigilant and sympathetic people. It has not done so. It has not awakened us."[8]

Seventy years later we have yet to wake up. Why is that? Why don't we care that much about a problem that has left millions of people dead or maimed?

Unless we are talking about the *Blues Brothers* movie,[a] car crashes tend to happen intermittently and in relatively small numbers. More often than not, though, they happen to somebody else.

Even though somebody dies on a US street every 13.5 minutes, on average, the typical individual driver has a 1 percent chance of dying every 50 years.[9] A 1973 *Traffic Quarterly* paper suggests why this is problematic: "Traffic accidents are among the most common events of our daily lives, and yet it their rarity that deludes the individual driver and baffles the scientist."[10]

Every time we complete a trip without dying ourselves or killing anyone else is another check mark in the success column. For many of us, there are few check marks on the side of failure. A 1971 highway safety book calculated that only one out of every *two billion* decisions made by drivers results in a fatality.[11] So, on an individual level, we turn a blind eye to these risks and costs because most of our available evidence tells us that it probably won't happen to us. While it's great that so many people haven't been personally impacted by all the road safety carnage, it leaves us without a public mandate to fix the problem.

We treat plane or train crashes very differently from car crashes. It's like the difference between a leaky faucet and a broken pipe. With a leaky faucet, we get used to a few drips here and there. With a broken pipe, we turn off the water completely until we figure out what is wrong and fix it.

Either way, we waste a lot of water … and a lot of lives.

---

[a] *Blues Brothers* (1980) destroyed 104 cars during filming (most of which were police cars). *The Fast and the Furious* franchise has totaled well over 1,000 and counting.

20 Do Better, Be Better

—

The three E's—*engineering*, *education*, and *enforcement*—as they've stood for the last century are clearly not enough. I could talk all day about how we just need to add other E's, such as emergency response, evaluation, engagement, equity, ethics, and empathy.[a]

Instead of adding more E's, I want this book to fundamentally change how we think—and what we do—about road safety.

I'm focused on traffic engineering because that is what I know we need to fix. I also know that most traffic engineers truly believe we aren't the problem. If only everybody followed the rules, we'd all be safe.

The science of road safety may no longer be in its infancy, but it's safe to say it remains nowhere near adulthood. When it comes to public safety, traffic engineers make more of a difference than police officers. Yet road safety professional is scarcely a real job.

To get a Professional Engineering license, you take the Professional Engineering exam. It includes a *breadth* section, which covers the gamut of civil engineering, and a *depth* section. Even though my undergraduate concentration was structural engineering, most of my professional experience leaned toward transportation, so I decided to take the Professional Engineering licensure test with the transportation depth. When I received my Professional Engineering license in 2003, I had taken a grand total of one college course specific to transportation. That's it. But I was professionally licensed to design any street or freeway in the state of Connecticut.

This isn't unusual. Most undergraduate civil engineering majors

---

[a] Prior to the 1965 Selma march, according to news reports at the time, Alabama's governor, George Wallace, said in his press conference, "There will be no march from Selma to Montgomery. It is not conducive to traffic flow on Route 80." He then authorized using "whatever measures are necessary to prevent a march," including blockading Route 80 with his own police and horses. But Governor, what about that traffic flow on Route 80?

take one or maybe two transportation-specific classes.[b] When you think about what civil engineers are responsible for, that seems insane. A 1964 *Traffic Quarterly* paper says, "The civil engineer has contributed significantly to the creation of man's urban environment.... The civil engineer has a vital role in shaping urban growth through expressway planning. He has control of forces which vitally affect the ultimate pattern of our cities."[1]

Another *Traffic Quarterly* paper from the same year goes into more detail because expressways are just the start when it comes to what civil engineers do, saying that they must take on "the responsibility of providing access to all parts of the urban area. This assignment must be viewed in its overall context. The location and design of major express highways and public transportation facilities are only a part of the assignment, though a major part. These must mesh into a proper configuration of arterial, collector, and local streets, and points of primary pedestrian destination."[2]

Is that it? Not even close. "A further breakdown must consider the question of vehicular parking, involving the manner in which they will be handled and the location of these facilities. The ease of pedestrian access to final destination from transportation-media termination points is particularly significant. Driveways and curb cuts, truck docks, and access points are all part of the transportation planning picture."[3]

Despite this range of responsibility, the executive director of the Institute of Transportation Engineers (ITE) said in a 1949 *Traffic Quarterly* paper that most traffic engineers receive training only in how to structurally build a road.[c] When it came to everything else, "traffic engineers matriculated from the school of hard knocks."[4]

Really!?!

A 1962 *Traffic Quarterly* paper tells us how to train the best traffic engineers and lays out the ideal syllabus.[5] Digging through that ideal syllabus, we see nothing specific to road safety. The paper then calls traffic engineering a "new-fangled Science," even though we were already neck deep in the construction of our interstate highway system.

---

[b] As we'll see in part 10, some undergraduate civil engineering students take no—that's zero—transportation classes.

[c] We'll also see in part 10 that traffic engineering education is better today, but not by much.

Going back to that 1949 paper, the ITE director presents Springfield, Massachusetts, as a city that is ahead of the curve. Led by a nonprofit group called Future Springfield Inc., the city hired a full-time traffic engineer—which was a rarity at the time—to lead the charge to build freeways in and around Springfield.

Unless you are from this area or visiting the Basketball Hall of Fame, you've probably never been to Springfield, Massachusetts. Within 10 seconds of searching online, I found Springfield on top of a list of worst cities to live in the United States.[6] Two minutes later, I found it listed among the worst cities in the country to drive across.[7] As for Springfield's road safety record?[d] It's terrible.[8]

So to all the traffic engineers out there, it's not your fault (despite the title of this book). How can we expect you to make the roads safer if you've never learned how?

The tragic disconnect is that traffic engineers believe they've been trained in road safety because they believe the manuals are steeped in a century of road safety knowledge.

They aren't.

The reality is that much of what we learn as traffic engineers has nothing to do with actual road safety outcomes whatsoever.

That's why traffic engineers don't take too kindly to suggestions about road safety.

That's why traffic engineers believe in things that diverge from common sense when it comes to road safety.

That's why others look at traffic engineering like a religious cult.

When it comes to courses specific to road safety, you'd be hard-pressed to find any at the undergraduate level. And classes specific to road safety for pedestrians or bicyclists, even at the graduate level, are few and far between.

During graduate school, I was fortunate to take an entire course dedicated to road safety. Even though it was one of the best courses I've taken,[e] I remember a lecture mentioning that bicyclists were safer in traffic that they were on off-street trails.

---

[d] In a report identifying the 200 least safe intersections across Massachusetts over a three-year period, Springfield boasts 23 of them (but has only 2 percent of the state's population). If you want to read more about the transportation problems of Springfield, take a look at Chuck Marohn's great book, *Confessions of a Recovering Engineer: Transportation for a Strong Town*.

[e] Thanks to Dr. Lisa Aultman-Hall.

How could that possibly be the case? It defies logic, right?

I asked that question and was told about "vehicular cycling" and a 1970s California study. It's probably the same thing that a generation of traffic engineers before me had been told.

You might remember my reasons for going back to graduate school in the first place. I wanted to understand **why** we did the things we did, so I dug up this old study and realized that it was based on one guy—an industrial engineer named John Forester[f]—biking through Palo Alto as fast as he could. He biked several hundred days in the street and said that he was never in danger once.

He then tested the off-street trail ... once.

In trying to maintain the same speeds he would've on the street, he had seven "incipient collisions" and humblebrags that he "was able to avoid only through a combination of expert understanding of traffic and expert bicycle handling skill."[9]

What does he conclude from this one ride on the path? That the risk of the off-street trail was at least 1,000-times greater than on the street.

Yup.

That paper, and Forester's subsequent books on the subject, changed what traffic engineers put in their manuals—and how traffic engineers design for bicyclists—for decades. A 1993 *Transportation Quarterly* paper tells us that it was about time: "It appears professionals are recognizing the mistakes of earlier bikeway planning practices that encouraged separated facilities, including sidewalk bikeways. They now appear to prefer facilities that put the bicycle where it belongs—on the roadway."[10] And many traffic engineers still buy Forester's statement that "13-year-olds on multilane streets with fast traffic" can ride a bicycle safer than they could on an off-street trail.[11]

Even if they could, why would we want them to? We can do better. We can be better.

David Byrne of the Talking Heads gets it. Here is what he says in his book *Bicycle Diaries*: "Maybe, for some, part of the thrill would go away if urban bicycling became safe. But that might be the price to pay if it means more people will start using bikes to get around. That thrill isn't really an appropriate thing to oblige schoolkids and

---

[f] He was the son of author C. S. Forester (whose 1955 book, *The Good Shepherd*, was turned into a Tom Hanks movie called *Greyhound*).

seniors to be burdened with. Living in New York used to be a lot more dangerous in general, but that's hardly something to get all nostalgic about."[12]

I don't want to wax nostalgic, but I've been working in this field for 25 years. After getting my undergraduate degree, I thought I knew enough to get me started. When the time came, I figured I could learn whatever else I needed to know by using those 1,000-page manuals. They would be more than enough.

I was fortunate to become disillusioned early and often.

This led me back to graduate school. My wife and I were about to have our first child, so one reason I chose the University of Connecticut (over what might be considered more prestigious transportation programs) was its great health insurance for graduate students.

I came to realize that a PhD is much more about having a great advisor than anything else. Luckily,[g] I ended up sitting in front of the person who would eventually become my Jamaican Dumbledore, Dr. Norman Garrick.

After growing up in Jamaica, Norman trained as a civil engineer specializing in pavements. He got his master's and PhD degrees at Purdue University in the mid-1980s and ended up as a professor at UConn, teaching and researching all about pavements.

In the years before I arrived at UConn, he was sick of pavement research and ready to quit his tenured position. Instead, he figured he should take his sabbatical first. Spending that time at Cambridge University in England transformed his mind-set on transportation. He returned intent on changing his career within civil engineering.[h]

Norman wasn't even considered the road safety professor at UConn,[i] but he taught me how to think differently about transportation engineering, like learning to experience cities and streets in person rather than though an AutoCAD monitor.[j] Our computer-simulated drivers don't run people over like real drivers do. And as

---

[g] Lucky is an understatement.

[h] That the work Norman used to do as a pavement engineer and what he does now both fall within the same discipline is one of the most useful things about having a civil engineering degree.

[i] That would be the one and only Dr. John Ivan.

[j] When I was a consultant, I was the AutoCAD king. Now, I'm thrilled that it isn't installed on my computer.

Jane Jacobs wrote in a 1958 *Fortune Magazine* article, "The best way to plan for downtown is to see how people use it today. You've got to get out and walk."[13]

Looking back, that is what we need to begin to understand road safety. Data can only tell part of the story. In one of the many human error books I read, Alina Tugend, a *New York Times* journalist, writes, "In developing systems to avoid errors, observation and feedback are crucial. We all make too many assumptions about how we and others act based on what we want to be true, not necessarily on what is true."[14]

It was this sort of thinking that I brought to my early research work when I was a graduate student at UConn.

The second research paper I ever wrote looked at on-street parking. It had been decided, decades before, that on-street parking was unsafe.

I would occasionally do consulting work in small New England towns whose Main Streets look like a Norman Rockwell painting. The one problem was always a state-owned highway cutting right through the middle of the town center. Wouldn't it be nice if that highway—just for the Main Street portion—was a slower street with on-street parking in front of the shops and restaurants? The DOT officials would say it was impossible because on-street parking simply wasn't safe. Dating back to the 1950s, the research seemed to agree.

But when I got out there and observed examples of slower streets with on-street parking, they didn't seem unsafe, and it turns out they weren't. My research paper looked at hundreds of streets and found that those with on-street parking and speed limits slower than 35 mph were indeed safer. They had 7 percent more fender benders but were killing or severely injuring three times fewer people.[15]

You might recall that one problem with the old research was that fender benders were treated the same as fatalities. That is one reason traffic engineers branded on-street parking as unsafe.

In trying to figure out my dissertation topic, I took on another question that was nagging me. When it comes time to buy a house, many people want a cul-de-sac because it's safer. Dating back to a well-known 1950s study by Harold Marks, the research seemed to agree.[16]

Yet every time I spent time in a place with lots of cul-de-sacs, it felt unsafe to leave the cul-de-sac without a car. When I spent time in a gridded street network, I could walk or bike around town without much of a problem.

My PhD dissertation looked at hundreds of thousands of crashes in two dozen cities over more than a dozen years. What I found is that the more compact, gridded street networks had slightly more fender benders but significantly fewer deaths or severe injuries.

Every time I presented my work to a traffic engineering audience though, I'd get a "question" that usually consisted of an older guy doubting my results because of the 50-year-old study by Harold Marks that few people alive have ever read.

I'll get more into this issue in part 6, but the gridded neighborhoods from the "renowned" 1957 Marks study didn't have much, if any, traffic control. In other words, the so-called dangerous gridded street networks that Marks studied in the 1950s were full of four-way intersections with no traffic lights, stop signs, or yield signs.

Seriously.

These are some simple examples of where we've gone wrong—and why. And if you now think this book will just be me rehashing my own old research papers, you'd be wrong. While much of my research looks at road safety, this book was a separate effort. Don't get me wrong; I'll mention my own work when need be, but it's only the tip of the iceberg.

Now that you've heard my origin story, I'm hoping you have a better understanding of why I do what I do.

But what keeps you up at night? Does it have anything to do with unsafe streets?

It might not be your job that provides that passion; you might find it somewhere else. If you get lucky like me, the passion and the job are one in the same.

So here it is, everything that you need to know—and a couple things you don't—to do something about it. Let's have some fun. Let's save some lives.

Part 2

Mismeasuring Safety

21   The Relativity of Safety

Is our transportation system safe?

Compared to Russian roulette—where you have a 16.7 percent chance of dying each turn—we are doing pretty darn well.

But I don't play Russian roulette very often. This means that my chances of dying on our streets are much higher than my chances of dying while playing Russian roulette. This has to do with what traffic engineers call *exposure*. Let's hold off on that for a moment and first ask, what exactly is "safe"?

To traffic engineers, there is no simple answer because, in their eyes, safety is relative.

In other words, an intersection is considered dangerous only if it sees significantly more crashes than other similar intersections given similar traffic volumes.

We determine that by collecting crash data for various intersections and then graphing those crashes against the number of cars coming through. This gives us a baseline off which to compare other intersections. If we plot our intersection of interest on that same graph and it falls too far above that baseline, it's dangerous. If it falls right along that baseline, it's normal. We interpret normal as safe.[a]

Is the baseline level dangerous in itself? Maybe, but it's irrelevant in the eyes of the traffic engineer. Danger is defined only in comparison to the status quo.

That's why we'd tell you that drinking and driving at night on a busy road without your headlights on is dangerous. But for the regular driver, traffic engineers want you to know that our streets are safer than they've ever been.

The recent US Department of Transportation "Beyond Traffic: 2045" report boasts that the "motor vehicle fatality rate has dropped by 80% over the last 50 years."[1] But even before this improvement, traffic engineers would still have bragged about how safe you were.

---

[a] If it falls below that baseline, that is good too.

This is from a 1947 *Traffic Quarterly* paper: "Is it surprising that in traveling a distance 360 times around the world one person should be killed in a traffic accident?"[2]

Go to any traffic engineering conference about road safety today and you'll see a slide showing a graph of our incredible improvements over the last 50 years or so. My own version of this goes all the way back to the year 1899, which is the earliest I could find a number for US road deaths.[b] For the next decade, the metric we call the road fatality rate—which is essentially the number of road deaths divided by the number of miles we drive—skyrockets to 45.3 deaths per 100 million vehicle miles traveled (VMT).[3] It then creeps lower and lower almost every year for next century. Since 2020, our road fatality rate has hovered between 1.3 and 1.4 deaths per 100 million VMT.[c]

I then show a picture of the infamous "Mission Accomplished" banner that was unfurled across the bridge of a US Navy aircraft carrier in 2003.

What caused this mind-boggling improvement in road safety? As a 1977 *Traffic Quarterly* paper says, "Engineers like to believe that they have contributed to the overall decline in the death rate and thus could easily fall into the same trap."[4]

What trap is that author talking about? The trap is that "almost any event in the past 50 years could thus be shown as having been related to a downward trend in death rates."[5]

For example, the road fatality rate was stagnant in the early 2000s and then started to drop from 2007 through 2009. These years happen to coincide with when I started researching road safety. Was I the hero the world was waiting for? Or maybe it was Steve Jobs for giving the world the iPhone in 2007? Or maybe we can give credit to Tay Zonday for singing his "Chocolate Rain" song on YouTube in 2007?

I'm sure the auto industry would be happy to take credit. And the police would like to remind us about all the *enforcement* they do. Maybe we finally made a good enough commercial that enticed people to start driving safer. I'm sure the medical field can make a strong case for these improvements being their doing as well.

No matter who we give credit to, the underlying assumption is

---

[b] Evidently, 26 people died on US roads in 1899.

[c] Although for the decade prior to the pandemic, it fell to between 1.1 and 1.2.

that there is something worth taking credit for. So again, let's ask the question, is our transportation system safe?

In other disciplines, safety critical systems are engineered systems "where a malfunction may result in a death or injury."[6] The threshold for reliability in such a system is defined as 1×10^9 per hour. In other words, more than one death or incapacitating injury every one billion hours is too many.

It's tricky to figure this out for transportation, but the American Time Use Survey estimates that Americans aged 15 and older spend an average of *1 hour, 11 minutes traveling each day.*[7]

For the sake of argument, let's just say that *every* American travels 71 minutes on a typical day. It's an overestimation, but let's overestimate it even further and assume that 71 minutes is for every day of the year. That puts us in the range of *143 billion hours of travel each year.*

In a system that meets the 1×10^9 per hour safety critical threshold, we'd want *fewer than 143 deaths or incapacitating injuries each year.*

Even if we ignore the incapacitating injuries and only consider fatalities, the US transportation system is *300 times worse* than the safety critical threshold for reliability.

Maybe the status quo isn't quite so safe?

22 Exposing Exposure

—

Traffic engineers try to measure road safety in terms of some sort of *exposure* metric. The idea is pretty simple. Let's compare two different streets that happen to have the same crash outcomes. How can we tell which street is safer?

We typically do so using vehicle miles traveled as exposure. The idea is to normalize the crash data so that we can make more of an apples-to-apples comparison.

Let's say that the first street had 100,000 miles driven on it and the other 200,000. This means that the second street—with 200,000 driven on it—is safer. Why? Well, it took twice as much driving to get to the same number of crashes on the second street. Hence, it's safer.

Things are a little more complicated than this because the relationship between crashes and traffic volumes isn't linear. Or instead of mileage as the exposure metric, we might use time spent traveling, number of trips taken, number of registered cars, number of licensed drivers, or even the number of streets crossed. The list goes on and on, but that's the gist of how traffic engineers measure safety.

The logic behind this strategy makes sense. The number of crashes (or injuries, deaths, and so on) is the numerator. The distance we travel (or time traveling, number of trips, and so on) is the denominator. The result lets us look at crash rates instead of the raw number of crashes. It seems like a fair way to compare safety.

Let's test it. Say you drove 100 miles today and didn't get into a crash. I drove 5 miles today and also didn't get into a crash. Who was safer? Traffic engineers would say that you are because you drove 20 times the distance with the same level of safety. From my perspective, that's crazy talk. Let me explain.

What if you and I drive this far as part of our everyday commute? Luckily, neither of us gets into a crash over the course of an entire year. Who is contributing more to making the roads safer? Again, you are. Assuming something like 250 working days, you drove 25,000 miles. I only drove 1,250. If I biked to work for some of those days, my mileage drops even more and, somehow, so does my contribution to safety.

What if you got into 10 crashes during those 25,000 miles but I only got into a single crash in my 1,250 miles. Who's safer now? Traffic engineers would look at the numbers and still say that you are. How's that? Well, you are averaging one crash every 2,500 miles while I am averaging one crash every 1,250 miles. Even though you had 10 times more crashes than me, your crash rate says that you are two times safer than I am.

But which of us would you rather be?

Let's blow this example out and consider road fatalities for two different cities. The first is Driveton. The houses in Driveton are nice and big, surrounded by giant yards. Land uses are spread out. Transit exists but isn't a great option. Walking and bicycling are not considered safe or practical for anything other than exercise or walking a dog around Driveton's individual neighborhoods. For the most part, the people of Driveton have little choice but to drive everywhere they need to go. Thus, the average Drivetonian driver drives 25,000 miles each year. Over the last 10 years, unfortunately, Driveton lost 150 of its citizens to fatal crashes.

The other city is Heresville. Houses here are nice but smaller than those in Driveton. They rarely have big yards, but there are parks and playgrounds sprinkled across the various neighborhoods. Heresville also has plenty of apartments, some of them located directly above coffee shops and other commercial uses. Land uses tend to be closer together, and transit is sometimes a decent option. Walking and bicycling are practical, especially when the weather is good, but not always that safe. The average Heresvillian driver drives 5,000 miles each year. Yet, over the last 10 years, Heresville also lost 150 of its citizens to fatal crashes.

So, which city is safer?

Traffic engineers would first multiply the vehicle miles traveled by each person by the number of people living there. Then we can divide the average number of fatalities each year (15 deaths per year) by all those miles traveled. Driveton's fatal crash rate ends up at 1.2 fatalities per 100 million vehicle miles traveled, right around the US average. For Heresville, it's at 1.5 fatalities per 100 million vehicle miles traveled. The fatal crash rate in Heresville is about 25 percent worse than Driveton.

If you are concerned about road safety, where would you want to live? The answer seems to be Driveton.

But what if I told you that 200,000 people live in Heresville while Driveton's population is 50,000? In the public health world, "fatalities per capita is a standard, widely accepted metric of risk at the scale of an entire society, and it is used to calculate annual or lifetime risk."[1]

So what if we treated road safety like any other public health issue and calculated our fatality rate using population as the *exposure* metric?

After all, dying on our streets is not good for your health.

Using that mind-set, Driveton's road fatality rate is 30 road fatalities per 100,000 residents per year, more than double the US average. In Heresville, it's 7.5 fatalities per 100,000 residents per year. This is at least a third better than the US average. Now, Driveton is four times more dangerous than Heresville.

But conventional traffic engineering metrics tell us that Driveton is safer.

Conventional metrics imply two ways to improve safety. The first is to decrease the numerator and the number of fatal crashes. The second is to increase the denominator and how much we drive. A 1971 *Traffic Quarterly* paper makes it clear that "since death rate is dependent upon motorization, it can be reduced simply by increasing the number of vehicles on the highways. The formula leaves no room for any other conclusion." Accordingly, a city like Driveton "could expect to lower its death rate purely by increasing motorization."[2]

Given these numbers, traffic engineers would tell you that Driveton is what we should emulate when building other places.

But where would you rather live?

Welcome to Heresville.

23 The Mirage of More Mileage

—

Everybody seems to want to push toward data-driven safety solutions. Few take a step back to think about *how* our safety measures influence that drive.

Given how we measure road safety, there are two ways to improve. One is to reduce crashes, injuries, and/or fatalities. The other is to increase how much we drive. The latter isn't real safety.

But *why* do we measure road safety in terms of mileage?

This story wasn't an easy one to track down.

In the early 1900s, road fatalities skyrocketed. When the Model T first went into production in 1908, the United States had 751 road fatalities. In the year that the last Model T rolled off the assembly line, 1927, nearly 25,000 Americans died on the roads.

A few years earlier, President Calvin Coolidge had gathered the first President's National Conference on Street and Highway Safety. Instead of treating road fatalities like the cost of doing business, as we do now, they were being recognized as a national tragedy. Americans were mad. As L. J. Aurbach describes in his valuable book, *A History of Street Networks*, "The shocking headlines about children, women, and seniors killed by cars sparked a nationwide grassroots safety movement in the early 1920s. Citizens protested and gathered in angry mobs. They held large memorial events and raised monuments to the dead. Autos were demonized in the press."[1]

By 1934, US road fatalities eclipsed 34,000 deaths in a single year. In 1935, J. C. Furnas wrote a piece in *Reader's Digest* titled "—And Sudden Death" that delved into the reality of traffic fatalities with nauseating details about organs being torn apart and skulls being crushed.[2] Americans would later become somewhat desensitized to scare tactics. But this was new, and the article struck a nerve across the country. Years later, a *Traffic Quarterly* paper tried to describe just how impactful it was: "The article was so shocking it was fascinating. Millions of reprints were distributed by civic agencies."[a, 3]

Car manufacturers were well aware that the number of road fatalities was bad for business. They had been trying to change the headlines for years. L. J. Aurbach's book mentions a 1921 car industry annual that brags, "Automobiles Now Twice as Safe: Ratio Fatalities per Car Halved in Five Years."[4] The denominator in this case isn't mileage; it's the number of cars. With car ownership growing exponentially, this rate made them look pretty good.

The Great Depression put a dent in car buying and tanked the "fatalities per car" metric. The National Safety Council then tried using "traffic fatalities per gallons of gasoline consumed," claiming "that increased gasoline consumption was the single factor that could explain the recent increase in traffic fatalities."[5]

Traffic engineers seemed happy to use this metric, but the auto industry realized it wouldn't suffice over the long haul. As cars became more fuel efficient, they'd seem less and less safe.

But after the *Reader's Digest* article, the auto industry knew it needed to change the narrative. That is where Paul Hoffman came in.

At the time of "—And Sudden Death," Hoffman was president of Studebaker.[b] Most auto industry executives were more concerned with the lingering effects of the Great Depression and dealing with the burgeoning auto workers union movement[c] than they were with road safety. The *Reader's Digest* piece was different. Hoffman says the article was "the spark that touched off a tremendous explosion of public interest and concern."[6] The next year, a national poll found that two of out of every three Americans were in favor of mandating that car manufacturers install speed governors. The tide had turned drastically since the failed speed governor referendums.

All the negative press, let alone these deaths, was bad for business.

Hoffman had been involved in road safety efforts before, becoming chairman of the Traffic Planning and Safety Committee for the

---

[a] Columbia University Library preserved two copies of the *Reader's Digest* issue. Perhaps due to its popularity, or maybe because somebody wanted it to disappear, "—And Sudden Death" has been torn out of both copies.

[b] He'd been working with Studebaker since 1911.

[c] The big dog they were fighting was the United Auto Workers union, which was founded in 1935, the same year that *Reader's Digest* piece was published.

National Automobile Chamber of Commerce in 1932, so he decided to write a book. His goal was to establish that cars were safe and getting safer. His challenge? Most metrics showed that road safety was getting worse. And worse. And worse.

But the auto industry knew that "different statistical metrics showed drastically different results," so the answer was obvious.[7] Just pick whichever one makes the auto industry look the best and would continue to make it look good.

In 1938, Hoffman published a preview[d] of his forthcoming book in the *Saturday Evening Post*.[e] He doesn't introduce the mileage-based metric as new. He doesn't present it as another option for us to consider. He simply pretends it's the way we've always measured road safety: "Our present highway accident rate is 15.9 deaths per 100,000,000 vehicle miles. The average American automobile could complete more than 240 trips around the earth at the equator before killing anybody. A lot of safety. Few countries have as low a motor-death rate as the United States, per mile traveled."[8]

He goes through almost the same spiel at the very beginning of his 1939 book, *Seven Roads to Safety*.[9] The death rate goes up slightly, but, magically, that still equated to an extra 10 safe trips around Earth's equator. Hoffman writes, "The highway accident rate for 1937 was about 16 deaths per 100,000,000 vehicle miles.[f] Try to visualize that. If the average American automobile were driven around the earth at the Equator, it would complete more than 250 round trips before it killed anybody. Sounds like a lot of safety. It is."[10]

Hoffman didn't have much else to say, so most of the rest of his 76-page book repeats what people like Thomas MacDonald, chief of the Bureau of Public Roads, and traffic engineer Miller McClintock[g]

---

[d] The title of this 1938 article, "The White Line Isn't Enough," wouldn't be out of place for an article about pedestrian or bicyclist safety today.

[e] Flipping through the full issue, it was impossible not to notice the nearly three dozen car-related advertisements, including one from Hoffman's company for an $875 Studebaker Commander (which is the same name of a recent Jeep model).

[f] We've dropped this fatality rate nearly 14-fold since then, but what he describes still sounds pretty good, right?

[g] *Life* magazine called Miller McClintock the "foremost U.S. traffic engineer" in a 1938 article, "The Traffic Problem: Its Best Solution Lies in Foolproof Highways which Reduce Driver Judgments to a Minimum."

had been saying for years about the three E's (*education*, *enforcement*, and *engineering*) of road safety.

Yet Hoffman became known as the "civic leader of the safety movement."[h]

Hoffman formed the Automotive Safety Foundation during these years, and as president of that group, he barnstormed the country, championing ever-improving road fatality rates, and talking to this new traffic engineering discipline about how the three E's have made us safe and will continue to make us safer if we buy in.

If you told Hoffman you are worried about the ever-increasing number of road fatalities, he'd say those numbers are concerning to the general public but they should be "wholly meaningless" to traffic engineers and to those who work in road safety.[11]

Hoffman makes the right point that we need to account for *exposure*. As a 1960 *Highway Research Board Bulletin* piece says, "Accident rates based on improper exposure values are misleading and can delay the proper correction of accident hazards."[12]

That's true. Hoffman just picks the exposure metric he likes best and sells it to us as best as he can. For the most part, traffic engineers bought what he was selling. We never thought much about the measurer when considering the measure.

Later when the US Department of Transportation was created in 1966, the US road fatality rate was presented using the same mileage-based metric that Studebaker CEO Paul Hoffman had imparted upon us in the 1930s. It's a denominator that had increased more than 12-fold since Hoffman wrote his book. So even if fatalities remained exactly the same each year, we'd seem 12 times safer. Even if road fatalities had gone up, way up, we'd still seem safer. And where does this important mileage data even come from? A 1968 *Traffic Quarterly* paper complains that "aside from some crude overall mileage estimates based on gasoline tax receipts and some sporadic and nongeneralizable survey data, we have no information on who drives what kind of vehicle how many miles per year on what kinds of road and under what conditions of weather, traffic, daylight, fatigue, intoxication, drug impairment, etc. And thus we have no way of

---

[h] Hoffman's most famous quote is, apparently, "A gentlemen is one who is too brave to lie, too generous to cheat, and who takes his share of the world and lets other people have theirs."

relating the number of violations and accidents in any population subgroup to the quantity and quality of its exposure."[13]

The way we collect mileage data is not much better today. Heck, it took us until the oil embargo in the 1970s before we realized that fewer miles driven might be a good thing. Yet the Department of Transportation and traffic engineers continue to measure road safety using the same ol' mileage-based metric.

24   Why Didn't the Chicken Cross the Road?

—

Part of the problem with **how** we measure road safety relates to **where** we measure road safety.

Where we measure is quite simple. Traffic engineers look at where the crashes take place. Why would we look elsewhere?

But let's think about Driveton and Heresville again. If a Driveton resident needs to drive all the way Heresville to go to the dentist, they run a greater risk of a crash than a Heresville resident who walks around the corner to their dentist. If that Drivetonian crashes during their long trip, there is a decent chance that the crash will happen outside of Driveton. Even though Driveton's design played a role in the crash—by forcing its resident to drive a long way to a dentist—the crash never gets attributed to Driveton. It may even get attributed to Heresville if it happened within those city limits.

The result? Many of the places that attract people from all over seem more dangerous than they are. Conversely, many of the places that send people all over, while attracting few, seem safer. Those dying on the streets in Heresville may not live in Heresville, but traffic engineers never take that into account. Instead, we only look at where crashes happen. A traffic engineer will look at this data and tell people how safe it is to live in Driveton and how dangerous it is to live in Heresville.

Are those living in Heresville—or any place with more people and jobs—less safe than those living in Driveton? If we solely focus on where crashes happen, we have no idea. Heresville may look guilty, but the evidence is circumstantial.

One of the few datasets where we can even begin to account for this is the US national fatal crash database, which collects the zip code of the driver.[a] This information lets us distinguish between

---

[a] It's called the Fatality Analysis Reporting System (FARS). The National Highway Traffic Safety Administration has been collecting detailed data on all fatal crashes since 1975, although the level of detail has improved over the years.

where crashes happened and where those involved were likely from. Looking at 24 years of data, I chopped up US zip codes into a rural to urban spectrum of 12 categories based on population density.[1] What I found out is that those living in the most urban places in the United States experience safety outcomes on par with the safest countries in the world. It gets worse and worse at every step in between until you reach the most rural places, which are on par with the most dangerous countries in the world. The difference? In the United States, you are six times more likely to die on the roads if you live in the most rural areas than in the most urban.[b]

Another question we rarely ask is, where are crashes **not** happening? Again, why would we?

When trying to explain the importance of ***exposure*** to my students, I show them a real picture of a bicyclist happily riding along on the Southeast Expressway, part of Interstate 93 running along the coastline outside of Boston. The guy is wearing flip-flops and riding his bike in the second of four expressway lanes headed south.

Why is he there? One reason could be that he looked at the crash data and realized that there have never, ever been any bicyclist crashes on that expressway. His conclusion? It must be the absolute safest place to ride a bicycle.

This isn't true, of course, but looking at the crash data without considering how many bicyclists use a facility could lead us to think so.

There aren't many motorcycle crashes in the snow, but that doesn't mean it's safest to ride a motorcycle in the snow.

We find a similar measurement problem when we try to assess pedestrian and bicyclist safety. Every city counts cars. When we assess vehicle safety on a street or at an intersection, vehicle exposure data is usually easy to come by.[c] Few cities count pedestrians or bicyclists. If we want to understand how safe they are on a street or at an

---

[b] A 2011 German study for Scheiner and Holz-Rau used license plate data to account for this discrepancy and found that "the risk of being killed or seriously injured in a road accident is considerably lower for" those living in more urban places "than for the suburban and rural population." Joachim Scheiner and Christian Holz-Rau, "A Residential Location Approach to Traffic Safety: Two Case Studies from Germany," *Accident Analysis & Prevention* 43, no. 1 (2011).

[c] According to a 1980 *Traffic Quarterly* paper, we've been counting cars since the early 1920s. John Hassell, "How Effective Has Urban Transportation Planning Been?," *Traffic Quarterly* 34, no. 1 (1980).

intersection, we usually have to stick with the raw number of times pedestrians and bicyclists get hit by a car.[d]

As a result, most pedestrian and bicyclist safety studies create fancy-looking hot-spot maps to highlight the most dangerous locations. In most cities, these hot spots simply correspond to downtown. Why? Because that's where most people are walking and bicycling. Without people walking and bicycling, we won't get any pedestrian or bicyclist crashes. And because these maps do not account for any sort of exposure, we don't get an apples-to-apples comparison.

What hot-spot maps also miss are the locations where a street or intersection is so damn dangerous that few people, if any, would attempt to walk or bicycle there. These streets and intersections don't have many pedestrian or bicyclist crashes, but it's not because they are safe places for people to walk or bike. It's because they are not.

As it stands, it's easy for very dangerous streets and intersections to be deemed safe; it's because we've scared away all people who want to walk or bike. It also means that we can improve pedestrian and bicyclist safety by making a street so dangerous that it forces people into cars. If we have no pedestrians or bicyclists, we don't have a pedestrian/bicyclist safety problem.

In the end, where we ***don't*** find crashes could be just as dangerous, if not more, than where we do.

---

[d] A moving vehicle would have to be involved for a pedestrian or bicyclist fatality to be eligible for the FARS fatal crash data. If you park your car, open your door, and kill a bicyclist, that—for whatever reason—doesn't count.

25 Don't Sweat the Small Stuff

—

We now know about *how* we measure road safety and a bit about *where*, but it would help to also think about *what* we measure.

I talked earlier about how traffic engineers studying safety often focus on all crashes instead of on those that lead to injuries or death. There are practical reasons for this. For one, having more data points helps with statistical significance. Plus, injury severity levels tend to be inconsistently coded.

It's also what we've always done.

However, I can't blame traffic engineers or the president of Studebaker for this. Instead, it goes back to the early accident research of H. W. Heinrich, a guy who worked for Traveler's Insurance. His 1931 book on industrial worker accidents found a proportional relationship between minor accidents and major accidents.[1] He put some specific numbers to this idea, and his 300–29–1 ratio became known as Heinrich's law or the *accident triangle*.[a] Heinrich said, "In a workplace, for every accident that causes a major injury, there are 29 accidents that cause minor injuries and 300 accidents that cause no injuries."[2]

This idea that accident severity is proportional implies that we can reduce severe injuries and fatalities by reducing the number of noninjury accidents.

These ideas seeped into traffic engineering. It was easier for us to focus on reducing all crashes and assume that doing so will simultaneously reduce fatalities and severe injuries at a similar rate. As a 1950 *Traffic Quarterly* paper argues, "For a truly comprehensive approach to the safety problem, we must understand that we must bring all accidents under attack and not simply fatal or serious personal injury accidents, because every accident has the same root causes."[3]

---

[a] The basic accident triangle looks like a pyramid with three layers. The base of the pyramid represents accidents with no injuries. The middle level refers to minor injury accidents. The top of the pyramid represents the major injury accidents.

This thinking turned out to be wrong. But it's the reason we declared our historic downtown streets dangerous and our new highways safe. The slow-moving downtown streets had lots of fender benders but few deaths. The fast-moving highways had fewer fender benders—especially when measured by miles driven—but far more deaths and severe injuries.

Thanks to the accident triangle's proportionality assumption, we figured that making our downtown streets look more like highways would make us safer. A 1957 manual, specific to urban areas, by the American Association of State Highway Officials (AASHO) says, "The accident experience on controlled-access highways is markedly better than noncontrolled facilities."[4]

Of course they are, because we're measuring the wrong thing.

This same 1957 AASHO manual then applies these crash rates to a cost-benefit analysis for the proposed 185 mile of freeways in Los Angeles. Guess what? Those freeways will pay for themselves in only seven years! How is that possible? Well, just look at all the money we'll save with fewer crashes.

We've inherited flawed infrastructure based on flawed thinking. Instead of fixing it, many traffic engineers still focus on the overwhelming majority of road crashes that result in no injury whatsoever. We do so even though the relationship is often the opposite of what Heinrich told us. As Sidney Dekker writes in his human error book, it's not uncommon to find a "negative correlation between number of incidents and fatalities."[5]

Traffic engineers have known this—and ignored this—for a long time. A 1963 manual by Hoffman's Automotive Safety Foundation complains that "most research to date on the relationship of accidents to traffic volumes has dealt with all accidents."[6] It praises a 1933 traffic signal study for differentiating by crash severity and disparages most of the research that came in between.[7]

A 1971 *Traffic Quarterly* paper also criticizes our overfocus on the "general reduction of accidental events," saying that "it appears wrong to measure safety achievement simply as a function of incidents that occur."[8] But why would this be a problem? The author writes, "Studies indicate, for example, that there may be no statistical relationship between rates of incidence and hazard control effectiveness. Therefore, safety approaches which are evaluated by rates … can be falsely praised or discredited."[9]

He goes on to say that if we do shift the focus to fatalities and severe injuries, all that we think we know about road safety may not mean much.

Exactly.

A focus on fender benders interests insurance companies. For traffic engineers, it compromises how we look at road safety. We not only end up with a misleading impression about what is safe, but we also believe we can make it safer by just "suppressing the small stuff."[10] A fender bender is the small stuff. Don't sweat it.

26 The Conflict Conflict

___

H. W. Heinrich of accident triangle fame claimed that 88 percent of workplace accidents are caused by human error (his theory of behavior-based safety). It would be easy to blame him for traffic engineers seizing on this human error thing. To Heinrich's credit, however, he did say that "no matter how strongly the statistical records emphasize personal faults or how imperatively the need for educational activity is shown, no safety procedure is complete or satisfactory that does not provide for the ... correction or elimination of ... physical hazards."[1]

This didn't stop people like Frank Bird[a] from doubling down on the accident triangle in the 1960s by assuming that both **near misses** and **unsafe acts** are also proportional to severe injuries.[2]

Near misses is a term that always bothers me. If I nearly miss someone, doesn't that mean I hit them? I'm going to use **near hits** from here on out.

So, Bird estimated 300 near hits for every major injury. He then takes it a step further, estimating 300,000 unsafe acts for every major injury. That's adding two new layers to the base of the pyramid.

Bird based these numbers on industrial accidents, not crash data. Yet the takeaway for traffic engineers was that we could assume proportionality in these relationships. Such thinking allowed traffic engineers to study safety without the need for crash data. To understand safety, we only need to count near hits or unsafe acts. Much of today's so-called safety research does just that, counting conflicts or bad behaviors and ignoring actual crash outcomes altogether.

While focusing on something like conflicts makes some logical sense, it can also steer us in the wrong direction. You may have heard of Hans Monderman, a Dutch traffic engineer who believed that conflicts were **good** for safety.[b] He removed all stops signs and signals

___

[a] No relation to Larry?

[b] I was lucky to speak to Hans Monderman and see him present his work in November 2007, less than two months before he died of cancer at the age of 62.

from his street and intersection designs, including for the entirety of his hometown of Bristol, England.[c] Monderman lays out his thinking: "A wide road with a lot of signs is telling a story. It's saying, 'Go ahead, don't worry, go as fast as you want, there's no need to pay attention to your surroundings.' And that's a very dangerous message."[3]

So instead of drivers waiting for a traffic light to tell them what to do, drivers have no choice but to respond to the other human beings in the street. How does that work? Who gets to go first? Monderman answers that question as well: "Who has the right of way? I *don't* care. People here have to find their own way, negotiate for themselves, use their own brains."[4] Interacting with other road users in this way leads to far more conflicts, but it also leads to people being careful and taking things a bit slower. As far as I know, every time he or one of his disciples[d] talked a city into trying this crazy idea, it led to fewer fatalities and fewer severe injuries.

Bird's assumptions also suggest that we can simply count how many "unsafe" acts people commit. Such counting is less common than conflicts as a proxy for crash data in road safety research, but it does happen. It is also why most cities—when it comes time to show their commitment to road safety—make a point of pulling over bicyclists for rolling through a four-way stop sign when no one is around. Will this make the streets safer? My Magic 8 Ball says, "Don't count on it."

---

[c] We call these stripped-down designs, which sometimes even remove the vertical separation of the curb, *shared spaces*. They are very different from pedestrian-only streets, which don't allow cars at all. Shared spaces allow cars, in some cases as many of 22,000 per day (such as a shared space intersection than Monderman designed in Drachten, the Netherlands), but drivers entering the space always have to watch out for everyone else. For a short time, shared spaces were called naked streets. But try Googling "naked streets," and you'll see why it changed.

[d] Including the late, great Ben Hamilton-Baillie, who I was also fortunate to meet and hear speak prior to his death from cancer in 2019 at the age of 63.

27   Conflating Congestion

Our stated transportation priorities for the last 100-plus years have always included improving road safety and reducing traffic congestion. But given our safety record, it's easy to wonder—as James Taylor did in his 1981 *Traffic Quarterly* paper—"How can one justify spending **any** money on relieving congestion or reducing delay when these same funds could be used to save lives and reduce serious injuries?"[1]

It's easy. We did so by conflating the two.

What do I mean by this? For much of the history of traffic engineering, we considered traffic congestion to be a proxy for road safety. As our good friends over at AASHO inform us in their 1957 manual on urban arterials, "Congestion breeds accidents."[2]

So how do we improve safety? We can do so by focusing on fighting off traffic congestion.

AASHO's 1957 manual gets a bit more specific about the causes of urban safety problems. It says, "Pedestrians and congestion, augmented by intersectional and roadside interferences, are chief causes of urban accidents."[3]

That's convenient. We can fight off both traffic congestion and pedestrians in the name of road safety!

Hey AASHO, any suggestions for how to do this? With fully controlled access, you say? What would that do for us? "With fully controlled access, there are no grade crossings and there is no warrant for the presence of pedestrians on the right-of-way."[4]

So let me get this straight. We should build expressways in our downtowns with no pedestrian infrastructure. That will rid us of both traffic congestion and pedestrians, and then we'll be safer. Really?

That is quite an argument. Was there any research supporting it?

Not at all. In fact, most of the research suggests the exact opposite. Even AASHO—in the same 1957 manual—points that out: "Accident rates increase with traffic volumes to a point where congestion becomes very acute when accident rates actually fall off."[5]

In the 1930s, several researchers found that crash rates increased as daily traffic increased.[6] They also found a threshold at which traffic congestion pushed the rates back down again. This threshold differed from study to study, but it was consistent.

Neil Arason suggests that we knew this to be true from the get-go: "In the 1910s and 20s, police often saw congestion as helpful in their ongoing efforts to slow traffic."[7]

Still, the prevailing sentiment—as described in a 1949 *Traffic Quarterly* paper—was that "if we eliminate or alleviate congestion, we shall simultaneously clear up related hazards or hindrances."[8]

Wilbur Smith—who along with William Eno is considered one of the grandfathers of traffic engineering—explains away the lack of supporting data in a 1947 *Traffic Quarterly* paper. Without anything to back it up, he suggests that traffic congestion in one place induces reckless driving elsewhere: "Delay and congestion in traffic, even though they do not produce accidents at the point of congestion, may cause accidents elsewhere."[9]

So we shunned pedestrians and built massive roads that we thought would solve traffic congestion **and** simultaneously improve road safety. We got it wrong on all accounts.

Traffic congestion was a problem long before cars and will remain a problem long after. A 1972 *Traffic Quarterly* paper dates it back to ancient Rome: "Ancient Rome tried to regulate vehicular traffic, with no greater success than its present-day counterparts, modern Rome included."[10]

Originally, cars were supposed to be the traffic congestion panacea. A 1909 article titled "Automobiles May Solve the Traffic Problem" lays out the case: "It looks as if one of the most vexing problems on municipalities—that of increasing the pavement space to prevent the congestion of traffic—will work out through the automobile. Every city of any size has had the question forced on its attention at one time or another, but the more general use of the automobile would allow putting off consideration of it for years."[11]

Longtime *Traffic Quarterly* editor Colonel Robert Goetz says in 1960 that "the problem of traffic congestion in large cities is a matter of grave concern. Congestion cannot grow worse for long without serious, damaging results to the important place traffic occupies in a functional city structure."[12]

In a 1956 piece, the Colonel even says that fixing traffic congestion is our patriotic duty.[13]

What will happen if we ***don't*** fix traffic congestion?

Here is Wilbur Smith again, this time in a 1949 *Traffic Quarterly* paper. If we don't get rid of traffic congestion in cities, he writes, "real estate values decline, and the whole plan of city growth is upset."[14]

This was a common refrain and remains so today. What shifted over the years is how we put numbers to the cost of congestion. Traffic engineers have always done that, at least to some extent, dating back to the 1840s when William Haywood wrote about the "many ingenious calculations" regarding "the loss of time consequent upon the inadequacy of the thoroughfares" in London.[15] What we do today honestly isn't much different from what John Wolfe Barry described in 1899. In making a case for "many more millions of expenditures" being spent on roads in London, he counted everyone on the road and compared their actual speed to what would be a free-flowing speed.[16] He then figured out the difference and multiplied that time by the average wage.

Although the basics remain the same, what changed is that we've made it national, recurring, and more systematic. In 1955, Paul Hoffman's Automotive Safety Foundation funded the Texas Transportation Institute (TTI) to conduct "economic impact research" on traffic congestion.[17] While the funding source changed, TTI does the same work today. Its current traffic congestion monitoring efforts started in 1982, and it has put out its "Urban Mobility Report" on a regular basis since 1987. Every time it is released, there's a slew of news stories about which cities have the worst congestion and how much this is costing us all.[a] In a recent report, the TTI estimated that traffic congestion cost Americans more than $190 billion in a single year.[18]

Traffic engineers rely on such numbers—extrapolated into the future—whenever they want to sell us a bigger and supposedly better road. They tell us that if we don't do something now, "traffic congestion will quadruple within the next 20 years."[19] This one was a 1993 quote from the Federal Highway Administration, which goes on to explain that "Americans will spend a stunning 8 billion hours each year sitting in traffic."[20] Sometimes the estimates don't seem all that scary in retrospect. A 1986 *Transportation Quarterly* paper implores

---

[a] Unsurprisingly, such a list of "worst" US cities would be similar to many people's list of best US cities.

us to build expressways in San Francisco or else "bottlenecks would reach 30 minutes" by the year 2000![21]

If we ask if we can afford all this, they tell us that we can't afford not to. They remind us that "traffic congestion undermines the economic productivity and competitiveness, contributes to air pollution, and degrades the quality of life in our metropolitan areas."[22]

And don't forget, traffic congestion is bad for road safety too.

Is any of this true?

Fortunately—or unfortunately, depending on how you look at it—the answer seems to be no.

Safety researcher Leonard Evans tells us in a 1993 *Transportation Quarterly* paper that "although congestion impedes mobility, it increases safety, as measured by serious injuries and fatalities."[23]

The research supports this, but the research also supports the opposite. That's too often the case in road safety research given discrepancies in what we actually measure. In this strand of research, most papers focus on total expressway crashes, typically using mileage as *exposure*. Other research considers crash severity but then uses that same mileage metric as a proxy for traffic congestion, but high vehicle miles traveled is *not* the same as high traffic congestion.

Some European research is beginning to look more comprehensively at road fatalities across urban areas, but the research doesn't differentiate between where the crash happened and where the person was from. Nor does it consider that traffic congestion not only slows speeds, but it entices people into lower *exposure*, and thus they might shift to safer modes or to more location-efficient housing.

It's no wonder we get different results.

The other major fear—that too much traffic congestion will kill a region's economy—is a research topic that suffered from similar problems. That traffic engineers continued to use this fear to build bigger and badder roads bothered me enough that I had to see for myself. A few years ago, I wrote a paper with Eric Dumbaugh that looked at 30 years of data for 89[b] of the largest metropolitan areas in the country.[24] Instead of traffic congestion stifling the economy, we found the opposite. In other words, more job growth and better per capita gross domestic product went hand in hand with more traffic congestion. Our takeaway: "There may be valid reasons to continue

---

[b] It was supposed to be the top 100, but we could only get the data we needed for 89.

the fight against congestion, but a fear that congestion will stifle the economy is not one of them."[25]

Leonard Evans, in that same 1993 paper quoted above, explains why this might be the case:

> Congestion is caused by people's desire to have access to amenities that require large concentrated populations; it is regulated by how much aggravation they are willing to pay for such benefits, not by the number of vehicles. When congestion reaches a certain level of unpleasantness, a natural balance arises between those willing and those not willing to suffer it. The tolerable level does not seem to have changed much over the centuries, nor for that matter, millennia.[26]

The reality is that people adapt. They switch to other modes like walking, bicycling, or transit. They work from home from time to time. Some may move away while others will move downtown and find a more efficient commute. As Leonard Evans explains, "Places that are already as congested as the human spirit can bear,[c] and there are a great number of them, will not become more congested."[27]

That is why our 100 years of trying to fix traffic congestion hasn't helped all that much. Leonard Evans also points out that "congestion today is not demonstrably worse in New York City than it was a hundred years ago, nor worse in London than 300 years ago, nor worse in Rome than 2,000 years ago."[28]

That is also why continuing to fight traffic congestion—even in the misguided name of safety—remains a fool's errand.

---

[c] If we ever get autonomous cars right, people can watch movies while they sit in traffic—and maybe the human spirit will be able to bear even more traffic congestion.

28   Aiming in the Wrong Direction

If we actually want to improve road safety, it would help if we actually measured safety outcomes.

Much of the early crosswalk research focused not on safety outcomes but on whether crosswalks improved pedestrian crossing habits. As AAA explains in their 1958 pedestrian planning guidebook, zebra crosswalks help make pedestrians more "satisfactory."[1] They may also lead to safety, but we have no idea because that isn't what we are measuring.

Here is a more recent example. The Colorado DOT (CDOT) spends more than $20 million every year on its pavement striping program. Why so expensive? Well, the goal is to make sure every pavement marking has enough retro-reflectivity, which means replacing striping every year in mountainous areas and every three years in urban areas. To make sure CDOT does this often enough, we pay for annual retro-reflectivity testing. CDOT hires drivers to drive tens of thousands of miles each year with trucks equipped to take millions of scans in an attempt to measure an average retro-reflectivity value for each road marking based on retro-reflectivity testing standards from groups like the American Association of State Highway and Transportation Officials (AASHTO) and the American Society for Testing and Materials (ASTM).

That's a mouthful, but CDOT does all this because better retro-reflectivity leads to better safety, right? Right??

That's a good question. But most research focuses on whether retro-reflectivity leads to better visibility. Researchers measure the ***detection distance*** and assume that "longer detection distances have a positive effect on vehicle-control measures and, consequently, crashes."[2]

That makes sense, but as we know, drivers adapt. Poor striping visibility may cajole somebody to drive slower and more carefully. Great striping may give people a false sense of confidence and tempt them to drive faster than they should or to attempt a drive in treacherous weather.

One of the few good studies to look at retro-reflectivity and safety outcomes was a 2006 National Cooperative Highway Research Program report.[3] Since retro-reflectivity fades over time, safety should improve soon after striping replacement and then get progressively worse until it gets replaced again. What this report found is that retro-reflectivity doesn't seem to matter when it comes to crash outcomes.

A former student at the University of Colorado Denver[a] looked at safety and high-contrast pavement markings for her master's report. She found that supposedly better markings were significantly associated with worse crash outcomes.

The actual link between retro-reflectivity and pavement markings may differ by context. It could matter more in mountainous or foggy places than in urban areas, but I could easily make an argument for the opposite. Either way, we don't know the answer because our "safety" studies on this topic don't look at actual safety outcomes.

Maybe we don't want to know the answer.[b]

In the late 1950s, a couple of midwestern states initiated edge-lining programs for their rural highways. An *edge line* is the white stripe used to demarcate the edge of the road, and, logically, they should help improve safety. Both Ohio and Kansas did what few departments of transportation would do today: they set up randomized control experiments, considered the gold standard of research methodologies. We don't see many randomized control experiments in transportation.

Ezra Hauer summarizes the results: "Contrary to what was expected, both studies found that between access points the number of accidents after edgelining was larger than what would have been expected without it."[4]

In Ohio, crashes increased 15 percent with edge lines.[5] In Kansas, they jumped 27 percent.[6]

Did the Ohio and Kansas DOTs change courses after seeing these results? Nope. They kept right on edge lining. In fact, both went ahead and initiated statewide edge-lining programs.

---

[a] Katrina Koberdanz.

[b] When I proposed safety and retro-reflectivity as a possible research project with CDOT, they asked that I instead create a better statistical model for predicting when pavement markings need replacing.

The point here isn't about whether retro-reflectivity or edge lining are safer or not. It's about what we measure and what that says about our goals and objectives.

In these cases, the primary goal is to increase roadway visibility. If better safety comes along for the ride, great. But our list of things to do begins and ends with making sure everyone can see the pavement markings.

Let's say we flipped this thinking and made improving safety the goal. Making sure everyone can see pavement markings may be one way to achieve that goal, but there is a long list of other things we could try. Increasing roadway visibility is just one.

Traffic congestion is also a great example of traffic engineers measuring—and caring about—the wrong thing.

Most people would say transportation is about moving people and goods from point A to point B. If transportation is about movement, we want to make sure that nothing—such as traffic congestion—hinders that movement.

But at its core, transport is about connecting people, goods, services, and activities. We want to make it easy and safe for people to access schools, jobs, shops, restaurants, hospitals, health care, concerts, social gatherings, parks, nature trails, and so on.

If transportation is about access, reducing traffic congestion is just one of many possible solutions.

Focusing on the wrong goal leads us astray. Mistaking traffic congestion as the goal is the reason that Elon Musk's Boring Company isn't selling anything that we haven't seen before. Musk's car tunnels are the *Stranger Things* version of the elevated roads proposed by Norman Bel Geddes with his Futurama exhibit at the 1939 New York World's Fair. We might as well invest in Lyle Lanley and his monorail.[c]

First thinking through what transportation is for helps set the right goal. We then find a subset of solutions well beyond using bigger roads to solve traffic congestion. Improved transit is just the start. It can now include thinking about land uses and where points A and B are located in the first place. Can people afford to live

---

[c] "I've sold monorails to Brockway, Ogdenville, and North Haverbrook, and by gum, it put them on the map!"—Lyle Lanley, a *Simpsons* character who Google describes as a "con artist and shyster"

in A, or have we pushed them out to C, D, or E? Do you need to be in a car to get there, or did we provide a range of safe and reliable options?

Road safety needs this kind of thinking. If we care about reducing injuries and deaths, our metrics need to follow that lead. The World Health Organization says that "road traffic injuries are a major but neglected public health challenge."[7] Fixing this public health problem means measuring it as a public health problem. No more focusing on fender benders. No more metrics that interpret more driving as more safety. No more measuring things like retro-reflectivity under an unproven assumption that it will make us safer. The old metrics have their place, but they don't tell us much about real safety. I'm sick of seeing people pretend that they do.

So if we want to get road safety right, let's start by measuring it right.

Part 3

Make No Mistake

———

29   The Human Error False Flag

When something bad happens, we want a clear and simple explanation of what went wrong to help us understand *why*. The inclination to blame human error is pervasive—across nearly all disciplines—because it simplifies things. It helps us feel better about the world we live in. The world is still safe as long as everyone follows the rules. People make mistakes from time to time, but don't worry, the system is sound.

Despite the title of this book, the point isn't to shift blame to traffic engineers. The problem is that we've worked under the assumption that what traffic engineers do borders on irreproachable. That assumption limits our ability to get better, but it's easier this way.

It is also easier for traffic engineers to cite the "fact" that human error accounts for 90-plus percent of our crashes than to consider what they themselves can change. These days, the number you hear most often is 94 percent. We can thank (or blame) the National Highway Traffic Safety Administration (NHTSA) for this so-called fact.

NHTSA conducted a crash causation study focused on what it calls the "critical reason." While it cautions us that the critical reason should not be interpreted as the crash cause, nor used to assign blame, it also defines it as the "immediate reason for the critical pre-crash event" and the "last failure in the causal chain of events leading up to the crash."[1] The infamous table from that study attributes each critical reason to one of four possibilities: drivers, vehicles, the environment, or something unknown. Add 'em up and you get more than two million critical reasons attributed to drivers. The other three categories have fewer than 150,000 combined. That equates to 94 percent on drivers and 2 percent for each of the other three categories. Look more closely at the driver estimate, and you will find a confidence interval of 2.2 percent in either direction. In other words, it's fair to attribute the critical reason to drivers for between 91.8 percent and 96.2 percent of crashes.

Many traffic engineers quote numbers toward the higher end of this range. During his presentation at a recent Institute of Transportation Engineers conference, Raid Tirhi, a senior traffic engineer from Bellevue, Washington, said that the 94 percent NHTSA number is wrong.

OK Raid, you've got my attention.

He next said that according to his own personal research, 99 percent of crashes are caused by "erroneous human behavior." Well, that wasn't what I was hoping for.

I also saw Nic Ward present at the Commercial Vehicle Safety Summit. He's a professor from Montana State University and director of the Center for Health and Safety Culture. I like his work, but one of the first things he said was that 95 percent of crashes "in whole or in part" can be attributed to human error. So how do we fix safety? He says we need to focus on the road user by improving the safety culture with *education*.

But just like how focusing on traffic congestion limits our solution subset, so does focusing on human error. Instead of trying to understand and fix a system that encourages human error, those behaviors become the problems we try to fix.

This type of thinking isn't Raid's fault, or Nic's fault, or even NHTSA's fault. Traffic engineers are not thinking this way out of villainy. They legitimately believe it. Traffic engineering is a discipline that has long been "misleading itself into believing that it has a narrow 'human error problem.'"[2] A 1950 *Traffic Quarterly* paper asks whether we've reached the limits of safety because "the human element is, of course, the great variable. Driver errors are involved in 80 to 90% of all accidents."[3] Given how bad humans are, that paper goes on to say that "perhaps the wonder is not that we have so many accidents but that the record is as good as it is."[4]

That's one way to put a positive spin on bad safety.

So why do traffic engineers like to blame human error? Passing the buck is part of the answer. The other is straight logic. As a 1969 *Traffic Quarterly* paper lays out, "The logic is deceptively simple. Roads, vehicles, and the environment are passive elements. By themselves they do not cause accidents. It can only be through human violation of the law, error, and lack of good judgment that accidents occur."[5]

In other words, the transportation system that traffic engineers design and build can't possibly cause crashes. It just sits there, quietly waiting for people to use it. If those people follow the rules that we give them, we'd be 100 percent safe.

Let's think about "those people" for a second. Who is it that uses our system? Who do we design it for? One problem is that the answer to these questions is not one in the same. Traffic engineers generally design for people without disabilities, who are clear-minded, not distracted, and not drunk or impaired in any way. But we also know that isn't who exclusively uses the transportation system. Isn't that disconnect a design flaw?

The characteristics of what we consider a normal road user date to a 1971 paper by automotive engineer John Treat.[6] In that paper, Treat and his coauthor, a physicist named Kent Joscelyn, wanted to respond to Ralph Nader's *Unsafe at Any Speed* and his 1965 indictment of the auto industry. Funded by NHTSA, Treat and Joscelyn tried to figure out what percentage of crashes could be attributed to vehicle failure. In doing so, they realized they would need to pass "judgement on whether what the driver did or did not do constitutes a 'causal factor.'" They decided on a "high but reasonable standard for defensive driving" from what would "reasonably be expected from a good, alert—but completely human—driver."[7] That sounds fair, right? The study looked at 600 crashes and considered three possibilities for blame—the human operator, the vehicle, and the environment—and found that the human was responsible about 90 percent of the time.

Treat continued to publish on similar topics throughout the 1970s, culminating with his seminal 1979 study of crash causation.[8] Treat and his coauthors dug into 2,258 crash records, including the accompanying police reports. This time they found that 92.6 percent of crashes had a human factor as the probable cause. These humans were a "deficiency without which the accident would not have occurred" and guilty of driving too fast, a lack of attention, as well as improper scanning and vision. Treat and his coauthors— including David Shinar—even blame crashes on "poor personal and social adjustment."[9]

Forty years later, Shinar was still trying to remedy the aftermath of Treat's "false assumptions that must be overcome to yield an effective crash countermeasures policy."[10] Shinar's biggest grievance? He says Treat's work was biased. Shinar says crash cause "is in the eye of the beholder" and makes it clear Treat was not an impartial beholder:

"The study team members were highly biased in their definition of the 'acceptable' human, environment, and vehicle standards for 'normal' and appropriate' versus deficient/causal." When it comes to vehicles or the environment, "the thresholds were very high … meaning tolerant of vehicle and roadway design deficiencies even if research had identified technologically feasible better designs."[11]

What about for humans? "The threshold was very low for human errors—meaning intolerant of errors even when they could be expected as part of 'normal' behavior," he says.

What does Treat's work mean when it comes to understanding crashes? It means we can blame any road user who isn't completely alert and attentive at all times. That's the case even though we know we've designed a system where it's close to impossible for road users to be completely alert and attentive at all times. Shinar cites his own papers from the early 1980s on this last point.[12] Interviewing drivers after passing warning signs under various conditions, Shinar and Amos Drory found that only "5 to 10% of drivers registered the sign."[13] So even though it's completely normal for 90-plus percent of drivers to be too preoccupied to notice a warning sign, those same drivers will be blamed for any crash related to whatever that sign was warning them about.

We'll see more examples later, but for now, here's a simple traffic engineering one. We design long and straight highways that we know make people sleepy. A 1998 *Transportation Quarterly* paper calls **highway hypnosis** the mental state "where a monotonous set of conditions lulls the vehicular operator into driving as if he is on 'automatic pilot.'"[14]

We also know that thousands of people die on our roads because they fall asleep at the wheel. One study found that this rate may be as high as 13 percent of all road deaths.[15] In addition to affecting those with sleep disorders[a]—which are more common than most would assume—falling asleep at the wheel tends to happen to younger people, night-shift workers, those driving long distances, drinking, or

---

[a] That 1998 *Traffic Quarterly* paper tells the story of an undiagnosed narcoleptic named Joe Piscipo (not the actor). Joe fell asleep and crashed eight different times during his four years at the University of Illinois in the 1960s. Even after his diagnosis, he had seven more such crashes before age 25. Amazingly, nobody was seriously hurt in any of these 15 crashes, and Piscipo ended up as chairman of the American Narcolepsy Association.

using drugs, or those who drive for their job. It even happened to Michael Doucette, a New Hampshire teen who was named "America's Safest Teen Driver" in 1989. The next year, he fell asleep at the wheel, resulting in a head-on collision with another car that killed both Michael and the other driver. This wasn't the middle of the night. This was at 5 p.m.

Traffic engineers could design highways differently, in ways that would help people ward off fatigue.[b] Instead, we blame the deaths on the people who fell asleep. These fatalities get chalked up to human error, just like almost every other road fatality. When we look at the data to try to figure out how to improve road safety, it's no wonder we fail to consider *engineering* solutions.

Today, nearly every crash causation study uses the same definition of human error as automotive engineer John Treat did in the 1970s. The percent blame shifts a little from study to study, but the definition of a "normal driver" being completely alert and attentive remains the same. Even the more recent "naturalistic driving studies," which continuously recorded drivers in real-world conditions, found that driver inattention contributed to 78 percent of all crashes.[16] That doesn't include all the other "errors" drivers make, such as speeding. That's just driver inattention. The follow-up study found drivers engaging in distracting activities nearly 52 percent of the time they are driving.[17]

If we want to save lives, traffic engineers might want to assume that all human error is engineering error. This isn't for liability purposes. I'm strictly talking about the need for traffic engineers to reconsider their work. After all, conventional traffic engineering is a petri dish for human error. We can do better.

---

[b] Traffic engineers like to add rumble strips along roadway edges. They are better than nothing (except when they place rumble strips where bicyclists ride). Some evidence shows that they don't improve overall safety because they lead to other crash types (such as when a dozing driver feels the rumble strips and jerks the steering wheel the opposite direction). I also saw a presenter give survey evidence that long-haul truckers are no longer scared of falling asleep because they know the rumble strips will wake them.

30 What Is Predictable Is Preventable

___

There once was an online ESPN series called *Detail: Kobe Bryant* during which Kobe would break down film of current NBA players. In talking about Ben Simmons's lack of an offensive game, Kobe said, "If it's predictable, it's preventable."[1]

That's exactly how I feel about road safety. Most crashes are predictable. Most human error falls into recurrent patterns. We know exactly when certain types of human errors take place. We even know why certain types of human error take place. But when that human error coincides with a crash, it's easy to blame the human error. Traffic engineers—and *engineering* in general—are off the hook.

Think about a signalized intersection with permissive phasing. Permissive phasing is when we give pedestrians the walk signal while we simultaneously allow drivers to turn directly into where we just told pedestrians they could safely walk. Of course, we tell drivers that they need to yield to pedestrians. We just do so while forcing drivers to look for gaps in multiple lanes of oncoming traffic—while simultaneously making sure the crosswalk is clear of pedestrians—so that they can speed through the gap in oncoming traffic and make the turn. It's a predictable recipe for disaster.

Studies suggest that between 4 and 7 percent of drivers never look toward the pedestrian before committing to the turn.[2] Why? Well, we've traffic engineered a situation in which the driver is only peripherally concerned about the most vulnerable road user, the pedestrian. We force the driver's attention elsewhere and hope for the best. There is a reason left-turning vehicle-pedestrian crashes outnumber right-turning ones by a factor of three to one. We've set them up to fail. We put up the walk symbol and tell pedestrians it's safe to go. If things don't go well, it's not like traffic engineers will take the blame.

One-third of San Francisco pedestrian fatalities took place in crosswalks when the pedestrian had the right-of-way.[3] We never stop to think *why* drivers hit so many pedestrians in such situations. If the driver does hit somebody, it's the driver's fault. Human error.

We also never stop to think *why* pedestrians may opt to cross a street in the middle of a block where they don't have to worry about turning drivers. But if the pedestrian gets hit, they were jaywalking; it's their fault. More human error.

In cases like these, the human error and the subsequent crashes are predictable. We might be able to prevent them with *engineering*, but it's easier to blame human error and hope that *education* or *enforcement* will fix the problem.

What traffic engineers fail to realize is that their definition of human error is a symptom of conventional traffic engineering. If we stop condemning road users and ask ourselves some basic questions, we might be able to help the cause. What conditions led to the road user error? What conditions might lead them toward a safer course of action?

Traffic engineers don't learn anything by blaming human error.

If we want to understand road user errors, especially those that lead to fatal and severe crashes, we need to dig deeper. Of course it would be easier if we could simply remind pedestrians to make eye contact before they cross the street, but that won't solve our real problems. Traffic engineers need to shift the attention and resources we waste on such reminders and focus on self-examination. What role did *engineering* play?

Although few and far between, some studies commit to a bit more self-examination. Helena Stigson and her coauthors conducted an in-depth assessment of fatal crashes in Sweden and found that more than 60 percent of road deaths were related to engineering deficiencies.[4] There is even a 1959 *Traffic Quarterly* paper that echoes this sentiment: "One phase of highway safety that should receive greater attention is the effect of the highways themselves.[a] Despite the fact that the driver is universally held responsible for a majority of accidents, it does not follow that highways cannot be made more accident-free regardless of the improvement of the driver."[5]

---

[a] I'll ignore that this same author goes on to wish that all roads could be limited-access expressways.

I don't believe traffic engineers are unwilling to change. The problem is that the approach we take to understanding crash causation—via human error—would never make us think we need to. But human error isn't the real cause, and it can't be separated from the context where the so-called errors occur. And that's a context designed by traffic engineers.

31   The Errors beneath the Errors

—

Books on human error spend a ton of time defining and nitpicking terms like *errors* versus *mistakes*, *slips* versus *lapses*, and *sharp* versus *blunt ends*. For your sake, and mine, I don't want to spend much time quibbling over terminology. To me, the most important difference is that between active and latent errors.

Almost all the human errors we talk about in traffic engineering are active errors. Somebody driving too fast is an active error. Somebody crossing the street when it's not their turn is an active error. Somebody pressing the wrong pedal is an active error. Somebody turning the wrong way down a one-way street is also an active error. And as the human error guru James Reason points out, somebody getting stung by a wasp and unintentionally running a red light would be an active error.[1]

Active errors may be undesirable, but they are not always intrinsically bad, nor do they always lead to crashes. In transportation, most active errors lead to nothing. Some cause unfathomable harm to people, property, and the built environment.

Latent errors refer to the transportation system itself. They are the long, straight, and boring highways. They are the permissive left turns across multiple lanes of traffic into a pedestrian phase. They are wide streets that entice speeds much faster than the speed limit.

Latent errors are harder to notice because they can lie dormant in our transportation system. But the hidden latent errors lead to the obvious active errors. Traffic engineers then treat active errors as random errors, never looking to the latent conditions that lead to their predictability—and to our underlying road safety issues.

James Reason describes the difference, again with an insect-related example: "Active failures are like mosquitoes. They can be swatted one by one, but they still keep coming. The best remedies are to create more effective defenses and to drain the swamps in which they breed. The swamps, in this case, are the ever present latent conditions."[2] In other words, the swamp is the underlying transportation system. The mosquitos are the road users. We aren't going to make road safety strides until we uncover the latent conditions that lead to the problematic road user errors.

Alina Tugend's book on human error makes the point that "latent errors pose the greatest threat to the safety of a complex system," but "it is generally difficult to quantify the contribution made by latent errors to systems failures."[3]

Why is it so difficult to quantify the latent errors? The Swiss cheese model of crash causation might help explain. Reason came up with this model to illustrate overlapping latent conditions. Imagine several pieces of Swiss cheese stacked on top of one another. The holes in each layer of cheese represent the latent errors, but we only see these holes when all the layers of cheese line up perfectly. If aligned, an active error can sneak through the aligned Swiss cheese holes and lead to a crash.

Let's consider the permissive left turn again in terms of Swiss cheese. Most of the time, the holes aren't aligned. There may be no pedestrian present, or the pedestrian may get to the other side before the driver gets a chance to turn. The opposing traffic may be light, giving the driver a better opportunity to scan for pedestrians. Whatever the reason, most permissive left-turn movements don't result in a death. But when the various latent design errors align just right, we have the perfect conditions for a driver to "accidentally" kill a pedestrian. The latent conditions led directly to an active failure. But since these latent error Swiss cheese holes align somewhat infrequently, the active error remains the obvious error. That's the error the officer writes down in the police report. That's the error that ends up in our crash database. That's the error we consider in our safety studies. That's the error we use when deciding which safety interventions to fund.

Even when traffic engineers provide a transportation system with lots of design holes, it still takes a bit of bad luck for the holes to align and provide a crash opportunity. That's why it's easy for the holes—the latent errors placed into the system by traffic engineers—to go unnoticed. But when it comes to road safety outcomes, it's the latent errors in the transportation system that are the most insidious. It's our refusal to deal with the latent issues proactively that keeps us spinning our wheels, never making any headway on real road safety improvements.

32   Tip of the Wrong Iceberg

—

Road user errors are the "tip of the causal iceberg" when it comes to understanding crash causation.[1] To paraphrase James Reason, drivers are the "final garnish to a lethal brew whose ingredients have already been long in the cooking" and "the root causes … were usually present within the system long before these active errors were committed."[2] Focusing on active errors lets us ignore the latent errors. It also lets us ignore the role that the traffic engineer played.

So if we really want to improve road safety, we need to look beyond human error.

But that's hard to do. Why? Well, it's just so damn easy to blame the road users.

Active errors are obvious. People fail to see or hear something. They fail to drive slow enough or walk fast enough. They fail to yield the right-of-way or seize the right-of-way. They fail to expect what we expect them to expect.[a]

When Uber's autonomous car killed Elaine Herzberg—the first recorded case of an autonomous car killing a pedestrian—we still somehow managed to blame the road users. In a press conference, the Tempe, Arizona, police shared video from the car's front camera of the moments leading up to the crash. The police's take? It was dark, and she was jaywalking. The police deemed the crash "unavoidable" and said no driver could be expected to see and react to the situation in time. The police blamed Elaine for her own death, never considering that the "eyes" of the autonomous car are LiDAR (for *l*ight-*d*etection *a*nd *r*anging), a laser-based remote-sensing technology. And guess what? LiDAR doesn't care if it's dark outside.

---

[a] Robert Sumwalt retired from chairman of the National Transportation Safety Board in 2021. Alina Tugend interviewed him and wrote that Sumwalt "believes that virtually all accidents can be traced back to human error somewhere in the system." If you include the errors of the humans that designed the system, he's right. Alina Tugend, *Better by Mistake: The Unexpected Benefits of Being Wrong* (New York: Penguin, 2011).

Uber eventually released video from the interior camera. Here we see the operator, Rafaela Vasquez, watching *The Voice* on her cell phone. Her job was to pay attention to the road, ready to take over for the autonomous vehicle technology at a moment's notice. She wasn't doing that, so when the National Transportation Safety Board (NTSB) released its report, it said the "probable cause"[b] of the crash was "the failure of the vehicle operator to monitor the driving environment and the operation of the automated driving system because she was visually distracted."[3]

Either way, we are quibbling about which road user to blame.

Instead, we could blame Uber and its vehicle design. The LiDAR system "saw" Herzberg walking 5.6 seconds before the crash. Though the car only needed 1.3 seconds to stop, it classified her as a false positive. Uber had previously decided that its system was getting too many false positives. To help ensure a smoother ride, the Uber engineers limited what would trigger a positive response and, in turn, what would trigger the emergency braking system. Also, could Uber reasonably expect Vasquez to stare at the road for hours on end? As Reason points out, "It is well known that even highly motivated operators cannot maintain effective vigilance for anything more than short periods; thus, they are demonstrably ill-suited to carry out this residual task of monitoring for rare, abnormal events."[4]

We could also blame the traffic engineers. **Why** were the crosswalks so far apart? **Why** was this street designed like a highway with a 45 mph speed limit? **Why** did they design what looks like a pedestrian path in the median of the street right where they didn't want pedestrians like Herzberg to cross?[c]

With almost every crash, we can find an unsuspecting road user to blame.[d] That's the case even when a series of latent design errors leads directly to the active errors. When filling out crash reports, it is insanely difficult for the police to identify anything other than human error. Human error is what gets coded into our crash data.

What's wrong with this system?

---

[b] Don't forget that NTSB tells us not to assume fault or blame with its "probable cause" statements.

[c] Since the crash, what looks like a pathway has been removed and replaced with vegetation.

[d] And if we can't find a road user to blame, we blame God.

It lets traffic engineers off the hook. We aren't able to learn from our mistakes. For traffic engineers, it means that, as Neil Arason said, "we rarely worried about a poorly designed road because, when things would go wrong, they would usually be viewed as the driver's fault anyway."[5]

We all love the idea of a data-driven approach to road safety. But if you dig through the crash data and see all the human error, it's hard not to walk away thinking that our problems need to be fixed by *education* and *enforcement*.

A few years ago, the Colorado Department of Transportation placed an advertising billboard high over Colfax Avenue in Denver reminding drivers to keep their eyes on the road, something that would be impossible to do while also looking at the billboard.

There is a great billboard safety study from a couple years ago suggesting that "removing the billboards was associated with a decrease of 30 to 40% in injury crashes and restoring the billboards was associated with an increase of 40 to 50% in injury crashes."[6]

That shouldn't be surprising.[e] We've known that was true for longer than I've been alive. A *Traffic Quarterly* paper from 1953 details a five-year study of Minnesota crashes and shows that crash rates increase significantly when you add more advertising signs per mile.[7]

But it's much easier to blame the drivers. We just need to get those bad apples off the road.

---

[e] "It must be the first of the month! New billboard day!"—Homer Simpson

33　Bad Apples

—

Early traffic engineers like Arnold Vey said traffic engineers would eliminate 70 percent of all crashes if everyone just let them do their thing.[1] It didn't take long for traffic engineers to realize they've over-promised and underdelivered. Instead of figuring out where we went wrong and how to do better, we shifted to the "bad apples" defense.

In the mid-1920s, Boston's public transit authority calculated that 27 percent of its drivers were causing 55 percent of crashes. So, the city's transit agency psychologically tested all drivers to determine which ones were "accident prone" and needed to be taken off the roads.

The basic idea was that we could distinguish between the good drivers and the bad apples. A 1969 *Traffic Quarterly* paper lays out the thinking: "It is argued that human errors are at the core of all accidents. If somehow humans making more than the normal share of errors could be detected, and if these 'high error' humans could be removed from the roads, a tremendously beneficial effect upon road safety might be possible."[2]

Akin to what Boston's transit agency found in the 1920s, traffic engineers trumpeted the idea "that about 10% of the drivers have nearly all the accidents."[3] Mathematically, it may be true, but as a 1955 *Traffic Quarterly* paper says, "What we have failed to recognize is that it is largely a different 10 percent each year. Remove every driver who had an accident last year and this year about the same number of people will have about the same number of accidents."[4]

This didn't stop traffic engineers from trying … and trying … and trying to find the bad apples. A 1973 *Traffic Quarterly* paper tells us that "it would be no exaggeration to say that a thousand papers have been published during the past 40 years in an attempt to measure the elusive phenomenon of the accident-prone driver, all with negative results."[5]

It seems obvious that some drivers drive more dangerous routes or at riskier times. Thus, we shouldn't expect all drivers to have the same number of crashes. Some overrepresentation is supposed to happen. Still, we blamed "the majority of accidents … to the nut behind the wheel" in 1968.[6] You heard the same complaints in 1992: "The 'nut behind the wheel' is usually perceived as the cause of most traffic accidents."[7] We still hear it today. In a 2018 paper published in the journal *Nature*, Jeff Hecht discusses conventional wisdom that artificial intelligence "would greatly reduce the toll of road accidents" by "replacing the human driver—'the nut behind the wheel.'"[8]

As James Reason says in his 2013 book, "They accept that there may be the occasional 'bad apple,' but they believe the barrel (the system) is in good shape."[9]

For us, the barrel is the transportation system, and it's not in good shape.

Don't get me wrong; removing those deemed to be bad apples from our streets isn't necessarily a bad thing. But we can't assume that is all we need to do to solve the problem. By putting too much energy into finding the bad apples, we overlook the problems with the barrel—that is, our transportation system. As Sidney Dekker says, "A 'bad apple' problem, to the extent that you can prove its existence, is a system problem and a system responsibility."[10]

In other words, getting rid of one bad apple doesn't help the conditions within the barrel that led to the apple rotting in the first place. Nor does getting rid of one bad driver get rid of the systematic conditions within the transportation system that led to the driver behaving badly in the first place.

34 Wishful Technological Thinking

—

Instead of removing the bad apples, why not remove the humans altogether? After all, aren't humans the weak link in the chain? If better *education* and more *enforcement* don't work like we hope, won't *technology* save the day?

That's the traffic engineering dream.

It also comes from the old-school automakers like Mary Barra, CEO of General Motors, who said as much on a *Freakonomics* podcast: "But I do think autonomous is an ultimate part of the solution, because when you look at the number of deaths that we've studied in the United States, 90 percent of them are caused by human error. And an autonomous vehicle is going to follow all the traffic laws and regulations. It's not going to drive drowsy or drunk or impaired. And it has A.I. technology that allows you to understand the light's going to turn green or red."[1]

It also comes from some of the progressive transportation voices such as Gabe Klein, who takes the safety benefits of autonomous cars as a given. In his book *Start-Up City*, he touts the additional benefits in "that cities will have the opportunity to reclaim a lot of land currently devoted to automobiles and their storage" and that "traffic signalization may become a thing of a past, and pedestrians should ultimately have priority and attain more freedom of movement."[2]

A few years ago, I saw Carlo Ratti of MIT's Senseable City Lab speak at an event in Sydney, Australia. His lab had recently released a study on the impacts of autonomous cars making traffic lights obsolete. Carlo talked about doubling traffic capacity and eliminating traffic congestion with autonomous intersections. He did so while showing a video depicting the simulated comparison of a conventional, congested intersection against an autonomous intersection with a never-ending stream of autonomous cars rushing through. Carlo then called this the "future of urbanism" and got a standing ovation from a crowd of hundreds.

I sat there thinking. Where do the pedestrians go?

That's the thing about this autonomous future of ours. Many of the benefits go away when you add one human to the mix. And given what we saw with Uber's autonomous car killing Elaine Herzberg, the humans will somehow still get blamed.

In terms of road safety, we also need to keep in mind that vehicle autonomy is not binary. Instead lies an autonomous spectrum, ranging from level 0 through 6, with level 0 being all human and level 6 being completely autonomous. Many current driver-assist technologies—such as lane change assistance or emergency brake assist—fall somewhere in the middle. The intent of these technologies? We want them to improve safety.

In a typical year, the United States sees more than 1.7 million rear-end crashes, with 500,000 injuries and 1,700 fatalities. The National Transportation Safety Board (NTSB) attributes 82 percent of fatalities and 94 percent of injuries to human error and suggests that "forward collision avoidance" technology would solve most of the problem.[3] Interestingly, the NTSB broke down related insurance claims of cars with this technology, but instead of solving 80 or 90 percent of the problem, related claims only dropped between 7 and 14 percent.

AAA studied how well the automatic emergency braking technology did with pedestrians, and according to a 2021 *Bloomberg News* video, "they do a really poor job even detecting adult pedestrians, only detecting adult pedestrians about 60% of the time and then when it comes to smaller pedestrians and children, they hardly detected them at all."[4]

AAA warned that this technology "proved to be completely ineffective at night" and ineffective with "a vehicle turning right onto an adjacent road with an adult crossing at the same time."[5] There is also evidence that such technologies are less effective detecting people with darker skin tones.[6]

I'm hoping it all improves over time. In the meantime, all this technology may make safety worse.

I remember driving from Denver up to the mountains for a ski trip more than a decade ago. Lane change assistance technology was new back then, and so was my neighbor's car. From the vantage point of the passenger's seat, he seemed to haphazardly change lanes, never looking over his shoulder nor using any mirrors. When I asked what the hell he was doing, he said that there was no need to look since the car would alert him when another car was in the way.

Now, if he uses that technology *in addition to* looking for himself, better safety might be a given. If he uses it *instead of* looking for himself, I'm not so sure.

AAA also studied lane change assistance and adaptive cruise control technologies. Researchers drove more than 4,000 miles, and, on average, the technology failed once every 8 miles.[7] If I'm putting my life in the hands of any technology, I'm looking for better odds than that.

Again, this technology will continue to improve. However, as Captain Sully Sullenberger once said, "Technology does not eliminate error, but it changes the nature of errors that are made, and it introduces new kinds of errors."[8]

Right now, I'm typing this on a laptop computer. It's easy to fix errors as I go, so I make a lot of them. Give me my grandmother's typewriter, and I'd be a lot more careful.

Back in 2013, Google was feeling pretty good about its autonomous car program and felt ready to let some of its employees try it out. Google told them that even though the car was "autonomous," they always needed to be ready to step in at a moment's notice. Google also hid cameras throughout the car. What it found were people texting, applying makeup, and taking naps at highway speeds. What it learned was that people begin to rely—and over rely—on technology quickly.

The safety benefits of safety technology are never as clear-cut as they seem.

NASCAR worked to improve safety with better helmets, seat belts, roll cages, fire-retardant uniforms, and softwall technology. These changes let drivers push the limits to a greater extent than ever before.[9]

Take away helmets and pads from football or hockey, and you'll see fewer big hits and maybe even fewer concussions.[10] Give them seemingly better equipment, and you could end up with more.[a]

Give somebody a backup camera on their vehicle, and they may stop looking over their shoulder.

---

[a] When my son Luke was in elementary school, he typically rode his bike like a maniac whenever he had his helmet on. Without a helmet, he no longer felt like a Marvel superhero and rode like a sane person. We still made him wear a helmet but couldn't help but wonder if it was safer.

Figuring out which is safer should be simple. But it's never that simple, especially when humans need to interact with the technology.

Sidney Dekker talks at length about this issue in his book and says the "human-technology interaction is an increasingly dominant source of error."[11] It's not just about substituting old errors for new errors. Like when switching from a typewriter to a computer, new technology can change the error game.

We assume that technology makes things easier on us. More often than not, it increases what is asked of us. I used to only look over my shoulder when backing up; now I do it while also trying to look at my backup camera.

Technology also asks more of our transportation system. Even though we fail at simple things like sidewalks, autonomous cars will expect us to maintain visible lane markings on every single road.

And when something goes wrong, we get a whole new set of problems. When my 16-year-old clothes dryer broke down, it was always one of two issues. Both of them, I could fix myself with a cheap part I found online and a YouTube video. With my new dryer, I would need both electrical engineering and computer science degrees to even consider cracking it open. There are way more pathways to break down, all hidden within an escape room of various black boxes. Similar things can be said for cars with more and more technology. They become more and more opaque.

This isn't just about keeping my dryer (or car) running; it's also about what should be simple functionality. As I mentioned earlier, when first jumping into a rental car it always takes me longer than it should to figure out how to turn the radio up or down. Radio volume may not be a big deal, but what if I have a similar issue when trying to switch my car into autonomous mode? What if I think I made the switch from cruise control to autonomous but didn't?

Technology influences us to create new errors. It mutates our old errors. It conceals what we would need to know to detect and fix our errors.

And it's not like we are going to retrain drivers like we might do for airline pilots when they switch to a new plane design.

Lucian Leape and Donald Berwick, doctors and foundational researchers of medical error, said that "the more complex any system is, the more chances it has to fail."[12]

Yet, as James Reason highlights, the mentality we have when it comes to transportation system safety is that "many systems designers view human operators as unreliable and inefficient and strive to supplant them with automated devices."[13]

Will it be safer? It's not as simple as we think.

35   Not So Simple

I usually love Malcolm Gladwell's work. His books.[a] His *New Yorker* pieces. His *Revisionist History* podcast. The point of each podcast is to reconsider something (or someone) that we think we know and examine why it (or they) may be misunderstood.

One early episode was called the "Blame Game."[1] The idea was to investigate the unintended acceleration crashes and deaths that were a fixture of media coverage in the early 2010s. Gladwell starts the episode with a 911 call from Mark Saylor, a California Highway Patrol officer who is barreling down Route 125 in Santee, California, telling the 911 operator that the accelerator pedal was stuck and the brakes weren't working. Then, you hear screams and a crash before the call cuts out.

Mark, his wife, Cleofe, his 13-year-old daughter Mahala, and his brother-in-law Chris hit another car, plunge down a ravine, and all died in a fiery crash.

It's a tough listen, not only because of this horrific 911 call but also because of where Malcolm lands in the end.

The press blamed the after-market, all-weather floor mats even though similarly tragic events happened in cars without such floor mats.

Malcolm then takes a trip to the General Motors Proving Ground in Michigan to test some of the cars most commonly associated with this problem. In every case, the brakes overpowered the accelerator. Even at 100 mph, it took longer to stop, but the car still stopped.

Over the course of the podcast episode, Malcolm talks with self-professed "car guys" from *Car and Driver* magazine as well as University of California, Los Angeles human performance professor Dick Schmidt.[b] Schmidt says this unintended acceleration issue

---

[a] His book *David and Goliath: Underdogs, Misfits, and the Art of Battling Giants* is my favorite.

[b] Unfortunately, Schmidt passed away before the podcast episode was released.

seems to happen to older drivers, shorter drivers, and those in unfamiliar cars (such as parking lot valets) and often at low speeds or when parking. Though not mentioned in the podcast, this group included an older California couple, Bulent and Anne Ezal, who were trying to park at a waterfront restaurant when their Toyota Camry accelerated through a fence and off a 70-foot-high cliff. Bulent survived; Anne, his wife of 46 years, did not. Unlike Mark Saylor, Bulent lived to tell his side of the story and remains adamant that his foot was "absolutely, positively on the brake."[2]

At the end of the episode, Malcolm concludes that "what the car guys want the rest of us to acknowledge is that driving is a complicated and dangerous act. It is not just the negligent or the reckless who make fatal mistakes. Ordinary people do under seemingly ordinary circumstances. Mark Saylor did nothing wrong, nothing. What happened to him could have happened to any of us and will happen again unless we can finally have an honest conversation about what a car is and what it isn't and what the responsibility of a driver is when things go awry. Cars do not have minds of their own. A car just does what the driver tells it to do."[3]

Despite adamantly saying that Mark Saylor did nothing wrong, this conclusion 100 percent blames him for pushing the wrong pedal (and for not realizing he could have put the car in neutral or apparently held down the push button ignition for 3 seconds).

Beyond the logical inconsistency, I have some thoughts. First, this conclusion ignores 100 years of blaming road users for anything and everything that happens to them on the roads. Second, even if it was a case of pushing the wrong pedal, it ignores that this setup could still be considered a design flaw. Does it make sense that STOP and GO are within three or four inches of each other and require the same basic action? Does it make sense that the setup is designed in a way in which older and shorter-than-average drivers are more likely to die? Does it make sense that we use our feet for these tasks even though our hands generally have far better gross and fine motor skills? In fact, in the mid-1930s, "it was called to the attention of the brake division at General Motors that a hand-operated brake would save distance in stopping because the hand is quicker than the foot. A finger-operated power brake would respond over twice as quickly as the foot-operated."[4]

Yet the National Highway Traffic Safety Administration (NHTSA) agreed with Malcolm's conclusion. Other than one case of "pedal entrapment," its report exonerates the cars and implicates the humans. Not one of the black boxes, which traffic engineers call event data recorders, indicated that any of the drivers applied the brakes until the last second.

Today's brakes are electronic. They aren't mechanical levers like in the old days. If there is a bug in the software, isn't it possible for both the car *and* the black box to miss that somebody is pushing the brake pedal? In other words, the act of pushing the brake pedal needs to be picked up by the car's software for that fact to be relayed to the car's mechanical system. If the software has a problem, it may not know the brake pedal was pressed. Neither would the black box.

Malcolm also interviewed Sean Kane, a consultant in this field, on this podcast. Kane tells us that modern-day cars have more than a hundred million lines of code written into their software. He is put in a bad light in the podcast, in part because his research was funded by a group of attorneys and in part because he implies a grander conspiracy and cover-up. In coming to my own conclusion, I happened upon a presentation by Phil Koopman, a professor of electrical and computer engineering at Carnegie Mellon University. Koopman makes it clear that black boxes "can record false information."[5] He also says that even if the software is perfect, computer chips have random faults on the order of one random error every 10,000 to 100,000 hours.

How often does your cell phone not work exactly as you expect it or as quickly as you would like? Why should cars be all that different?

Yes, pressing the wrong pedals can happen. One 1976 study suggests that it accounts for three out of every 10,000 crashes.[6] The most horrific case might be a 2003 crash in Santa Monica, California, where an 86-year-old named George Weller drove through a farmers' market, killing 10 people—including a baby named Brendon—and injuring 63 others.

The NTSB report for the Santa Monica farmers' market crash focused a lot on the need for event data recorders. Although this recommendation makes logical sense, car black boxes are not airplane black boxes. According to AAA, "The latest digital flight data recorders capture more than 700 pieces of information and the plane's position, while the companion cockpit voice recorder stores the last two hours of flight crew communications."[7]

In cars with black boxes, the recorders usually gather about 15 data points—such as speed, acceleration, and braking—for several seconds before and after an airbag deploys or an instance of extreme vehicle deceleration. If the vehicle manufacturer chooses to install an event data recorder, all that's required is that the manufacturer mentions it in the owner's manual. Car black boxes still aren't required, and even if they were, there isn't a lot of consistency in this world.

Sure, it's possible that all these people pushed the wrong pedal. It's also possible that—like cell phones—cars may not always do exactly what we expect as quickly as we expect. That's why some people install cameras to record their feet and pedals while they drive.[8] Either way, blaming these errors on the humans that make them doesn't much help the safety cause.

Trying to fix this with technology—or even autonomous cars—*may* help the cause. Since the harm of a malfunctioning autonomous car outstrips that of a malfunctioning cell phone, you would presume a higher reliability standard. At the same time, cell phones don't need to worry as much about simple things like rain, fog, snow, or sun glare. It is one thing for an autonomous car to perform in perfect conditions; it will be a challenge for it to do so while dealing with all that nature sends its way. Take dirt as an example. When it comes to conventional cars, dirt isn't usually a big deal. When it comes to autonomous cars, it could be. Will we need to legislate that people keep their cars clean? Will car washes become more ubiquitous than gas stations ever were?

When it comes to technology, don't forget that those who design these technologies are human as well—humans who make errors and errors that contribute to crashes. These designers realize they can't foresee every possible situation, so they tell us we need to be ready to take over at a moment's notice. As James Reason suggests, it's a catch-22 because humans just aren't meant for that: "Human supervisory control was not conceived with humans in mind. It was a by-product of the microchip revolution. Indeed, if a group of human factors specialists sat down with the malign intent of conceiving an activity that was wholly ill-matched to the strengths and weaknesses of human cognition, they might well have come up with something not altogether different from what is currently demanded."[9] He was talking about nuclear power plants, but it holds for our current iteration of so-called autonomous cars.

In 2019, Elon Musk claimed Tesla would have "feature-complete full self-driving this year."[10] Tesla started selling "full self-driving capability" for a hefty extra fee even though its 2021 quarterly US Securities and Exchange Commission (SEC) filing admitted that its cars may never achieve full self-driving. That didn't stop Musk from calling this feature "autopilot" or Tesla owners from treating their cars as if they are fully autonomous. Four months after the SEC filing, police arrested a California guy for sitting in the back seat of his Tesla while traveling across the San Francisco–Oakland Bay Bridge on Interstate 80.[11] Less than a month before this arrest, two Texas men died when their Tesla drove itself off the road.[12] Nobody was behind the wheel.

Before turning on driverless mode for the first time, Tesla makes you agree to "keep your hands on the steering wheel at all times" and to continuously "maintain control and responsibility for your vehicle."[13] Tesla seems to have a better black box than other cars, and it claims to know if you've kept your hands on the steering wheel.[c] If you didn't—like those guys in Texas—the crash is your fault. Apparently, Musk isn't to blame for making you think that it's full self-driving. Apparently, Tesla also isn't to blame for the fact that all it took was placing a 2-inch strip of black tape onto a 35 mph speed limit sign to trick its car into accelerating up to 85 mph.[14] The National Highway Traffic Safety Administration didn't seem bothered by any of this, but it did recently open an investigation into Tesla for its autopilot repeatedly steering cars into emergency and first responder vehicles sitting on shoulder.

Car manufacturers will follow Tesla's lead and make you agree to taking over the manual controls whenever there is a problem, which keeps the blame on the road users. But as autonomy continues to improve, drivers will drive less and less. As James Reason tells us, "One of the consequences of automation therefore, is that operators become de-skilled."[15] So when "manual takeover is necessary something has usually gone wrong; this means that operators need to be more rather than less skilled in order to cope with these atypical conditions."[16]

---

[c] My wife's Subaru often reminds me to put my hands back on the wheel when my hands are already on the wheel. You can also buy weights that attach to the steering wheel and trick the car into thinking your hands are on the wheel.

When something does go wrong, the situation will present a far greater burden on the so-called driver than we've ever been accustomed to. In his 1984 book *Normal Accidents: Living with High-Risk Technologies*, Charles Perrow says that "we shall see time and time again, the operator is confronted by unexpected and usually mysterious interactions among failures, saying that he or she should have zigged instead of zagged is only possible after the fact. Before the accident no one could know what was going on and what should have been done."[17]

Perrow then gives an example of two ships about to safely pass each other in the night. One of the captains thinks he needs to take over manual controls, but by doing so, he proceeds to ram directly into the other ship. The decision of whether to take over manual controls isn't an easy one, and even if we aren't out of practice when it comes to driving, we'll see entirely new types of errors on our roads.

If we ever reach full autonomy, driving will become "easier," which may coincide with an exponential increase in vehicle miles traveled. It could be exacerbated by policy decisions, such as whether we continue to own our own cars or whether they become a shared resource. If the latter, I'd go to work while the car serves others until I need it again. If the former, would it park and take up space downtown all day? Or could I have the car circle the block again and again, all by itself, while I sit in a meeting or grab a quick dinner? Or maybe I'd send it all the way home and have it come pick me up later? Whatever the case, none of these scenarios is safety neutral.

The future of cars will probably have more technology and may very well be autonomous. That doesn't mean we'll be any safer.

36   I Wish I Knew

_____

It's easy to come in after a crash and tell somebody that they should've zigged instead of zagged. Or that they should've paid more attention to the car directly in front of them. Or less attention to those inside their own car. They should've paid more attention to the side of the road where they missed the warning sign. Or the bicyclist. Or the pedestrian. Or the motorcycle. They should've been going slower. Or faster. They should've had two drinks instead of three. Or eight hours of sleep instead of five. Whatever circumstances led to the crash, we can figure out a way to tell them how they should have acted differently.

Here's the thing, though. Most errors weren't considered errors at the time. People were usually acting rationally given what they knew then. Of course they "would have acted differently if they had known the seriousness of the situation." But as Sidney Dekker tells us in his field guide to human error, "The point is, they did not."[1]

Knowing how things turned out creates a bias. We don't put ourselves in their shoes. We never ask _why_ they did what they did. We never think about what they knew at the time. Or what they might have been thinking. Or what uncertainty they faced. Or even how improbable a crash seemed to those that ended up being involved.

We thus overestimate our ability to predict what would have prevented the crash. We reverse engineer the outcome. We start with the crash and then look to see where those involved went wrong. By working backward instead of in the order things happened to those involved, we oversimplify things. As Dekker said, "You oversimplify causality because you reason backwards … which is the opposite from how people experienced it at the time."[2]

Several of the human error books that I read discuss the terrible 1988 Clapham Junction train crash in London, England, which killed 35 people and injured hundreds. They do so because the chairman of the postcrash investigation, Lord Anthony Hidden, said in his report, "There is almost no human action or decision that cannot be made to look flawed and less sensible in the misleading light of hindsight. It is essential that the critic should keep himself constantly aware of that fact."[3]

In other words, there is almost nothing they did that doesn't look flawed or causal in hindsight. But to paraphase James Reason, hindsight doesn't equal foresight.

Hindsight bias gets in the way of better understanding human error. Instead of trying to figure out *why* somebody did what they did, we focus our attention on telling them what they should have done differently. That wastes our time and resources. It also leads to a counterfactual argument, a rhetorical double contraction. As Dekker explains, "Saying what people could or should have done does not explain the reasons behind what they in fact did. Such questions and remarks are, literally, counterfactual—counter the known facts of the accident. What people didn't do (but could or should have done) is not as pressing as finding out why they in fact did what they did, which is where a 'human error' investigation needs to direct its resources."[4]

Whenever we say what the road user should or shouldn't've done, we are using a counterfactual. That situation never actually happened. It's like an alternate timeline. But when it comes to road safety, traffic engineers focus on that parallel universe. The same goes for the media when they talk about road crashes. It's logical to think counterfactuals will help us better understand human error. They won't.

In transportation, people constantly make decisions—both good and bad—that never result in a crash. Most of these decisions are never considered bad decisions. They were decisions that seemed reasonable at the time. They only become bad decisions when the Swiss cheese slices align and lead to a crash. We use hindsight and declare them, after the fact, to be bad decisions. Then we come up with a counterfactual to tell them that they never should've made the bad decision that they didn't know was a bad decision at the time the bad decision was being made.

Wait, what?

Let me explain it another way. The reality is that, in transportation, most bad decisions lead to perfectly fine outcomes.[a] In such

---

[a] Conversely, some good transportation decisions can sometimes lead to really bad outcomes.

cases, we rarely know we made a bad decision. Pointing out a bad decision, in hindsight only after a bad outcome happens, doesn't much help the cause.

For example, a lot of highway off-ramps have pedestrian crosswalks placed right where the driver needs to merge onto the arterial street. In most places, the chance of seeing a pedestrian in such a crosswalk hovers right around the point of almost never. Traffic engineers also ask drivers to look over their left shoulder toward the traffic they need to merge with instead of toward the crosswalk in front of them. Given the low chance of a pedestrian being present, it isn't usually a problem. If there is a pedestrian and the driver runs them over, we use hindsight to tell the driver they made a bad decision. They should've seen the pedestrian and stopped to let them cross. Yet this bad decision is one that most drivers in the same situation make every day. And in something like 99.99 percent of cases, that same bad decision doesn't lead to a bad outcome, nor would it be considered a bad decision to the person making it.

Using hindsight to come up with a counterfactual may be common practice, but it covers up the real problems. To me, this example is a clear design problem. The responsibility to do better needs to be placed in the lap of traffic engineers. Instead, the hindsight and counterfactuals help us feel like we have a simple human error explanation. To quote Chris Farley in *Tommy Boy*, this lets us "feel all warm and toasty inside." Even though something bad happened, we can go about our lives because there is a simple fix that we can hang our hat on.

But it's not that simple.

I'm not saying human errors aren't important, but we need to know what causes them. The problem isn't just a road user randomly doing something wrong; it's about how the transportation system leads them to such failures. As Paul Hoffman said in his 1939 safety book, "On certain kinds of roads, it is known, certain kinds of accidents can happen, do happen."[5]

Yet our process repeatedly casts the road users as the bad guys. Doing so implies that traffic engineers are the good guys. It allows traffic engineers to slap a "fancy guarantee" on the transportation system when all we might have sold you was a guaranteed piece of shit. If you want me to use hindsight bias to come up with a counterfactual, I won't. I don't have spare time.

37   Why and Why Not?

___

In January 2020, 14-year-old Taylor Crepeau was making his way to middle school in Vancouver, Washington. Andrew Friedt, on the day before his 18th birthday, woke up early to walk with Taylor. While trying to walk across 112th Avenue, 19-year-old Tyler Pickner, driving a green Chevy pickup truck, hit them on his way to work. Both Taylor and Andrew died at the scene before help arrived.

I heard about this crash the morning it happened and tried to follow the news as it unfolded. The initial media reports said the teenagers didn't use a crosswalk. The media also reported that the boys were wearing dark clothing. The takeaway was that it was an unfortunate accident, but these kids were clearly at fault.

If only they had used the crosswalk.

If only they had worn brighter clothes.

I purchased a copy of the police report, and it came to the same conclusion.[a] Does the report mention that 112th Avenue is designed like a 60 mph highway but signed at 35 mph with bad lighting, sidewalks that disappear for long stretches at a time, and nearly a half mile between crosswalks?

Nope.

Traffic engineers would look at the police report and think these road users needed more **education**. If the traffic engineers ask themselves **why** these road users did what they did, they might come to a very different conclusion.

Before Toyota got into the car manufacturing business in the 1930s, it was a looming company.[b] Sakichi Toyoda developed the **five whys** technique to tackle stubborn looming problems that wouldn't go away.[c] The basic idea is to dig down to the root cause of a problem by asking "Why" five times. Instead of treating symptoms

___

[a] It seems like these police reports should be free, right?

[b] Not looming, as in impending, but Toyota was a company that made actual looms.

[c] Stubborn problems that won't go away sounds a lot like road safety.

with quick-fix solutions, the five whys lead to countermeasures that help solve the underlying problem.

Traffic engineers still need to know **what** happened, but if we don't ask **why**, we'll never figure how what role the designs might've played in the why and how things could be designed differently. Put yourself in their shoes. Why did they make this so-called error? Keep asking *why*, digging deeper with each *why* until you begin to wrap your head around the underlying issues. It's hard to say how many *whys* we need, but the mind-set shift is easier than you think.

In March 2021, 24-year-old best friends Christine Yan and Isabelle Zhang were bringing food and clothes to an expectant mother in Lakewood, California. They crossed Bloomfield Avenue at an unsignalized crosswalk and never made it to the other side.

Let's ask the first *why*. Why didn't the driver stop? Given a five-lane arterial, the design of the road might suggest to the driver that such crosswalks are window dressing. If it was a real crosswalk, it would be signalized on a road like this one.

Second, why else might the driver not stop? The speed limit sign says 40 mph, but the design speed seems much higher. We know design influences drivers more than a speed limit sign, and at higher speeds, it would be difficult to see and react to potential pedestrians quickly enough.

Now let's put the third *why* on the pedestrian. Why did they cross there? Well, the crosswalk with zebra striping suggested that it would be safe.

Fourth, why didn't Christine and Isabelle walk to the nearest signalized crosswalk? First off, an extra third of a mile while carrying food and clothes doesn't sound like much fun. Even less fun is walking on a sidewalk shoved between a fast road and a concrete wall. Add in needing to cross multiple alleys just to arrive at a signalized crosswalk that isn't painted with the zebra striping like the unsignalized crosswalk, and their choice seems reasonable.

Here's the fifth *why*. Why did traffic engineers invite these young women to step out into the road to negotiate 100 feet of asphalt while simultaneously enticing drivers to drive at a speed at which it would be extremely difficult to stop for them in time? This intersection is also a block away from an elementary school.

Here is my bonus *why*. Why do we allow this?

This last question borders on unanswerable, but there is nothing in the crash report to suggest that traffic engineers should consider doing anything differently. We treat it as a tragic accident instead of a systemic failure. We all need to start with a fundamental rethinking of how we understand human error. Blaming crashes on the road users involved is easy. It lets traffic engineers go back to business as usual. The takeaway is that if road users simply followed the rules, our design is inherently safe. It's not. It's about time we question this and ask, ***why not?***

38   Cold, Wet, and a Little Embarrassed
—

One of the early pioneers of traffic engineering, Miller McClintock, said in the 1930s that "engineering skill, taking no account of financial limitations, could eliminate 98% of the highway accidents."[1] In fact, we "could build highways so nearly accident-proof that drivers would have to be very ingenious to cause them!"[2]

But as McClintock suggested, we can't achieve that because it costs too much. Paul Hoffman also tells us that we could "make all houses fireproof; but we don't" because it costs too much.[3]

You hear similar rhetoric today. Traffic engineers will tell you they can solve all the world's problems if you only gave us enough money to do so. The trillions of dollars we get every year isn't enough, so we need more money if you really want us to make these problems disappear.

But until we transform how we think about human error—and start asking ***what and why*** instead of ***what and who***—it's a bottomless money pit. It's a never-ending distraction from uncovering the latent conditions created by traffic engineers in the first place.

Sidney Dekker's field guide to human error does a nice job of differentiating between how we used to think about human error and how we should think about it now.[4]

Under the old view, we ask what happened, who did it, and what they should have done differently. Under the new view, we ask what happened, why it made sense for them to do what they did, and why there is a gap between what we want people to do and what they actually do.

Under the old view, the system is safe, and people are the problem. We count and categorize what they did wrong and use that list to try and control their behaviors and attitudes with ***education*** and ***enforcement***. Under the new view, people aren't the main problem; what they do is the consequence and a symptom of deeper issues. People's actions can help lead us to these deeper issues embedded within the system itself, many of which can be controlled with ***engineering***.

Under the old view, identifying the human error ends our crash investigation. Under the new view, that is where we start our crash investigation.

I'm not at all saying that people shouldn't be held accountable for their actions. I'm fully aware how some of this may come across, but personal responsibility is always going to be a vital part of road safety, especially for police, courts, and insurance companies. Traffic engineers, however, need a different perspective. We need to take the errors that people make and use them to help us understand what we can do better. Errors can't be viewed in isolation from the system we provided; they go hand in hand with the system. Understanding this connection is what we need if we want to improve safety. Fortunately, the human error research suggests that "there is no evidence that a systems approach dilutes personal accountability."[5]

Focusing on the system also makes it easy to blame traffic engineers for the mess we're in. But give us a chance. Yes, it's fair to say we have a vested interest in perpetuating the myth of human error. And yes, I found a 1976 article in something called *Traffic Engineering* magazine cautioning traffic engineers against buying into the myth that most crashes are the product of human error.[6] At the same time, I honestly believe that most traffic engineers don't know any better. They aren't trying to do a bad job. They've just been doing what we've all been taught to do. Unfortunately, focusing on human error throws a monkey wrench into our data-driven approach to road safety. It cuts us off from the feedback loop. It's been doing so for generations and still does so today. As a result, traffic engineers haven't yet been given the opportunity to identify the real problems and learn from our mistakes.

Focusing on human error is kind of like peeing your pants. At first, you feel better because you solved what seems like the most pressing problem, and as Billy Madison famously said, "You ain't cool unless you pee your pants." But now we've reached the point in the history of traffic engineering where focusing on human error should make us feel cold, wet, and a little embarrassed. Billy isn't going to come save the day.

Part 4

I Feel the Need for Speed

———

39   Disconnecting Speed from Safety

—

When it comes to speed and safety on a particular road, traffic engineers only care that everyone's speed is in the same ballpark. Actual speed—and whether that ballpark hovers in the 30 mph range or the 60 mph range—is irrelevant to how we define safety.

Back when we treated fender benders the same as fatal or severe injury crashes, it was easy to argue that this perspective makes sense. It doesn't. Speed kills. Yet traffic engineers design as if speed doesn't matter. It does.

So, how in the world did traffic engineers come to believe that speed is negligible in road safety?

Usually, this story begins with David Solomon and his research paper[a] from the mid-1960s suggesting that actual speeds matter far less than uniformity of speed.[1] But the infamous Solomon curves were just the culmination of 50 years of nudges in that direction.

A 1913 *Scientific American* article, forecasting the future of driving in cities, said that "the automobile was too valuable to be checked by speed limits, so all driving restrictions would be removed and 'if a man walked at the street level it would be at his own risk.'"[2]

Moving fast has always been the goal. The Will Ferrell movie *Talladega Nights: The Ballad of Ricky Bobby* kicks off with this opening title card: "America is all about speed, hot nasty bad-ass speed.— Eleanor Roosevelt, 1936."

I'm guessing Eleanor Roosevelt never said that. But it's hard to argue with the sentiment, and traffic engineers never wanted to be the ones standing in the way of this fast-moving future. The chief traffic engineer for the state of Illinois, Frank T. Sheets,[b] said as much in the late 1930s: "Speed will be limited only by the capacity of the driver and vehicle."[3]

Throughout those early years, traffic engineers were keen to

---

[a] Funded by the US Department of Commerce for some reason.

[b] It sounds like a fake name.

remind us that there is no relationship between driving fast and the accident-causing "bad apples" that we've long been trying to get off the roads. This was despite extensive research to the contrary, including a 1956 study saying that "it appears that faster drivers have more accidents than slower drivers."[4]

Yet traffic engineers treated such findings like a conspiracy theory that needed debunking. So we put lots of effort into trying to disconnect speed from safety.

Take a 1957 *Traffic Quarterly* paper about how we focus too much on speed as a problem in road safety.[5] If someone asks you how traffic engineers decided that safety outcomes have little to do with speed, you can point to this paper and tell them that back in the 1950s, we asked a bunch of college kids from UCLA to self-report how fast they drive. And guess what? Its findings "suggest that individuals … who report consistently higher driving speeds than average have, in reality, traffic records free of accidents as often as other drivers."[6]

Or take a 1958 *Traffic Quarterly* paper, which concluded that high speeds must be safe because most road fatalities reportedly happen at 40 mph or less.[7] Most road fatalities don't involve 10-year-old drivers either, but that doesn't mean they'd be safer. Some of the wildest papers would use a cost-benefit analysis to make it seem like we can't afford not to focus on higher speeds.

But based on such studies, the federal government released its own report on road safety in 1959. It included a graph detailing how crash severity increases with higher driving speeds, but it counterintuitively concluded that "moderately high speeds are safer than are slow speeds or excessively high speeds."[8]

The report acknowledges that this result is specific to rural highways, but that didn't stop traffic engineers from applying it everywhere.

That's where David Solomon and his curves come into the picture.[c] What Solomon did in the mid-1960s was show that driving fast isn't just safe, it's much safer than driving slow.

Solomon graphed crash "involvement rate" by speed. He first showed that the world was safest when all traffic speeds were within 10 mph of one another. He also found that, for example, driving 20 mph is more than 200 times more deadly than driving 70 mph.

---

[c] The Solomon curves look like mirror images of the Nike swoosh.

Why might we want to question those results?

Well, Solomon focused entirely on high-speed rural highways with no pedestrians or bicyclists.

How did Solomon get his speed data? He asked drivers how fast they were going, even after a crash.[d]

As for those slow-moving cars? Most of them were pulling in or out of businesses or driveways.[e] The rest were slowing down due to traffic.

Despite decades of research to the contrary, the Solomon curves continue to rent space in the minds of traffic engineers.[9] Many traffic engineers still contend that speed doesn't kill. As long as most drivers are within 10 mph of one another, they will say that speed saves. It saves time. It saves money. It saves lives.

Really?!?

[d] While Solomon acknowledged that drivers might underestimate their speed, he viewed this supposed limitation as quite "inconsequential." However, research by White and Nelson (1970) showed that biased responses from drivers about their estimated speeds could artificially manufacture Solomon's results. S. B. White and A. C. Nelson, "Some Effects of Measurement Errors in Estimating Involvement Rate as a Function of Deviation from Mean Traffic Speed," *Journal of Safety Research* (1970).

[e] Research by Fildes and Lee showed that removing the turning vehicles flips the results. Brian Fildes and Stephen Lee, "The Speed Review: Road Environment, Behaviour, Speed Limits, Enforcement and Crashes," Monash University Accident Research Centre, Clayton, Victoria, Australia, September 1993.

40 What's Up with That?

North of Salt Lake City, the Legacy Parkway parallels Interstate 15 up to the Wasatch Weave interchange where these highways come together. It's a four-lane, controlled-access highway with a wide, grassy median and more than its fair share of safety problems.[1]

So how did the Utah Department of Transportation (UDOT) respond?

It increased the speed limit from 55 mph to 65 mph. It said the speed limit jump will "eliminate the safety risk"[2] on the Legacy Parkway.[a]

Well, you don't say.

UDOT conducted speed studies up and down the Legacy Parkway. It found that most drivers were going much faster than the 55 mph speed limit. Channeling the ghost of traffic engineers past, the safety director for UDOT said, "We decided to raise the speed limit to a speed that is closer to what drivers are actually driving. In doing so, we hope to eliminate the safety risk of speed discrepancy, which can happen when you have a significant difference between the speed most drivers are actually traveling and those who are driving the posted speed limit."[3]

But wouldn't raising the speed limit indicate to drivers that they should drive even faster? Despite mountains of evidence suggesting that drivers treat speed limits as their low-end baseline, UDOT doesn't see it as a problem.

So how do traffic engineers fix a speeding/safety problem? They raise the speed limit.

Traffic engineers use what we call the 85th percentile speed. The 85th percentile speed is whatever speed 85 percent of drivers are

---

[a] The other safety problem UDOT was supposedly dealing with was that, as the Utah Highway Patrol tells us, "wrong-way crashes are too common on Legacy Parkway." The DOT still proposed increasing the speed limit. Ginna Roe, "Wrong-Way Crashes Are Too Common," 2KUTV, October 21, 2019.

traveling slower than. If we have 100 drivers on the road and rank them in order from fastest to slowest, the 15th fastest driver would give us our 85th percentile speed.

Traffic engineers will then look 5 mph faster and 5 mph slower to see what percentage of drivers fall into different 10 mph ranges. According to David Solomon and his curves, the magnitude of the speed range doesn't matter as long as we get as many drivers as possible into that 10 mph range.

What I found interesting was how common this thinking was before Solomon published his paper. A 1963 manual by the Automotive Safety Foundation (ASF) says that "accidents are not related as much to speed … as to the spread in speeds."[4] ASF does so while also admitting that "none of the data specifically supports the theory that speed-zoning encourages a more uniform flow of traffic and that the reduction in accidents results from this effect."[5]

Lucky for the ASF—and eventually UDOT—Solomon's flawed study came out the next year.

In the case of the Legacy Parkway, the 85th percentile speeds ranged from 65 mph to 75 mph. Based on that and what it deems *engineering judgment*, UDOT originally proposed raising the speed limit to 70 mph. After community pushback, it settled for 65 mph.

According to the *Manual on Uniform Traffic Control Devices* (MUTCD), this slight adjustment is acceptable.[b] The MUTCD specifies that speed limits "should be within 5 mph of the 85th percentile speed of free-flowing traffic."[6]

But hold on a second. Do traffic engineers really base speed limits on however fast drivers may be driving? To quote Jack Nicholson in *A Few Good Men*, you're goddamn right we do.

---

[b] The MUTCD is supposed to provide standards for traffic signs, signals, and markings. Why it thinks it needs to add its 2 cents on setting speed limits is beyond me.

41   Reasonable and Prudent

—

Why do traffic engineers crowdsource speed limits? The National Transportation Safety Board (NTSB) lays out today's conventional thinking:

The use of the operating speed, more specifically the 85th percentile speed, is based on the assumption that the majority of drivers:

(1) are capable of selecting appropriate speeds according to weather conditions, traffic, road geometry, and roadside development; and
(2) operate at reasonable and prudent speeds.[1]

Reasonable and prudent, you say? Even if that assumption was valid, this approach still hinges on the idea that more speed means more safety. Again, this thinking was common well before Solomon's study. For instance, a *Traffic Quarterly* paper from 1957 was able to foretell Solomon's research, as well as our current approach to setting speed limits: "In some states they argue that 90 percent or so, at least, of motorists are reasonable, decent, human beings, so, if a speed limit is mooted, they time the traffic, and make the limit that speed which was not being exceeded by 85 percent of motorists, *even if it means raising the limit*. This is known as the eighty-five percentile limit."[2]

The rest of that *Traffic Quarterly* paper philosophizes about the teachings of Saint Thomas Aquinas and why we can ignore unjust laws—like speed limits that are too low—because they don't represent the common good.[a] This idea—that low speed limits repressed drivers—was all the rage back then. In discussing speed limits, a 1959 *Traffic Quarterly* paper further connects it to what we continue to believe about drivers: "While perhaps well-intentioned, many of our motor vehicle laws are inappropriate, as the letter of the law makes everyone a violator. Generally speaking, I believe that drivers,

---

[a] Seriously.

as interested parties on the spot, have the intelligence and judgment, if not deadened by stimulants, to reckon with traffic situations."[3]

You might be thinking that this is the point where I start advocating for lower speeds limits. But that type of thinking misses the point, and not unlike the human error conundrum, it deflects responsibility from traffic engineers.

We've long known that drivers generally select their speed based on the design of the road, with the speed limit serving as the baseline. A 1963 manual by the Automotive Safety Foundation hits on this point—"motorists govern their speed more by traffic and roadway conditions than by indicated speed regulations"[4]—as does traffic engineering pioneer Wilbur Smith[b] in a 1948 *Traffic Quarterly* paper: "Except where unusually strict enforcement can be effected, motorists are inclined to fox their own driving speeds, regardless of posted limits."[5]

So Wilbur, what does it look like in practice when drivers "fox" their own speeds? "Almost without exception, the posted values are exceeded appreciably by the average driver—even at the risk of arrest."

But Wilbur, can't we just make the speed limits lower? No, "the hoax of attempting to frighten motorists into driving at unreasonably low speeds by fixing absurdly low speed limits should be abandoned."

OK, what should we do Wilbur? "This indicates either that calculated limits are too low, or that most drivers are incapable of determining safe and reasonable limits."[6]

We've spent the years since then assuming that the former was true ("limits are too low") and the latter was false ("drivers are incapable of determining safe and reasonable limits").

When it comes to speed, this combination traps us in a vicious cycle.

We measure the 85th percentile speed and set a higher speed limit. Drivers perceive this new limit as their low-end target speed, and they drive a little faster. We measure the 85th percentile speed again. It's higher, so we set an even higher speed limit. So on and so forth. As the NTSB makes it abundantly clear in a 2017 report, "Raising the speed limit to match the 85th percentile speed may lead to higher operating speeds, and hence a higher 85th percentile speed."[7]

---

[b] And coauthor Charles Le Craw Jr.

When it comes to safety, this approach hasn't gotten us very far. The reality of the situation is that we're setting everyone up to fail.

Lower speed limits on their own will help, but only to some extent. Since drivers select their speeds based on the design of the road more so than on the posted limit, we need street designs intended to self-enforce these speeds.

The good news? It's a problem that traffic engineers can do something about.

42   Lukewarm Chicken

—

Let's say we went to 100 restaurants and measured the temperature of their cooked chicken. We then put these chicken temperatures in rank order and realize that the 85th hottest is 150 degrees Fahrenheit. For the sake of argument, let's assume that these restaurants have the appropriate "intelligence and judgment" and that they are "reasonable and prudent"[1] when it comes to cooking chicken. So instead of requiring 165 degrees for cooked chicken, we change the rule to 150 degrees.

Sounds delicious, right?

In all honestly, safety isn't what we're after with the 85th percentile approach. Our goal was to increase mobility, and our thinking was that the best way to do so was through high speeds. We slowly but surely manufactured a hypothesis where "pace" was the only thing that mattered. Raw speed? Raw chicken? Who cares?

Then along came Solomon's study, and it seemed too good to be true—because it was. But nearly six decades later, some traffic engineers still argue that pace is the only thing that matters.

We should know better by now. Pace matters, but when it comes to safety, so does speed. The 2017 National Transportation Safety Board (NTSB) report on speeding and safety was quite clear on this: "The NTSB concludes that the MUTCD [*Manual on Uniform Traffic Control Devices*] guidance for setting speed limits in speed zones is based on the 85th percentile speed, but there is not strong evidence that, within a given traffic flow, the 85th percentile speed equates to the speed with the lowest crash involvement rate on all road types."[2]

And if the 85th percentile approach was in fact based on sound engineering principles, it wouldn't make any sense for some states and municipalities to overrule speed studies with statutory speed limits for certain road types. Yet they do so regularly, and such statutory speed limits often stand in the way of cities trying to improve safety with lower speeds. Under Oregon law, a commission appointed by the Oregon Department of Transportation gets to review all speed limit changes, including for city-owned streets. But the MUTCD tells us that it's OK for statuary speed limits to supersede what supposedly sound engineering principles dictate, so there you go.

Traffic engineers argue that we need to let drivers vote with their feet as to how fast is fast enough. And after tallying the votes, we tell 15 percent of drivers that hey, we're sorry, but you voted for a number that is too high. Right off the bat, we have a 15 percent failure rate.

We then tell the police to go catch these monsters who voted for too high of a number. The police use techniques like high-visibility enforcement, where the basic idea is to have conspicuous enforcement in high crash and high crime areas. This data-driven approach makes sense, but it inevitably leads police to set up shop in low-income, minority neighborhoods.

Some states make enforcing speeds impossible via loopholes based, in principle, on the 85th percentile speed. California has a 1959 law intended to protect drivers from speed traps.[3] If traffic engineers haven't conducted a speed study recently enough, police officers can't use a radar gun. If they used a radar gun on a street without a recent speed study, the speeding ticket gets thrown out. If they estimated the speed any other way, it often gets thrown out as well. So at any given time, there are thousands of miles of streets in California on which police officers can't enforce speeds. As the *Los Angeles Times* explains, "Some streets ... had not seen speed enforcement for more than a decade. Drivers had grown accustomed to speeding with no citations."[4]

When it's time to do a speed study, the results can put cities in a catch-22 situation. Do they do a speed study that will force them to raise the speed limit in order to enforce that new higher limit? Or do they forgo enforcing speeds and keep the speed limit as is?

In 2009, a 60-year-old Los Angeles woman named Victoria Ramos was run over and killed in a crosswalk on Zelzah Avenue. Zelzah Avenue is a four-lane arterial near Cal State Northridge with a speed limit of 35 mph. A few months earlier, 19-year-old student Johoney Lobos was killed when trying to cross the same street. She was also in a crosswalk. The city did a study and found that 68 percent of drivers speed here. Given a high crash rate as well, one might expect some drastic physical changes in the near future. Instead, the city raised the speed limit to 40 mph so it could resume ***enforcement***.

Over the next decade, Zelzah Avenue became one of the "city's most dangerous streets for pedestrians and bicyclists," with 13 pedestrian or bike crashes over three years along a short two-mile stretch of road.[5] So the city did another traffic study and found an 85th percentile speed of 48 mph. They rounded down but still raised the speed limit again, this time to 45 mph.[6]

Few would've been surprised if the traffic engineers did another speed study and next put up a 50 mph sign. Fewer would be surprised if that again didn't fix the safety problem.

In this case, it took the passage of a statewide bill granting local jurisdictions in California a little extra leeway in deviating from the 85th percentile speed rule.[7] So while Zelzah Avenue is signed at 40 mph again, we still have a lot of work to do.

43   Be Careful What You Wish For

——

In the early 1950s, citizens of the small township of Paraclifta, Arkansas, successfully lobbied the Arkansas Department of Transportation (DOT) to "improve" their Main Street. What does that mean to a state DOT? It means turning their Main Street into a highway.

After the changes, the townspeople clamored for the DOT to install a traffic signal to slow down traffic.

In a bizarre 1956 *Traffic Quarterly* paper, Fred Herring, a 31-year veteran of the Arkansas DOT—and award-winning AASHO engineer—explains why a traffic light is not needed on Paraclifta's Main Street.[1]

He explains how "human relationships are the key" when traffic engineers need to go to a public meeting and tell people that we know more than they do. His paper reads as if he is teaching a robot how to interact with humans for the first time.

Fred then uses this newfound power of "human relationships" to convince the townspeople that "speed control is basically an enforcement problem."[2]

Fortunately for everyone, "the meeting was friendly and agreeable."[3] Unfortunately for everyone, you can now only find Paraclifta on websites dedicated to ghost towns.

This wasn't an *enforcement* problem.

This problem wasn't caused by the lack of a traffic light.

It was a design problem.

When structural engineers design a building, they calculate the "design load" and then select a beam that can handle approximately three times that maximum load.

When traffic engineers design a road, they select the *design speed* and then use that to set *minimums* for all the cross-sectional features and for the curvature of the road.

The design mentality in civil engineering disciplines seems similar across the board. In my undergraduate civil engineering courses, whether we were designing a beam, culvert, footing, or road, it felt like the same basic process. We figured out the minimum size and then made sure our design was bigger. Bigger was better. Bigger was safer.

Unfortunately, design speed isn't the same as a design load. Where we get into trouble is when we treat them like they are.

There are a few important differences.

In structural engineering, hitting the design load is one that we—fingers crossed—will likely never experience. In transportation, hitting the design speed is one that we see hundreds or thousands of times every single day.

In other words, in structural engineering, failure of a design load is obvious. The structure collapses. In transportation engineering, it's hard to recognize failure when it comes to design speed. Most traffic engineers would say the only possibility of failure would be a design speed that is too low.

In structural engineering, choosing a bigger beam than needed may cost more money but should be at least as safe, if not safer. In transportation engineering, selecting a higher design speed changes our design and, in turn, how people use the road. The result can be worse safety.

So what exactly is design speed? And where did it come from?

Today, the American Association of State Highway and Transportation Officials (AASHTO) defines *design speed* simply as "a selected speed used to determine the various geometric design features of the roadway."[4]

That sounds pretty innocuous, right? Especially when it goes on to say that design speed should be a "logical" selection "with respect to the anticipated operating speed, topography, the adjacent land use, modal mix, and the functional classification."[5]

And once we select the design speed? "All of the pertinent roadway features should be related to it to obtain a balanced design."[6]

It all makes logical sense.

But prior to the fourth edition of AASHTO's Green Book[a] in 2001, AASHTO used the words "maximum safe" in place of

---

[a] We'll talk a lot more about this design manual—and others—later. For the time being, the "Green Book" refers to *A Policy on Geometric Design of Highways and Streets* by the American Association of State Highway and Transportation Officials (AASHTO).

"selected" in their definition of design speed. More specifically, it said that "design speed is the maximum safe speed that can be maintained over a specified section of highway when conditions are so favorable that the design features of the highways govern."[7]

And more than a decade before the first edition of the Green Book, AASHTO used the exact same wording in its 1973 manual specific to urban areas and arterial streets.[8]

The roots of this definition date back to the 1930s.

Joseph Barnett first mentioned design speed in a 1936 Highway Research Board[b] paper and defined it akin to how we might for the design load of a beam:[c] "The maximum reasonably uniform speed which would be adopted by the faster driving group of vehicle operators, once clear of urban areas."[9]

AASHTO reeled things in a bit when it officially adopted the idea in 1938 by excluding the reckless drivers in their first definition of design speed, saying, "The maximum approximately uniform speed which probably will be adopted by the faster group of drivers but not, necessarily, by the small percentage of reckless ones."[10]

Still, the sentiment that design speed represented the uppermost limit stuck around for at least the next six decades. Even when the folks at AASHTO changed the definition in the 2001 Green Book to get rid of the phrase "maximum safe," it did so "in order to avoid the perception that speeds greater than the design speed were 'unsafe'" and sidestep liability concerns.[11]

They didn't change anything else.

In fact, for another decade, AASHTO continued to tell us that "every effort should be made to use as high a design speed as practical to attain a desired degree of safety."

And AASHTO kept this doozy in right on up until 2018: "The selected design speed should fit the travel desires and habits of nearly all drivers expected to use a particular facility."[12]

---

[b] The Highway Research Board is now known as the Transportation Research Board.

[c] This link to design load wasn't just implied; Barnett explicitly told us so in his 1939 report, *Toll Roads and Free Roads*: "Design loads in geometric highway design are the traffic density, the type of traffic and design speed." Bureau of Public Roads, US Department of Agriculture, *Toll Roads and Free Roads* (Washington, DC: US Government Printing Office, 1939).

Add all this up, and I'm not surprised that traffic engineers still treat design speed like our max design load. So we end up designing our roads to placate and protect those that think they are in the next *Fast and Furious* movie.

This isn't the road to safety.

44 Designing for Speed

—

When it comes to selecting a *design speed*, traffic engineers will tell you that making the right choice will improve safety. But if you asked traffic engineers to quantify the safety difference between one *design speed* selection and another, they'd have no idea where to start.

Before selecting a design speed, AASHTO asks us about functional classification. That is, we need to know if the road is going to be a freeway, an arterial, a collector, or a local street. We also need to know if it's in an urban or rural location. Then, it's nice to know something about the terrain. If it's mountainous, we refer to a mountainous table to find an appropriate design speed.

And that's it.

Many traffic engineers couldn't tell you what the relationship is between design speed and actual driving speeds. If the average traffic engineer knew that more than 50 percent of Americans routinely exceed design speeds set at less than 55 mph, they'd probably see it as a safety issue and wish they had used a higher design speed.[1]

And what happens if we use a higher design speed?

As Ezra Hauer says in his underappreciated 2019 paper, "The higher the design speed, the faster will drivers go, and the longer is the required sight-distance. If roads had to be designed for a speed that is exceeded only extremely rarely the required sight-distance would be so long that all roads would have to be nearly flat and straight—an impractical proposition."[2]

Sight distance refers to the length of roadway visible to a driver. The underlying thinking is that more sight distance is safer because it gives drivers more of a chance to perceive and react to problems. The *stopping sight distance* equation, for instance, lets traffic engineers estimate how much distance drivers would need to see something in the road and stop their car in time not to hit it. Design speed is the key input value. Other than how steep the road is, all other factors— such as the time it takes to see the problem and push the brakes or the relative amount of tire-pavement friction[a]—are assumed values that don't often change.

Selecting a higher design speed directly leads to higher sight distances. So we design flatter and straighter streets.

If people behaved exactly the same on streets with short and long sight distances, it would make perfect sense that a flatter and straighter street would be safer.

When we put in a bigger beam, the wind doesn't care. The design load remains the same.

When we install a bigger drainage system, the rain doesn't care. The peak flow remains the same.

But when we build a bigger street, people know. They may not realize it, but they know.

They drive faster.

They pay less attention.

Safety doesn't remain the same.

How about instead of treating the design speed as something close to the uppermost limit, we set it as a ***target speed***?

Then we use design to make it hard to drive faster than that target speed. If traffic engineers do a good enough job, the speed limit almost becomes irrelevant. Sure, we can set a speed limit that matches the target speed, but we've fundamentally flipped how we do business. Instead of letting drivers tell us how fast they want to go, we design the road to tell them what to do.

If we do this well, enforcing speeds starts to become unnecessary. Why? Because the street is designed to be self-enforcing.

If we don't do this well and the 85th percentile speed far surpasses the target speed, we can look to the traffic engineers to see if their designs failed us.

---

[a] The assumed tire friction factors have little to do with actual tire friction. Instead, we base them on driver comfort—in terms of what will keep drivers from reducing their speed. The result is overly conservative and facilitates higher speeds.

45 Above Minimum

—

When a driver speeds and a crash happens, we blame the driver. It's another case of human error. Add up all that speeding, and speeding becomes a contributing factor in about 30 percent of all fatal US crashes.

In 2003, a 34-year-old off-duty police officer named Edward Belvin was speeding in his 1998 Mercedes at 2 a.m. along US Route 1 in Linden, New Jersey.[1] Somehow, he managed to jump his car over a raised median into a head-on collision with oncoming traffic, where he killed himself and five Latin American immigrants coming home from their shift at a Houlihan's restaurant.[a] The National Transportation Safety Board (NTSB) later estimated that Belvin was going between 48 and 62 mph on a six-lane stretch of road signed for 40 mph.[2]

Given the posted speed limit, he was speeding. Should we consider his speed unreasonable? Maybe, but probably not. The NTSB did its own speed study and found a median speed on that stretch of road of 62 mph. As we know, traffic engineers like to focus on the 85th percentile speed.[b] That speed on that stretch of road was 72 mph. If 85 percent of drivers are going slower than 72 mph, 15 percent of drivers are going faster than 72 mph. The NTSB found that 6 percent of drivers were going faster than 80 mph.

Given the NTSB estimates, Belvin was going slower than the majority of other drivers on that road. Did his speed violate the law? Yup. Should that speed be considered human error when most drivers are going even faster?

It turned out that Belvin was also insanely drunk. His blood alcohol content was between 0.326 and 0.351, more than four times the

---

[a] This included 33-year-old Jorge Alpuin, the driver, as well as Wilman Rojas, 31, Marlon Alexis Castro, 23, and 20-year-old siblings Juan and Olga Chavez.

[b] The speed at which 85 percent of traffic is slower than.

legal limit.[c] This level is problematic, and it's not like everyone else being wasted too would change our minds about that.

Speeding is a little different though, right?

Most people would look at this situation and say that we need more *enforcement*. The NTSB agreed and made enforcing the speed limit one of its main recommendations.

Most traffic engineers would look at this situation and say to raise the *speed limit*. Given 85th percentile traffic engineering guidelines, a 70 mph speed limit is more appropriate than 40 mph.

More discerning traffic engineers might say that we got the *design speed* wrong. In trying to be conservative, traffic engineers probably selected a design speed of 60 or 70 mph even though they want people driving 40 or 50 mph.

In this case, they'd all be wrong.

The original plans show that the traffic engineer selected 45 mph as the design speed but still built a road that handles 70-plus mph without a problem.

Why?

Every traffic engineering guidebook gives us the "minimum" size needed for every feature on the road. The word *minimum* implies that bigger is better.

But the word *implies* doesn't do justice here. AASHTO's Green Book specifically tells traffic engineers that when sizing the various elements that comprise a road, "above-minimum design values should be used, where practical."[3]

So even if we pick an appropriate design speed—like somebody did here—we still end up building roads as if the design speed is much, much higher.

It's not enough to question the traffic engineers' choice of design speed. We also need to understand that the factor of safety mentality permeates deeper, to the point where we end up in situations where the 85th percentile guidelines tell us to set the speed limit higher than the design speed.

Setting the speed limit higher than the design speed—which we long defined as the "maximum safe speed" of the road—doesn't make

---

[c] In the days following the crash, a police spokesman told the *New York Times* that alcohol wasn't thought to be a factor. Ronald Smothers, "Officer's Blood Alcohol Level Was Over Limit in Fatal Crash," *New York Times*, June 14, 2003.

any sense whatsoever.[d] In some cases, we disregard common sense and do so anyway. In other cases, we don't because as a Federal Highway Administration report states, "officials are concerned with the potential liability."[4]

Where does that leave us? With big roads and low speed limits just begging us to go faster.

---

[d] That's why AASHTO redefined design speed in 2001.

46   The Fundamental Physics

Police pulled over a 19-year-old guy from Maine for going 126 mph in a 65 mph zone. His reason? "I was a little late for work."

Not long after, New Hampshire police caught a 45-year-old man going 150 mph in a 65 mph zone. He had a Corvette but didn't have a valid driver's license.

In a season 13 episode of *The Simpsons*, 10-year-old Bart gets arrested for joy-riding in a police car. In the courtroom scene at the end of the episode, Lisa says, "Motion to declare a writ of boys-will-be-boys." The judge replies, "Motion granted."

The "boys-will-be-boys" defense is, we hope, pushing things too far today. But we still don't take speeding seriously. Driving at more than double the speed limit may be ridiculous, but we all speed. When we do, we're generally not worried about the safety implications. We may be worried about getting a speeding ticket. But with most tickets, it's not like we'd consider ourselves a criminal or a menace to the safety of others. In fact, I bet you have a friend who swears that driving slow is more dangerous than driving fast. That person might even be you.

But speed matters.

That seems obvious, but we've somehow lost sight of the fundamental physics.

When things move faster, they cover more distance in the amount of time that it takes for someone to react. Or, as a 1990 *Transportation Quarterly* paper tells us, "A fast moving vehicle leaves less time for drivers to make evasive maneuvers."[1]

When the driver does react and start braking, that braking also requires more distance.

At 30 mph, the ***stopping sight distance*** is around 200 feet. At 55 mph, it's closer to 500 feet. At 80 mph, it exceeds 900 feet.

The research likes to point out that there are fewer crashes at higher speeds than at slower speeds. But driving a mile on an interstate is not exactly an apples-to-apples comparison to driving a mile on an urban street. Yes, the latter results in more crashes per mile driven. But if there is a crash, as discussed in a 1985 *Transportation Quarterly* paper, "the potential for injury severity at higher impact speeds increases exponentially."[2] A 1982 study looked at 10,000 crashes and found that a "driver crashing with a change in velocity of 50 mph is twice as likely to be killed as one crashing with a change in velocity of 40 mph."[3]

Even then, that wasn't new news.

In a 1958 US government paper called "Physics and Math for Drivers," Frank Palmer asks, "Did you know that the 3,200-pound car you are driving will develop enough kinetic energy at 20 miles per hour to lift a 1,000-pound elevator 3 floors?"[a]

Frank then goes on to tell us that "should you accelerate a vehicle from 30 to 60 mph and strike a fixed object, the force of that impact would not be twice but four times as hard."[4]

That's exactly why there are a hundred different versions of a graph highlighting the risk of a pedestrian dying when they are hit by cars at various impact speeds. And Frank is right: the risk increases exponentially. In most iterations, researchers suggest that "5% of pedestrians struck by a vehicle at 20 mph are fatally injured. This likelihood increases to 45% at 30 mph, and 85% at 40 mph."[5]

Few studies account for sedans becoming an endangered species, the victim of natural selection and bigger, heavier SUVs. However, one study did argue that obese pedestrians are safer, saying that "it is not implausible that a greater proportion of torso and extremity fat may protect against injury."[6] Another study looked at crashes more generally and found that the risk of dying only increased at the very high and very low ends of the body mass index spectrum.[7]

I used to see these pedestrian casualty percentages at different speeds attributed to Transport for London or the UK Department for Transport. Now, you sometimes see the European Transport Safety Council cited or, in most cases, nobody at all.

This led me down a rabbit hole to find the original study and eventually to the 1986 dissertation of Finnish researcher—and forensic crime fiction novelist—Eero Pasanen. When I spoke to

---

[a] So, it's not surprising that a pedestrian hit in a crosswalk may end up very far away from that crosswalk, something a driver who hit that pedestrian may not willingly share with an investigating officer.

him about his work shortly before his death in 2013, he was sur-prised—and disappointed—to hear that his results were pretty much everywhere. Given that his data came from 1970s England and completely ignored the underreporting of slower-speed vehicle-pedestrian crashes with little to no injuries, he considered his own research obsolete, or as he said, "not anymore too relevant."

Accounting for the underreporting issue should lead to slower-speed crashes seeming even safer than they do already. Accounting for the shift in vehicle size should shift the results in the other direc-tion. But Eero warned me that this may not turn out to be the case due to improved emergency medical care and that cell phones help improve response time. It's awesome to have improved emergency response and medical care, but they can complicate studying these relationships over time.

OK, but what can we do with all this information?

Well, we can stop taking the driver's word for it when it comes to self-reporting speed. That's been a problem from the very begin-ning. In 1896, 44-year-old Bridget Driscoll was run down by Arthur Edsall in Surrey, England, becoming the first pedestrian to be killed in a car crash. Even though a witness said the car was moving at "a reckless pace, in fact, like a fire engine, as fast as a good horse could gallop," Edsall said he was only going 4 mph, maybe 4.5 mph. He also said he rang his bell and shouted for her to get out of the way. The jury agreed with Edsall and deemed it an "accidental' death." The coroner hoped "such a thing would never happen again."[8] The coroner wasn't right on that front. The United States alone has sur-passed 870,000 total pedestrian deaths since the advent of the car.

We can stop equating the danger of a car with that of a bike (or a galloping horse). An average American adult male (weighing close to 200 pounds) would have to bicycle at 120 mph to create the same kinetic energy as a typical sedan at 30 mph. When compared to a typical SUV at 30 mph, he'd need to reach 140 mph to create the same amount of kinetic energy. Even on an 80-pound e-bike, he'd have to be going over 100 mph to come close to the kinetic energy of a typical sedan. Nobody is saying that bicyclists can't hurt pedes-trians, but treating them as if they are just as dangerous as drivers isn't helping.

We can also stop perpetuating the myth that speed doesn't matter. Traffic engineers have mostly acknowledged that speed and injury severity go hand in hand. But I still hear traffic engineers promoting the idea that higher speeds lead to lower crash involvement. It's true that it's a more difficult relationship to isolate, but the overwhelming majority of the research points in the same direction. One of the better road safety researchers, Rune Elvik, dug into 98 studies on this topic. His meta-analysis found that

> when speed goes down, the number of accidents or injured road users also goes down in 95% of the cases. When speed goes up, the number of accidents or injured road users goes up in 71% of the cases. While it may to some extent be possible to offset the impacts of higher speed by introducing other road safety measures, a reduction in speed will almost always improve road safety.[9]

A common refrain from researchers is still: correlation, not causation. And when it comes to speed-related safety research, traffic engineers like to repeat this refrain. For instance, a 1965 *Traffic Quarterly* paper[10] says that "the records simply show that certain vehicles were traveling at certain speeds when the accidents occurred—a collation only, and under specific conditions; not a cause."[b]

Elvik's analysis, however, tells us that when it comes to speed and safety, it's more than fair to place blame on speed because "the causal direction between speed and road safety is clear."[11]

So feel free to tell your "friend" that speed matters and that driving slow is not more dangerous than driving fast.

---

[b] This 1965 paper then tries to connect this point to doctors and the idea that those seeking out medical care are more likely to die (because they are already sick) than those who don't. So he cautions us against plotting "deaths against doctor's care"— or speed against crashes—because they might lead us to the wrong conclusion. Maybe speed is a preexisting condition?

47 Common Knowledge

—

It's easy to think drivers just need more **education** or **enforcement** when it comes to speed. But trying to get folks to choose to do something for their own good—or for the good of others—always seems to be an uphill battle.

It doesn't help that people have long been entranced by speed. When it comes to cars in the United States, this goes back to Anderson Cooper's great-uncle, William K. Vanderbilt II. A 1908 *Cosmo* magazine article titled "Owners of America" said, "To William K. Vanderbilt, Jr., nothing seems worth while but automobiles. It is a matter of wonder that he has not been killed, for a more reckless driver and a more daring, speed-mad rider has never been known."[1]

Even then, we knew that speed wasn't great for safety. In discussing Peter Norton's invaluable book, *Fighting Traffic*, Neil Arason writes, "Speed was regarded as a great danger and safety campaigns were common, starting in the late 1910s…. As almost everyone agreed that speed was a chief problem, police imposed speed limits of 10 mph in the cities."[2]

In fact, the first speed limits date to 1901, when Connecticut created a statewide maximum of 12 mph and 8 mph in cities.[3] To be more specific, those were the first speed limits for cars. Exactly 200 years earlier, in 1701, in Boston, it was "Ordered, That no person whatsoever Shall at any time hereafter ride or drive a gallop or other extreme pace within any of the Streets, lanes, or alleys in this Town on penalty of forfeiting three Shillings for every such offence, and it may be lawfull for any of the Inhabitants of this Town to make Stop of such horse or Rider until the name of the offender be known in order to prosecution."[4]

Drivers want to get where they are going, but the reality is that high speeds in cities don't even help that much.[5]

In a review of the research on this topic, an Organisation for Economic Co-operation and Development report sums it up by saying,

"Mathematically, higher speed leads to reduced travel time. However, the effects of speed in reducing travel time are generally overestimated by road users and, at least in urban areas, the time savings are often small or negligible because of intersections and delays at traffic lights."[6]

Jeff Speck's fantastic *Walkable City Rules* book compares two 15-mile commutes. In the first example, the driver approaches the city at 45 mph. Once they reach the city, they continue at 45 mph until they get to work. In the second example, the driver also approaches the city at 45 mph but then slows down to 25 mph once they get into the city. The difference in arrival time is only 48 seconds.[7]

In the early 1940s, the insurance company Aetna made the same point in an advertising campaign: "Stop Saving Seconds, Save Lives Instead!"[8]

But somehow we turned hundreds of years of common knowledge into a giant question mark. Worse yet, we shifted the conversation so that speed seems negligible in safety.

Take the celebrated Martin Wagner and Walter Gropius urban design studios at Harvard University during the 1940s, where one of their stated goals was to increase traffic speed. To them, the idea that we should slow cars down was laughable and contrary to the very nature of cars. In a 1943 article in *Architectural Forum*, they wrote, "It seems preposterous that the railway system, with less than one tenth of the energy potential of the automotive traffic, has its own right of way, whereas the automobile is made to share the road confusedly with pedestrians and horses and must be bridled and policed on an obsolete net of streets by stop lights and speed regulations which contradict the very purpose of its invention."[9]

Safety wasn't even in the conversation.

Safety also wasn't a part of the conversation when the United States dropped the national speed limit to 55 mph at the beginning of 1975. The rationale? Saving oil.

We also saved thousands of lives.

The National Safety Council estimated that we saved 40,000 lives between 1973 and 1979, half of which the council attributed to a lower speed limit.[10] In the mid-1980s, the Transportation Research Board conducted a more thorough study that focused on roads where the speed differences were significant and found that lower speeds helped save between 2,000 and 4,000 lives each year. The end of the report concludes that "Congress should be aware of the substantial benefits of the 55 mph speed limit" and that we should probably keep it around.[11]

Congress didn't see saving lives as a priority, and the United States started allowing 65 mph on certain roads in the late 1980s before repealing the 55 mph law entirely in 1995. A 2009 study "estimated that higher speed limits across the United States led to 12,545 excess deaths" and "36,583 injuries in fatal crashes" in the ten years after Congress repealed the national maximum speed limit law.[12]

A related problem is that initially trucks were allowed to carry more weight to offset the slower 55 mph speed limit. But when the speed limit went back up, cargo weights didn't do back down.[13]

Another problem is that we like to leave speed up to "personal responsibility." But sadly, more than 35 percent of those who die in speeding-related road crashes are not the people who were speeding.[14] I may be crazy, but I see a big difference between allowing your lack of personal responsibility to kill yourself versus your lack of personal responsibility killing innocent people.

So[1] what are we going to do about it?

The National Transportation Safety Board's recent speed report suggests five approaches.[15]

First, they want to lower speed limits. Cool.

Second, they want to use "data-driven approaches for speed enforcement" in combination with their third approach, automated enforcement. OK.

Fourth on the list is what they call "intelligent speed adaptation." This term refers to things like onboard warnings when the driver speeds, but also includes using technology to limit car speeds in particular locations and on specific streets. Sounding better.

Last, they say we need to do better when it comes to exercising "national leadership," which basically means we need more funding and more **education**. I guess.

What do I find most interesting about this list? They don't say a damn thing about street design. All that's suggested for traffic engineers to do is set lower speed limits, but that's it. Traffic engineers can otherwise go about business as usual.

A 1955 *Traffic Quarterly* paper speaks to our attitudes about speed: "First off, some way must be found to persuade everyone to adopt a realistic attitude towards speed. We Americans pride ourselves on our common sense, yet nothing could be less realistic than our attitude toward speed."[16]

It's hard to disagree with the sentiment. But this definition of "everyone" needs to include traffic engineers as well. The mind-set of a traffic engineer is where we most need a more realistic attitude toward speed.

As is, traffic engineers typically design our streets and intersections in hopes of reducing traffic congestion. Reducing speeds? That isn't even a secondary consideration. In fact, we are more than happy to leave the choice of an appropriate speed up to the whims of drivers. We've even used the argument that we need to increase speeds for the sake of the environment. One 1978 paper, for example, claimed that increasing speeds by 10 percent could result in 6 to 7 percent less fuel consumption.[17] The authors didn't mention that it would probably result in at least 6 to 7 percent more deaths as well.

The traffic engineering mind-set toward speed has cost tens of thousands of people their lives. It still does.

Part 5

Designing Time

———

48 Forecasting Overkill
—

Some people fear heights. Others fear public speaking. Some people fear clowns. Others fear going to the dentist.

Just about everyone fears traffic congestion.

Unfortunately, traffic engineers turned this near-universal fear of traffic congestion into our most important design factor.

When it comes to sizing streets during the design phase, traffic engineers focus on capacity. We are theoretically trying to accommodate traffic levels during the peak hour by estimating the design capacity, also called the design hourly volume. The design hourly volume typically equates to the 30th highest hourly volume. The 30th highest hourly volume means that if we take all the hours in a year and rank them in order by traffic level, there are only 29 hours over the course of a year with more traffic than the design value. It also means there are 8,730 hours in a typical year with less traffic than the design value. For intersections, we take it a step further and estimate the traffic during the peak 15 minutes of a typical peak hour.

The basic idea is simple: we want to make sure we can handle the expected traffic during those peak times. As the American Association of State Highway Officials (AASHO) says in its 1957 manual specific to urban arterials, "Design capacity must be sufficient for the design-hour volume."[1]

On one hand, that makes perfect sense. Of course we want enough capacity to handle the traffic we expect. On the other hand, what is the street like during the rest of the day? Or the rest of the year? And how does designing for peak traffic impact road safety?

Before I get to those questions, let's look at where this 30th peak hour idea comes from. It first appeared in a 1941 study by the Bureau of Public Roads that analyzed one year of data from 89 permanent count stations spread across 43 states (about two counters per state).[2] According to Frederick Cron's 1976 paper, the bureau graphed that data but was worried that "the 115th hour might be frequently congested."[3] They wanted a sensible "cutoff point," as designing for the very peak is "not economical," so they figured that "the 30th highest hour was a good compromise for designers to use in planning the width of new roads."[4] A few years later, in 1945, AASHO jumped on board and has recommended the 30th peak hour as the *design load* ever since.

Although traffic engineers design for the 30th peak hour, we don't actually count a whole year's worth of traffic to figure that number out. Instead, we base it on the average daily traffic. We do so even though AASHO tells us in both its 1957 and 1973 urban arterial manuals that daily traffic volumes have "little direct application to geometric features of urban design."[5]

Nevertheless, our typical design process takes a daily count and applies an averaging factor to estimate the design hourly volume. In the 1957 manual, AASHO suggests that the 30th highest hour of the year is between 11 and 20 percent of the average daily traffic in rural areas. In urban areas, they tell us that it's between 7 and 18 percent of the average daily traffic, even though the "determination of future traffic volumes is much more difficult in urban than in rural areas." Why is it more difficult? Well, AASHO doesn't know because "insufficient data are available to account for this difference."[6] Yet, by 1973, AASHO narrowed it down even more and told people that the 30th peak hour usually falls between 8 and 12 percent for urban areas (but the report doesn't provide much more in the way of detail).[7] Forty-five years later, it remains the same. The 2018 AASHTO Green Book continues to say that we can apply "a representative percentage (usually 8 to 12 percent in urban areas)."[8]

If you think that multiplying the daily traffic by a simple percentage to get our design load feels like a gross oversimplification, you aren't alone.

A 1980 *Traffic Quarterly* paper tells us these estimates "grossly simplify the actual relationship of peak-hour volumes to daily travel volumes."[9] And even though these estimates "could result in an appropriate facility and system design," it is the traffic "engineer who 'picks' a suitable peak hour ratio to establish the design-hour volume required at the facility design stage."[10]

In other words, the actual multiplicative factor that a traffic engineer uses comes down to *engineering judgment* and whether we are talking about an urban, suburban, or rural area. If that all seems a bit arbitrary, wait until you see what comes next.

The 2018 AASHTO Green Book tells us that "values should then be projected to the design year, usually 20 years into the future."[11]

AASHO decided on this way back in 1945. The projection then got codified into law with the Federal-Aid Highway Act of 1956 that, according to a 1957 *Traffic Quarterly* paper, required that "we must look ahead nearly 20 years—to 1975—and locate, plan, and design the kind of highways that will be needed to handle future traffic volumes."[12]

In other words, we have no intention of sizing our streets based on the 30th busiest hour for *this* year. What we want to do is design them for the 30th busiest hour for 20 years from now. Since the DeLorean ran out of plutonium, this process often resembles a giant guesstimate.

During my undergraduate days and when I first started working as a consultant, I was told to assume that car traffic grows 3 percent each year.[a] Mathematically, this means that today's 10,000 cars in 20 years' time becomes over 18,000. It's not that different than the magic of compounding interest.[b]

It's hard to pinpoint the origin of the back-of-the-envelope estimate I was given, but I can promise you it isn't uncommon. It is not even egregiously high. A 1957 *Traffic Quarterly* paper suggests we should assume a 5 percent compounding traffic increase.[13] Under than assumption, 10,000 cars would turn into more than 26,000 cars in 20 years' time. A 1958 *Traffic Quarterly* paper then argues that 5 percent is far too low. That data, looking at the New York metropolitan area between 1925 through 1957, suggests that 6.4 percent is more appropriate.[14] Over 20 years, this estimate turns 10,000 cars into more than 34,000 cars. If we followed this logic from 1925 to 2025, we'd turn a road with 10,000 daily cars into one with nearly five million cars.[c]

The thinking behind these growth factors, as described in a 1967

---

[a] I've never seen a traffic engineer use this simplified growth factor method for estimating future walking, bicycling, or transit use.

[b] The internet likes to attribute the following quote to Albert Einstein: "Compound interest is the eighth wonder of the world. He who understands it, earns it; he who doesn't, pays it." Who knows if he actually said these words, but they probably shouldn't apply to traffic engineering.

[c] The highest suggested growth factor I came across was 9 percent in a 1965 British manual.

Traffic Quarterly paper, is to apply today's travel behaviors to the expected future population. For example, "A method much used in the past to obtain an estimate of the future trip generation of a zone is to multiply the present trips by a *growth factor*, which may be the product of the ratios between the future and the present population, the future and the present car density, and the future and the present car utilization."[15]

And since we expect incomes to rise in our growing populations over time, we characteristically predict that car ownership—and driving—will rise as well. The end result often boils down to an unassuming growth factor percentage.

So all that work we did earlier to try to figure out the 30th busiest hour doesn't much matter in the face of almost any traffic growth rate compounded over 20 years. Yet we continue to size our streets and intersections for that compounded future even though we've long known it to be a problem. Here is a quote from a Federal Highway Administration representative at a 1976 session at the Transportation Research Board conference: "80% of the level of effort is focused on 20-year forecasts.... We have got to analyze the problems that are on the street today. I would recommend that [we] ... not commit more than 20% of planning resources to the long term."[16]

The same complaint could be said today.

The good news? These models continue to improve, and the growth factors they provide now at least seem a bit more reasonable than the earlier ones. The Colorado Department of Transportation's most recent traffic projections show average 20-year growth factors around 1.3 percent and a maximum at 3.2 percent.

The bad news? Instead of designing for the amount of vehicle traffic we deem appropriate, we think it's our job to accommodate whatever future traffic these growth factors predict. What's worse is that we size our streets for a busy rush hour 20 years into the future while simultaneously designing it for the speeding we expect when the street is empty.

This process isn't about safe streets. It never was. It was never supposed to be.

49 An Origin Story for the High-Injury Network

___

The third design factor—in addition to design speed and design capacity—is the ***design vehicle***.[a]

The main idea is to design streets for the biggest and least maneuverable vehicle that may use it. This goal often leads us to design streets and intersections for Optimus Prime instead of Bumblebee, even if Optimus Prime doesn't come along all that often.[b]

What do traffic engineers do next? The process is essentially the same as it was in 1957 when the American Association of State Highway Officials said that "a design check should be made to insure that the largest vehicle expected can negotiate the designated turns."[1]

In terms of a design example, larger design vehicles lead to larger turning radii. Larger turning radii lead to … well, let's hear what AAA has to say about it in their 1958 book about pedestrian safety: "The radius should not be greatly increased beyond this value [12 feet], because long radii encourage turning at too high speeds and extend the crosswalk length, increasing the hazard for the pedestrian."[2]

Use conventional guidelines and equations and you'd be hard-pressed to end up with a turning radii as small as what AAA

---

[a] It is worth noting the difference between the design vehicle and the control vehicle. The design vehicle is the biggest and least maneuverable vehicle that frequently—not occasionally—uses the street. The control vehicle, however, is the biggest and least maneuverable vehicle that seldom comes along; when it does, it can do so slowly and can even venture into an adjacent lane to make the turn. Traffic engineers have a lot more leeway than they let on when it comes to selecting a design vehicle.

[b] In case you care but don't know, Optimus Prime and Bumblebee are two of the more popular Transformers toys that Hasbro started making in the 1980s. Transformers are robots that usually turn into vehicles. Optimus Prime transformed into a big Freightliner semitruck while Bumblebee turned into a small, yellow Volkswagen Beetle. One of Optimus Prime's iconic lines—"no sacrifice is too great in the service of freedom"—could easily be the tagline for our approach to building big roads (or for most any car company).

suggests.[c] Interestingly, smaller turning radii rarely preclude bigger vehicles from using a street; they just need to do so more slowly. But we can't have that because it might reduce capacity.

It always comes back to capacity.

So now let's ask, what's the problem with traffic engineers focusing on capacity? And what does that have to do with safety?

It has everything to do with it.

It's the reason the street "needs" to be four lanes instead of two. Or six lanes instead of four. Or even eight lanes instead of six.

It's the reason all those lanes still aren't enough and the reason we need to add multiple vehicle turning lanes as well.

It's the reason these lanes need to be 12 feet wide instead of 10 or 11.

It's the reason bike lanes magically disappear at large intersections.

It's the reason we add slip lanes to facilitate right-turning cars slightly before they get to the intersection (which eliminates the need to include these right-turning cars in our intersection capacity calculations).

It's the reason we then make these pedestrians hustle across half of the huge street, only to be told to hang out here on the median while cars get another turn.

It's the reason we assume that pedestrians walk faster than we know that at least 15 percent of pedestrians can actually walk.

It's the reason we can't give pedestrians their own time to cross and instead tell them it's their turn while simultaneously telling drivers to turn directly into the same crosswalk.

This list could go on and on.

The reality is that when we combine the conventional traffic engineering approach to our three main street design criteria—design capacity, design speed, and the design vehicle—traffic engineers end up building every city's high-injury network.

What do I mean by that?

The fundamental idea behind a high-injury network is to identify the streets where most of our serious injuries and fatalities take place.

---

[c] In practice, the "effective" turning radius often ends up much larger than the design value. The design value refers to the radius of the centerline of the road and not the corner itself. Some traffic engineers don't know that and use it to design the corner instead. Add in a bike lane or on-street parking, and drivers can easily take such corners much faster than the designer intended.

Most cities run this analysis and find that 50 percent of their serious injuries and fatalities happen on only 5 percent of streets or that 60 percent happen on 6 percent of streets. Whatever the numbers, the resulting map inevitably provides an unflinching look at our most deadly streets.

I've now seen a few dozen of these high-injury network analyses done for various cities. Every time, the map ends up highlighting the exact big roads where state departments of transportation have emphasized the conventional, capacity-based traffic engineering design approaches described here.

One of the better papers I've seen recently dug into more than 62,000 pedestrian deaths that happened in the United States across 16 years.[3] Instead of simply identifying hot-spot corridors, the researchers focused on design characteristics. What they found is that "nearly two-thirds (63%) of all hot spot corridors were roadways with 1) three or more lanes, 2) speed limits of 30 mph or higher, and 3) high traffic volumes."[4] In fact, more than 97 percent of their hot spots had three or more lanes, and 70 percent of them had five or more lanes.

Traffic engineers continue to tell ourselves—and the general public—that road safety underpins how we design streets. As Gabe Klein writes in his book *Start-Up City*, "Traffic engineers often use 'safety' as a defense when they widen roads or remove crosswalks on busy roads. Pedestrian overpasses and underpasses, which can compromise personal safety in other ways, are often justified using the same logic."[5]

But when we consider our three main design criteria—design capacity, design speed, and the design vehicle—very little of the process has much to do with making our roads safer. What's worse is that the combination of our main three design criteria has systematically led to the creation of the deadliest streets in most every city.

It was a recipe for disaster. It is a recipe we still use today.

50 It's a Tradition

—

In a season 5 episode of the filthy (but incredibly well-done) car-
toon *Big Mouth* on Netflix, Andrew's father, Marty Glouberman,
says, "Some things are wrong, but if they're wrong long enough, they
become tradition." Mr. Glouberman was talking about the mascot
of the town's middle school—the Scheming Gypsy—but the same
sort of thinking applies to many of our traffic engineering design
"traditions."

Let's go back to the AASHO 1957 design manual on urban arterials. In that manual, we learn that "lane widths should be not less
than 11 and desirably 12 feet." In fact, it suggests that up to "13 feet
are desirable."[1]

Ezra Hauer traces the origin of this particular design recommendation to a time well before the AASHTO Green Book existed. As he
says in a 2000 paper, "The historical roots of the lane-width standard
go back to the period of 1938–1944 when seven 'geometric design
policies' were written by the Committee on Planning and Design
Policies of the American Association of State Highway Officials."[2]

Eventually, the committee assembled these seven policies into
its 1954 guidebook on rural highways and the 1957 guidebook on
urban arterials. In its 1954 rural guidebook, for instance, AASHO
says, "No feature of a highway has a greater influence on safety
and comfort of driving than the width of the surface.... Ten- to
12-foot lane width are now standard and the tendency is toward the
larger value."[3]

This same mentality carried over to the 1957 urban guidebook:
"In the interest of safety, efficiency, and ease of operation, lane widths
of 11 to 13 feet are desirable on any arterial highway, the larger values providing the additional freedom and ease of operation consistent with high traffic volumes."[4]

In defense of this design recommendation, AASHO cites one
study done by Asriel Taragin in 1945.[5] AASHO describes Taragin's
study in the 1954 guidebook, saying that "hazardous conditions

exist" when things are too narrow. AASHO goes on to say that "from this and similar studies it has been concluded and generally accepted that lane width of 11 feet and preferably 12 feet should be provided."[6]

AASHO comes to this conclusion even though Taragin never looked at road safety outcomes. In fact, he never looked at lane width either. His study only looked at relative vehicle placement in terms of overall roadway width. He found that trucks have to shift slightly more to the right when there is oncoming traffic on a 22-foot-wide road than they do on a 24-foot-wide road. His takeaway? This maneuver leads to "hazardous traffic conditions."[7] As Hauer points out, AASHO literally copied Taragin's conclusion and pasted it onto lane widths.[8] They did so for the next four decades, finally removing it for the 1994 edition of the AASHTO Green Book. But even then, AASHTO made the same recommendation without referring to any research whatsoever.

In one of the more comprehensive studies on this subject, Robert Noland accounted for all sorts of confounding factors and found that it is "not possible to support the engineering hypotheses" that wider lanes equate to safer streets. In many cases, Noland found that narrower lanes meant fewer injuries and fatalities. Noland then says that "these results are quite surprising as it is general practice to improve the safety of roads by increasing lane widths."[9]

Should these results be surprising?

They shouldn't be when we consider that wider lanes typically mean higher speeds. A 2001 street design study found that "only lane width was a significant variable, explaining about 25 percent of the variability of the speeds."[10] This isn't news. Even the *Highway Capacity Manual*[a] warns us that using a 10-foot-wide lane instead of a 12-foot-wide lane would reduce free-flowing speeds by 6.6 mph.[b]

Noland's results shouldn't have been news either, because the research has long told us that wider lanes maybe weren't the safest. In a 1953 study looking at crash data from thousands of miles of roads across 15 states, Morton Raff admitted that when it comes to lane

---

[a] Published by the Transportation Research Board, the *Highway Capacity Manual* has long been the traffic engineer's holy book when it comes to measuring the capacity of a street or intersection.

[b] The *Highway Capacity Manual* makes it seem like reducing speeds would be a bad thing.

width, "the evidence is confusing." Narrower widths, he concludes, do *not* seem to have "a detrimental effect on the accident rates."[11] Raff wasn't alone; a 1954 California study by D. M. Belmont found that 11-foot-wide lanes are safer than 12-foot-wide lanes.[12] Hauer points to a 1978 National Cooperative Highway Research Program report, which also found that narrower roads were safer.[13] But because the results were "inconsistent with the expectation," the authors of that report regrouped their results to make the widest category appear safest.[14] By doing so, they turned an 11 percent difference in safety outcomes into a nothing burger.

Ask AASHTO about any of this, and they will say that they've always allowed for narrower lanes. That's true. It's also true that "there has always existed a conventional wisdom that narrower lanes result in higher crash frequencies."[15]

At best, the research is inconclusive. At worst, wider traffic lanes lead to higher speeds and less safety.

But, as we have seen with other aspects of traffic engineering, safety concerns aren't prioritized in design.[c] Traffic engineers design for capacity, and all our manuals tell us that narrower lanes reduce capacity.

Flipping through my old undergraduate traffic engineering textbook, it says that "traffic flow tends to be restricted when lane widths are narrower than 12 ft."[16] When my textbook then explains how to calculate level of service, it presents an "adjustment factor for restricted lane width."[17] In other words, the capacity of the lane drops about 10 percent when going from 12-foot-wide lanes to 10-foot-wide ones.

To my surprise, the first edition of the *Highway Capacity Manual*—from 1965—says pretty much the exact same thing with pretty much the exact same numbers.[18] When discussing intersection level of service, the capacity continues to increase up to a 16-foot-wide lane. Why we would want a 16-foot-wide lane? I'm not sure, but it

---

[c] In a 1957 *Traffic Quarterly* paper, Ralph Fisher acknowledges that "there is almost universal agreement" regarding the safety benefit of shoulders but also advocates that we "find the space for additional lanes by ... doing away with the shoulders." Whether shoulders contribute to safety is beside the point. The point is that this traffic engineer fully believed that shoulders helped safety but was willing to set them aside for sake of capacity. Ralph L. Fisher, "Designing Roads for High Capacity," *Traffic Quarterly* 11, no. 2 (1957).

apparently has the most capacity (or at least the most capacity shown in this table).

Despite the lack of evidence, the 1965 *Highway Capacity Manual* says that narrower lanes "adversely affect driver comfort and increase potential hazard."[19]

As a result, our design criteria focus on providing as much capacity as we can, and wider lanes seem to give us more capacity. Instead of thinking about wider lanes in terms of a trade-off with safety, we told ourselves that wider lanes help improve safety. We initially did so by citing a study—conducted before almost any of us were born—that didn't consider safety outcomes. Slowly, over time, this thinking—that wider lanes are safer—became accepted fact, to the point where we no longer need to cite any research. The inertia of what came before becomes almost too big to question. After all, it's been in our 1,000-page guidebooks for decades. It's easiest to sit back and assume that whoever wrote them knew what they were doing.

Given all that, it's not at all surprising we do what we do. It's tradition.

51　One-Way Conflicts

The promise of more capacity also steered traffic engineers down a one-way street.

Instead of tearing up the whole city and starting from scratch—which as Wilbur Smith says in a 1949 *Traffic Quarterly* paper is "obviously impractical"[1]—we can transform our streets into one-ways. How will that help? In addition to increasing capacity and speed, Wilbur tells us that it will also "reduce accidents."[2]

He then offers some evidence from a case study in New Haven, Connecticut, where the city flipped 10 miles of streets to one-way streets for seven months in 1948. What did Wilbur learn from that experiment? He found that traffic volumes went up, and drivers drove faster. Wilbur doesn't present data on safety, but he does give us an early version of what we still see in our guidelines about one-way streets: the conflict point diagram.

A *conflict point diagram* identifies the locations within an intersection where vehicle paths cross. The underlying assumption is that fewer conflict points equate with better safety. Wilbur Smith's diagram shows 16 conflict points for an intersection of two-way streets and one for an intersection of one-way streets.[3] Given our assumptions about what conflict points mean for safety, one-way streets must be safer.[a]

One problem with these conflict point diagrams is that like some HGTV shows, they give us an unrealistic flip. The two-ways streets in these diagrams have four lanes total. The one-way streets have two lanes total. It's not an apples-to-apples comparison.

The more fundamental problem I have with the conflict point mentality is that it doesn't consider actual safety outcomes. We just assume that having more conflict points is bad for safety. Maybe

---

[a] Today's conflict point diagrams include merging or diverging movements. Add those in and the two-way intersection jumps to 32 conflict points whereas the one-way intersection ends up with five conflict points. Given our assumptions, one-way streets seem even safer.

that's true, but given how counterintuitive transportation outcomes can be, maybe not.

We make the same assumption today. And when we do look at real safety outcomes, we are quick to explain away whatever results don't hit the mark.

Louisville, Kentucky, flipped 33 miles of streets into one-ways in the 1940s, and road safety got worse. The traffic engineers blamed the worse safety on increased driving and not the one-way streets.[b]

Traffic engineers also assumed that one-way streets will lead to the extinction of head-on collisions. The AASHO 1957 manual on urban arterials says that one-way streets help safety most of all "by eliminating the hazardous head-on type of accident."[4] This assumption, however, relies on people never making the mistake of driving the wrong way. You may not watch the 1987 movie *Planes, Trains and Automobiles* every Thanksgiving like I do, but you might remember Steve Martin and John Candy making that same mistake.[c]

Denver is home to a downtown full of one-way streets, and I see a similar scene play out there almost every day. Occasionally, the scene plays out differently than the movie and somebody dies.

We blame this death on the driver rather than the design even though the 1957 AAHSO manual admits that one-ways streets may cause "some confusion to nonfamiliar drivers." They then assure us that "advantages generally overweigh disadvantages"[d] when it comes to one-ways.[5]

Are there any other safety disadvantages worth noting?

A 2004 Institute of Transportation Engineers report admits that "one-way pairs … seemed to elicit red-light running behavior."[6] And given the more circuitous routes precipitated by one-way streets, people drive farther and turn more often.[7] Both add unnecessary exposure. That exposure also tends to be at higher speeds on those streets.

AAA's 1958 pedestrian report casually says one-way streets "are also often beneficial to pedestrians," but "despite the above-mentioned

---

[b] Even though the one-ways might've led to the increased driving.

[c] "How would he know where we're going?"—Del (John Candy) in *Planes, Trains and Automobiles* (1987)

[d] In the same 1957 AASHO manual, we're told not to worry about any "opposition" to one-way streets because it'll all disappear "after a trial period."

advantages of one-way streets, there can be no categorical claim that they are always safer for pedestrians."[8] In fact, "the ability of the one-way street to handle more vehicle traffic at higher speeds stands as a fundamental hazard to persons on foot."[9] An interesting 1964 *Traffic Quarterly* paper about suburban town centers also talks about how well-designed streets can "unite the community on a higher level."[10] But when we design one-way streets for high capacity and high speeds, we end up with highway-like streets that create barriers instead of bonds. Streets represent the building blocks of our cities, and any city street that "stands as a fundamental hazard to persons on foot" doesn't seem like building block we want to start with.[11]

Still, the current iteration of the AASHTO Green Book continues to suggest flipping two-ways to one-ways in urban areas where pedestrian safety is a concern.

So when it comes to safety, I'm not philosophically against one-way streets. There are examples of narrow one-way streets that encourage slow driving and active transportation, but they are few and far between. Most are the wide and fast variety—focused on capacity and speed—that cut through our cities.[e]

The more fundamental problem relates to how traffic engineers deluded themselves into believing that one-way streets were safer. It had little to do with empirical crash outcomes. We assumed that some other thing—like fewer conflict points in this case—would automatically improve safety. Unfortunately, transportation is rarely that simple.

---

[e] It's sad that one-way streets were supposed to solve our capacity problem, but they didn't even do that well. In a 1959 *Traffic Quarterly* paper, Bellis tested the same set of origins and destinations and found that the one-way network equated to 35 percent more driving and caused "more harm than the benefit intended." W. R. Bellis, "Increasing City Street Capacity," *Traffic Quarterly* 13, no. 1 (1959).

52   Inconvenient Evidence

One thing I love about transportation is how counterintuitive things can be. If we have too much traffic on a highway, for example, it's perfectly logical to think we need more lanes. But time and time again, the seemingly simple solution fails. Instead of the same people doing the same thing at the same time of day, people adapt to what we've put in front of them.

Let's say I always leave for work at 5 a.m. to beat the traffic. Now that new highway lanes have been added, leaving at 7:30 a.m. with some extra sleep under my belt and time for a cup of coffee sounds pretty, pretty good.

Or maybe I used to take the train every day because of all that traffic on the highway. Now, somebody told me that a ton of money has been spent trying to fix the traffic problem, so I'll try driving again.

Or perhaps I used to take back roads to work. Instead of the highway, the back roads were a little farther distance-wise but got me there quicker. If the traffic problem has been fixed, I might go back to the highway.

Or possibly I used to get a ride from a coworker, but now we drive separately again.

All these scenarios add up to what we call ***induced demand*** and, in turn, a giant fail.

Whenever induced demand comes up, so does Anthony Downs. Downs lays out the above thinking in a 1962 *Traffic Quarterly* paper and takes it a step further: "In pure theory, only a road or system of roads wide enough to carry every commuter simultaneously at an optimal speed would be sufficient to eliminate all peak-hour congestions. It is obvious that no such roads are practical unless we convert our metropolitan areas into giant cement slabs."[1]

For years, I imagined Downs as a lone voice in the wilderness, shouting about induced demand and the law of peak-hour traffic congestion. But this issue was discussed repeatedly well before

Downs's 1962 paper.[a] For instance, a 1956 *Traffic Quarterly* paper by the chief engineer for the Arkansas Department of Transportation complains that "one might assume that an expected annual compounded traffic increase of 3 or 4 percent might be applied to the present traffic and thus arrive at a logical and satisfactory answer for twenty years from now. To follow such a procedure is the height of folly. Much has been learned of induced traffic on modern thoroughfares and much remains to be learned."[2]

Other papers from the 1950s were even quantifying induced demand, something we rarely do even today. The "after" traffic on the Eden Expressway in Illinois, for example, was shown to be between 32.2 and 62.2 percent higher than anyone would have expected.[3]

Even *Peppa Pig* gets it. In one episode, Mr. Bull wants to remove the old "*wibbly-wobbly*" road and replace it with new one that is wide and straight. At first, Mr. Bull plans to tear down the veterinarian's office, but Dr. Hamster asks if the road can go around the building instead. That confuses Mr. Bull. "Around the building? Around? But then the road wouldn't be straight. Busy people can't waste time driving around things."

After deciding that the place for sick pets is worth keeping, the work crew tunnels under. But once the new road is done, Peppa asks why the cars still aren't going that quickly. Mr. Rabbit, the traffic engineer, wonders aloud, saying, "Hmmm. There are more cars using this road than we had planned for." And the narrator explains that "the new road is so nice and straight that lots of cars have come to use it." The solution? "We'll need a bigger road!"

And everyone falls onto their backs and laughs uncontrollably, which, if you've seen *Peppa Pig*, makes a lot more sense.

Others kept trying to downplay induced demand. Russell Singer, in a 1964 *Traffic Quarterly* paper, tells us it's a myth: "Myth No. 2: Building freeways and providing parking space is self-defeating because such facilities merely attract more traffic, causing greater congestion and leaving the city worse off."[4]

One 1971 *Traffic Quarterly* paper confronts the "conventional

[a] Including as far back as 1928 when a Los Angeles official said that "a newly ... widened street immediately becomes glutted by the access of cars that hitherto have reposed more in their garages than they have utilized the streets." Brian Ladd, "'You can't build your way out of congestion.'–Or can you? A Century of Highway Plans and Induced Traffic," *disP - The Planning Review* 48, no. 3 (2012).

wisdom" of induced demand and this "spiral theory of more high-ways leading to more cars" by saying that induced demand won't matter once we've acquired enough land for highways (and parking).[5]

Despite the empirical evidence,[b] most traffic engineers drank the Kool-Aid. What was once conventional wisdom became whatever the opposite of conventional wisdom is. It got to the point where traffic engineers could no longer figure out why our "investments" weren't working out.[c] As a 1986 *Transportation Quarterly* paper says, "In spite of these 'investments,' traffic congestion and the physical condition of the commuter transportation system continue to deteriorate."[6]

Why am I talking so much about induced demand in a book about safety? One reason has to do with seemingly logical underlying assumptions. When empirical outcomes don't jive with what we assumed, we treat them as an inconvenience that can be explained away. And since transportation outcomes tend to be counterintuitive, that happens way too often.

Remember the edge-lining experiments from the 1950s? The results suggested that edge lining didn't help safety: "Total accidents, including those at access points, increased by 1% and the number of persons killed and injured increased by 16%."[7]

Yet states kept right on edge lining, because it makes logical sense that edge lining should help safety. A few years later, the American Association of State Highway Officials recommended edge lining nationwide.[8]

The 1963 design manual by the Automotive Safety Foundation couldn't "explain some of the apparent contradictions in certain data."[9] For instance in terms of sight distance, "when substandard features are a common occurrence, drivers apparently compensate for them and the accident rate for that stretch of road may be less than average."[10]

Yet one of our basic design criteria remains sight distance

---

[b] And despite that Russell Singer worked for AAA.

[c] A higher-up at an unnamed state department of transportation recently told one of my former students that induced demand remains "controversial and unproven." After my student suggested otherwise and asked about DOT leadership often using the phrase "you can't build your way out of congestion" at public meetings, the higher-up responded by claiming that that phrase has nothing to do with induced demand and that we still need to add vehicle capacity. Yup.

because, for example in the case of uncontrolled railroad crossings, we want to make sure the driver can see if a train is coming and stop in time. Accordingly, the AASHTO Green Book states that "sight distance is a primary consideration at crossings without train-activated warning devices."[11]

AASHTO also refers us to the *Highway-Rail Crossing Handbook*, which tells us that "adequacy of sight distance is critical at passive crossings."[12]

Traffic engineers use a combination of math and physics along with some assumed values to calculate sight distance. This calculated sight distance becomes our minimum design value.[d] Anything bigger is better.[13]

Why do we do that? Well, we do it for safety. A 1987 report from the National Cooperative Highway Research Program (NCHRP) says we can "improve safety conditions at the crossing" with "improved sight distances and a wider roadway."[14]

As for how much of a safety benefit will we get, the report doesn't know: "Techniques to accurately estimate the change are not available and so no estimate is presented."[15]

So even if we don't know how big of a safety benefit we get, it's at least a benefit, right?

Right?!?

A 1968 NCHRP report brushed this question off, and instead the authors tried to come up with "a logical explanation for the nonexistence, or existence as a very minor variable, of sight distance in predictive equations."[16] Even though the data says that sight distance—our critical design criteria—doesn't matter much in terms of safety, "this does not seem logical: sight distance should be one of the most important variables." They never found that logical explanation other than to say that "common sense indicates that there is a minimum value which should be provided at all crossings."[17] So that is what we did.

Researchers finally tested different sight distances at the same railroad crossings in 1996. When they increased sight distance, drivers

---

[d] Some states consider sight distance at railroad crossings so important that they've set laws establishing minimums far beyond our calculated minimums. Even if our equations tell us that 69 feet is enough, Illinois wants "a distance of not less than 500 feet in either direction from each grade crossing." Illinois Compiled Statutes, "Sec. 18c-7401. Safety Requirements for Track, Facilities, and Equipment."

could see farther down the tracks. Drivers felt safer, so drivers drove faster.[e] The researchers concluded that increased sight distance, the very thing we say is so "critical" at such crossings, "resulted in no demonstrable net safety benefit."[18]

I'm not trying to get you fired up about induced demand, railroad crossings, or even sight distances. I'm pointing out that traffic engineering outcomes are often counterintuitive and that traffic engineers have a history of overlooking the disconnect between what we think leads to better safety and what actually does.

When traffic engineers assume that some design criteria will lead to better safety, they develop mathematical equations to calculate how much of that design criteria is needed. Do these calculations give us actual safety benefits? Maybe, but it's tough to tell because these design criteria are not as based on empirical crash outcomes as any of us might assume.

So when it comes to safety, traffic engineers don't know as much as *you* would think. Hell, traffic engineers don't know as much as *traffic engineers* would think.

---

[e] It is worth noting that this study didn't look at actual crash outcomes either, only at driver behavior.

53   Unclear Zones

—

Design details matter, but it would take another whole book to cover them all or to get into all the safety benefits of basics like sidewalks and curb extensions. What I'm focused on here are the fundamentals behind our designs. Why do we end up with the wider streets and big intersections that make up our high-injury network? Traffic engineers may tell you these designs prioritized safety, but as we now know, capacity is king.

You need to step beyond the street to see where traffic engineers did prioritize safety. Even there, whose safety they prioritized isn't what you'd hope for.

In its 1963 roadside design guidebook, the Automotive Safety Foundation (ASF) said that "a high percentage of accidents involves vehicles which leave the traveled lanes. These usually involve only one vehicle and include those that overturn or strike an object near the road. Stonex points out that this type of accident accounts for between 30 and 35 percent of all fatalities."[1]

The Stonex referred to here is Ken Stonex, the General Motors Proving Ground engineer who would eventually testify at the congressional road safety hearings. At the GM test track, he says that 72 percent of crashes were run-off-the-road crashes. His takeaway? "Stonex suggests that since it is inevitable that some vehicles will leave the road, the roadside should be designed to minimize the consequences of the accident."[2]

ASF then references a Virginia study, which found that around 25 percent of run-off-the-road drivers hit a tree. Its solution was simply the "removal of all trees within 15 feet of the roadway."[3] What happens when we cut down all these trees? Well, fewer drivers hit trees. Does this kind of solution help overall safety? The same 1963 ASF book makes it clear that we really didn't know: "Little research has been done to determine the likelihood of an accident in terms of proximity to a fixed object…. Past findings have related only to the frequency with which such objects have been struck."[4]

Nevertheless, AASHO forged ahead with the "clear zone" concept in their 1967 manual on highway safety, a manual that became known as the Yellow Book.[5] AASHO introduces the idea by looking at 507 run-off-the-road fatal crashes that happened in 1965, lamenting about how these "occur even on the safest facilities yet devised."[6] The remedy? A *Traffic Quarterly* paper explains the basic idea underlying the 1967 AASHO design guidance: "The goal is to protect the vehicle that runs off the normal traveled surface—for whatever reason. The 'forgiving roadside,' as it has appropriately been called, provides a traversable area, free of solid obstructions that will allow a driver to recover control of the vehicle and stop or get back on the traveled way."[7]

A "forgiving roadside" sounds pretty good, right? It seemed right in line with what Ralph Nader was thinking[a] with his own congressional testimony: "Even if people have accidents, even if they make mistakes, even if they are looking out a window, or they are drunk, we should have a second line of defense for these people."[8]

But what does the clear zone concept mean in practice? Here is AASHO in 1967: "Corrective programs should be undertaken at once to eliminate from the roadside or to relocate to protected positions such hazardous fixed objects as trees, drainage structures, massive sign supports, utility poles, and other ground-mounted obstructions that are now exposed to traffic."[9]

At the time, 30 feet was considered an appropriate clear zone distance: "For adequate safety it is desirable to provide an unencumbered recovery area up to 30 feet from the edge of the traveled way."[b]

AASHO chose this distance because when it looked at those 507 fatal crashes, "80% of the vehicles in 'ran-off-the-road' accidents did not travel beyond this limit."[10] What sort of safety benefit might we see? AASHO didn't have an answer then—and it still admits that clear zone dimensions remain based on "limited empirical data" that is "neither absolute not precise"[11]—but my undergraduate textbook suggests: the more, the better.[12] A 5-foot clear zone will give us a

---

[a] Nader also talked about the benefits of using technology (at the time, radar) to force cars to automatically brake for pedestrians and about how vehicle fronts could be designed to help save pedestrians. We only recently got around to the former (I believe a 2010 Volvo was the first vehicle with such technology) and continue to go in the wrong direction on the latter.

[b] AASHO also reminds us that "design standards more liberal than the minimums prescribed will often increase safety."

13 percent reduction in "related" crashes. If we gave 10 feet of clear zone, we'd see a 25 percent decrease. Twenty feet would give us a 44 percent drop.

That's pretty impressive. It's hard to think of much else we could do that would be better for safety. Neil Arason's more recent safety book makes a similar point: "Another concept that will bring us nearer to the goal of zero deaths is that of the clear zone. On rural highways, single vehicle, run-off-the-road crashes comprise a significant portion of fatal crashes, making it important for roadside environments to be free of rigid objects or hazards."[13]

In fact, why stop with rural highways? Aren't we basically building the same big roads in cities? We should really be doing this clear zone thing everywhere.

So that is what AASHO did, and that is what AASHTO continues to do. In their words, we should apply the clear zone "wherever practical." That includes cities. Now, AASHTO at least acknowledges that the rural clear zone distances "may be unlikely in an urban setting" due to more limited rights-of-way. Yet even in urban areas, "the clear roadside concept is still the goal of the designer."[14]

AASHTO long defined the clear zone as the space we give "errant vehicles" to recover. In its more recent corrections to the most recent *Roadside Design Guide*, the clear zone is defined as "the unobstructed, traversable area provided beyond the edge of the through traveled way for the recovery of errant vehicles. The clear zone includes shoulders, bike lanes, and auxiliary lanes, except those auxiliary lanes that function like through lanes."[15]

In other words, the clear zone includes shoulders, which makes some sense, as well as bike lanes, which doesn't make any sense. Why would we intentionally design our major roads to place bicyclists in the exact same place as where we want errant vehicles to go?

It's worse than that in practice. I'm currently looking at the new, well-regarded typical section for an urban arterial that was recently put out by the Florida Department of Transportation. It labels both the bike lane and the sidewalk as "Clear Zone." So instead of protecting vulnerable road users with trees or other fixed objects, we deliberately remove those protections and overlap the limited space we give them with that for "errant vehicles."

Here's the funny—and sad—thing. Much of the evidence suggests that urban clear zones don't help the cause.

Eric Dumbaugh's excellent paper analyzes two sections of the same Orlando, Florida, corridor.[16] Controlling for traffic volumes and crash severity, he found the one with the well-engineered clear zone to be far more dangerous than the one without.

In my own safety paper on street trees in the clear zone, "the results suggest that—at least with respect to street trees in urban areas—the benefits of adhering to the clear zone concept seem overblown."[17]

Imagine a street with massive trees or other fixed objects right up against the edge of the driving lane. How fast are you driving? If you drove the exact same way on a street with a big clear zone, a clear zone would probably be safer. If a big clear zone makes the road feel wide open and you drive faster or pay less attention, the supposed safety benefit isn't as clear-cut. Even in 1967, AASHO warns that "a completely clear roadside within right-of-way lines could be very unattractive and monotonous and might create other safety problems."[18]

I like to think of so-called forgiving roadsides as the "prevent defense" of traffic engineering. We think we've got this huge lead and just need to run out the clock. On paper, the prevent defense looks like the safest option. But what is the one thing we should know about the prevent defense? It prevents you from winning.

The clear zone mentality is one of the biggest missed opportunities in urban areas. We could be proactive and use trees or other fixed objects help slow speeds and serve as protection between cars and more vulnerable road users. Instead, we play a prevent defense with a seemingly forgiving roadside and hope for the best.

In the 1950s and 1960s, we hypothesized that rural solutions like a big old clear zone would make cities safer. At the same time, we knew that our urban areas were safer, even when measured on a per mile basis like traffic engineers love to use.

A *Traffic Quarterly* paper uses 1959 data to compare the road fatality rate of urban areas, overall, to rural interstates with full access control.[19] I probably don't need to tell you, but the urban areas were safer. And then you have a 1963 *Traffic Quarterly* paper, which found crash injury rates for those living outside of urban areas to be double that of urban dwellers.[20] When looking at those aged 5 to 24, it is more than five times more dangerous to be outside of our urban areas. Even AASHO's 1957 manual on urban arterials admits that the fatality rate of urban areas is less than half that of rural areas.[21]

But little of that mattered because our assumptions and theories for what makes things safer too often trumps the empirical evidence.

Remember James Reason, the kingpin of human error research? He discussed some theories about that kind of thinking in his most recent book, writing that "planners are heavily influenced by their theories: where these are inappropriate, systematic sources of bias will be introduced into the planning process."[22]

This can lead transportation planners to "unconsciously fill in the bits missing from the current evidence in accord with their theories." In addition to seeking "confirmatory evidence for the soundness of the plan," we may also "disregard information that suggests the plan may fail." Because we "tend to have a simplistic view of causality," it's also easy to blame the trees instead of trying to understand their real role in human behavior and road safety.[c]

> It is time, it seems to me, to start thinking realistically in regard to our urban transportation system; time for highway planners to stop trying to build rural highways in cities and start developing urban roadways that are practical and economically feasible.

That previous paragraph could've been written today, but it was actually written in 1960 by Bill Marston, the traffic engineer for the City of Chicago at the time.[23] He was right in 1960, and if he were alive today,[d] he'd still be right.

But when it comes to street design, we keep spinning our wheels on road safety because we remain tied to assumptions and theories that didn't pan out quite like we thought they would. We hitched our wagons to these assumptions and theories, often despite a lack of evidence or even when faced with evidence to the contrary. As Sherlock Holmes once said, "It is a capital mistake to theorize before one has data. Insensibly one begins to twist facts to suit theories, instead of theories to suit facts."[24] Twisting facts to suit theories led us to create design guidelines where much of what we think we do in the name of safety makes things worse.

These design guidelines—often disconnected from any real

---

[c] Or account for all the other benefits of trees in cities.

[d] Marston died in 2008 at age 98.

safety outcomes—remain what we work with today. Their existence allows us to gloss over the harm happening to real people currently on our streets while we sit at our desks calculating how to provide enough road space for theoretical traffic counts off in some far-distant future. We then ask a somewhat rhetorical question: how can these deaths possibly be our fault if we followed the guidelines to perfection?

54 The Fuzzy Math of Urban Freeways

—

When traffic engineers force one-way driving onto streets with plenty of wide lanes, big sight distances and corner radii, and a massive clear zone, city streets will start to look like freeways.

Sadly, that was exactly what we were trying to do. Even worse, we thought it would make us safer.

Wait, what?

Traffic engineers thought the math was simple. According to a 1964 *Traffic Quarterly* paper, the United States averaged 5.3 fatalities for every 100 million miles driven. But on interstates, the fatality rate was only 2.8 fatalities per 100 million miles. Turning our major streets into freeways—"particularly in our fast-growing metropolitan areas"—is going to be key when it comes to the "safety of future travel."[1]

In a 1965 *Traffic Quarterly* paper, Rex Whitton—appointed by President John F. Kennedy to lead the Federal Highway Administration—compares the interstate crash rate of 2.8 fatalities per 100 million miles driven to that on whatever he considers to be "older" roads, which stood at 9.7 fatalities per 100 million miles driven.[2] He concludes that "it is well to look at design feature of the interstate system.... Some of these design features are adaptable to a degree to lessor roads and streets."[3]

What does that mean in terms of design? Rex tells us that it means "wide lanes, easy curves and grades, and long sight distance." In terms of safety, Rex says that "head-on crashes, opposite direction sideswipes, and angle and pedestrian collision are virtually eliminated by such features."[4] Thanks to Rex, all these so-called safety improvements now qualify for federal aid. And yes, Rex wanted to see this happen in our cities.

The origin story of traffic engineers turning city streets into freeways usually starts with the 1965 and 1966 congressional hearings on road safety prompted by Ralph Nader's 1965 book *Unsafe at Any Speed*. In those hearings, Ken Stonex—that same engineer from the

General Motors Proving Ground test track—uses a similar logic as Rex Whitton to conclude that "what we must do is to operate the 90% or more of our surface streets just as we do our freeways … [converting] the surface highway and street network to freeway road and roadside conditions."[5]

Such quotes make for a good origin story. Heck, it's one that I often repeat myself. In reality, traffic engineers were already gung-ho on freeway'ing up our city streets well before 1965. Look again at the AASHO 1957 manual on urban arterials and it's obvious that major streets just aren't good enough. Here AASHO is salivating over the idea of at-grade expressways in cities: "A major street serves both arterial and local traffic, but an expressway, for the most part, separates the two. An at-grade expressway has some characteristics of a major street but it embodies many features of a freeway."

So, 1957 AASHO manual on urban arterials, what exactly is an at-grade expressway?

"An expressway at grade is a surface facility practically free from roadside interference and on which the crossing or entrance of traffic from the less important streets is eliminated."

That sounds interesting, but what do you mean by *eliminated*?

"All minor cross streets should be terminated."

Terminated? I guess we could do that, but how much space do we need for one of these at-grade expressways?

"For at-grade expressways with frontage roads to be provided in areas with a grid of city streets, right-of-way widths corresponding to either a half or a full block are needed."

A full city block? That seems like a whole lot. What about people who live there?

"Most cities have blighted areas slated for redevelopment. Where they are near general desire lines of travel, arterial routes might be located through them in coordination with slum clearance and redevelopment programs."

Oh, you want us to strategically kick certain people out of our cities. Is there anything else we should worry about?

"Pedestrians and vehicles at intersections at grade interfere with one another and present a serious problem in the development of at-grade expressways."

Slum clearance *and* pedestrian clearance? Can you remind me why we don't want the remaining pedestrians to easily cross the street?

"One-half of urban fatal accidents involve pedestrians."

Well, that sounds bad, but shouldn't we design to help pedestrians instead of trying to eliminate them?

"Pedestrian movements and congestion, augmented by intersectional and roadside interference, are the chief causes of urban accidents."[6]

This will keep us from having to fear traffic congestion too! Why didn't you say so? I'm sold.

Part 6

A Bird's-Eye View

_____

55 Not If You Leave Your Cul-de-Sac

—

It would make sense to assume that newer cities, built with all the knowledge that traffic engineers continue to accumulate, should be our safest cities. But that is not the case.

It's the older cities—mostly built before traffic engineers existed—that tend to be safer.

There are lots of reasons for this disconnect, but when we look down on the world, some differences become obvious. The older cities seem more compact, with a connected network of streets. The newer cities appear sparse, with a disconnected network of streets branching out like a tree.

In its 1963 street design manual, the Automotive Safety Foundation (ASF) acknowledges that "general street layout may also figure in accident experience."[1] If we have any doubts about what a proper street layout should look like, "there is considerable evidence" that "a street system which is properly laid out with collector and arterial streets to serve through-movements ... and discontinuous local streets" will "produce lower accident rates"[2] than the older style. In other words, traffic engineers want tree-like street networks instead of the gridded networks of yesteryear.

When it comes to "considerable evidence," the ASF readily admits that "much of the supporting data is taken from studies which deal more with intersections than with street network design."[3]

They first mention a 1953 *Traffic Quarterly* paper, which found that the more advertising signs per mile, the more crashes.[4] This study also found more crashes at four-way than at three-way intersections. We never heeded the sign results but were willing to generalize the intersection-level findings to transform the way we build cities.

They next point to a 1957 *Traffic Quarterly* paper by Harold Marks, who found eight times more crashes in the gridded neighborhoods and 14 times more crashes at four-way intersections than at T-intersections.[5] Though the wheels of tree-like street networks were already in motion, this study set that in stone for traffic engineers.

I'd even have traffic engineers reciting the 50-year-old Marks results when I presented my own research.

Unlike the signage study, Marks at least looked at street networks. But he only considered crashes within the residential neighborhoods (the small tree branches) while ignoring crashes on the collector and arterial roads (the medium and large branches). Given that tree-like street networks force traffic—and crashes—onto those bigger roads, excluding those crashes seems unfair.

What's even more unfair? The gridded street networks Marks studied had no stop signs or even yield signs.

Wait, what?

Marks doesn't mention yield signs because they hadn't been invented at the time of his data collection.[a] In terms of stop signs, Marks tells us that "each of the tracts is protected with the normal types of intersection traffic control, placed according to nationally accepted practices."[6]

What were those practices at the time?[b] Stop signs at the intersection of two residential streets were not likely unless there was "high speed, restricted view, or serious accident record."[7] In fact, Marks says that "a STOP sign is essentially a device used to correct an inherent design weakness." He then pines over the "built-in traffic safety" of "many of our limited access subdivisions which do not have a single STOP sign and have very few accidents."[8]

When later comparing the four-way and three-way intersections, Marks reminds us that "only uncontrolled intersections, those without STOP signs, were included in this comparison in order to isolate the effectiveness of the basic design feature, unaffected by traffic control devices."[9]

No wonder Harold Marks got the results he did. Does that mean that a gridded street is more dangerous than a cul-de-sac? Maybe.

---

[a] That didn't happen until Clinton Riggs installed the first yield sign in the United States in Tulsa, Oklahoma, in 1950. Riggs got some help developing a prototype from a Tulsa city engineer named Paul Rice, but they installed it without getting official permission. Yield signs weren't in the 1948 *Manual on Uniform Traffic Control Devices* but were added to the 1954 version in a short addendum. Even then, yield signs were considered experimental and only used in a handful of cities (none of which were Los Angeles, where this study took place).

[b] Stop signs were also yellow with black lettering back then. What you'd recognize as a stop sign today wasn't typical until 1954.

Does it also mean that gridded cities are more dangerous than cities with tree-like networks?

Harry doesn't know.

But it didn't stop him from issuing a call-to-traffic-engineering-arms because "design inadequacies come home to roost with the traffic engineer who is the man on the firing line when poor planning results in accidents."[10] Given how easy it is to blame crashes on those involved, this last part never came to fruition. Still, traffic engineers were happy to cosign everything else.

Were they right about street network design?

If you never leave your cul-de-sac, sure, it's probably safer.

But in studying more than 230,000 individual crashes in more than 1,000 census block groups spread across 11 years of data for 24 cities for the sake of my dissertation, I found that things aren't that simple.[11]

After separating out the fundamental street network design elements—compactness, connectivity, and configuration—we'd expect 70 percent fewer road fatalities in the more compact neighborhoods as compared to a conventionally sparser one when all other variables are held constant.[c] Do the same for street connectivity, and you'd expect slightly worse safety outcomes with more connectivity at the same level of compactness.[d]

But that's why things aren't that simple. When I accounted for configuration in terms of how disconnected street networks are built in practice—spread out and supported by big, arterial roads—the compact *and* connected street networks have between 27 and 44 percent fewer road fatalities than conventional sparse, tree-like street networks.[e] If I were able to account for where the crash victims lived

---

[c] Speed seems to be one of the biggest differences. Sparser street networks with intersections separated by greater distances tend to lead to faster driving. By the way, we'd also expect about 30 percent fewer fender benders in the more compact neighborhood when we hold all the other variables constant.

[d] Given my results, I'd like to see more US examples of compact networks completely connected for pedestrians and bicyclists but intentionally disconnected for cars. There are limited examples of such designs (the Village Homes neighborhood in Davis, California, is an interesting case; cities like Portland, Oregon, and Berkeley, California, strategically place car diverters within their grids a well) but not quite enough to study the safety impacts … yet.

[e] They also saw between 4 and 9 percent more fender benders, but if fender benders are still our focus, we're missing the point.

instead of just crash location, these differences would likely have been more drastic.

My follow-up research to the safety work found loads of other advantages for those living in compact and connected street networks. They drove far fewer miles.[12] They walked and biked much more often.[13] They also had drastically lower rates of obesity, diabetes, high blood pressure, and heart disease.[14]

You may say that healthier people choose to live in these places, and there's some truth to that. At the same time, you can't ignore that most Americans live where street networks are designed to thwart any sort of active transportation. That can't be helping the cause.

We treat these underlying street network design as a given. A given that we designed and left unexamined for generations.

56 What's Your Function?

___

Why oh why did we completely change the way we build the bones of our cities over the course of the 20th century?

I can let traffic engineers off the hook here—to some extent—because they were late to the game. The Automotive Safety Foundation manual didn't come out until 1963. Two years later, the Institute of Traffic Engineers[a] published *Recommended Practice for Subdivision Streets*, where traffic engineers finally put their ceremonial stamp on disconnected networks with cul-de-sacs and curvilinear streets.[1]

But even before the prevalence of cars, planners like Charles Mulford Robinson campaigned against gridded street networks by separating city streets into two types, "the busy major traffic thoroughfares and the quiet minor residential street."[2]

We didn't find our influential example until 1929 in Radburn, New Jersey. Radburn was one of the earliest tree-like, hierarchical street networks in the United States. To guys like Charles Stein, one of the planners behind Radburn, the idea of a gridded street network was "as obsolete as a fortified town wall."[3] Based on Clarence Perry's "neighborhood unit" concept, Radburn consisted of massive blocks of cul-de-sacs (30 to 50 acres) supported by a network of larger roads.

Road safety was not their main motivation. Perry, for instance, never mentioned safety as a rationale until well after Radburn. Prior to that, he sold his ideas under the guise of patriotism and as a way to get away from immigrants.[4] The Federal Housing Administration (FHA) liked the sound of that and jumped on board in 1935 when it started publishing a series of technical bulletins that were explicit about what constituted "good" and "bad" street network design.

The FHA sought Radburn-esque plans of large blocks and cul-de-sacs surrounded by big roads and highways. It lambasted gridded street networks, calling them a "safety issue" despite a lack of such

___

[a] It didn't change its name to the Institute of Transportation Engineers for another 10 years.

evidence. It filled the pamphlets with paired images of connected and disconnected street networks. Next to the connected street network was the word *"BAD"* written in all capital letters. And sure enough, right there on the disconnected network was the capitalized word *"GOOD."*

Who cares what the FHA has to say about street networks? I hear you, but keep in mind that the FHA played a role in guaranteeing the mortgages[b] on more than 22 million properties over the next 25 years.[5] If the proposed street network—or person—didn't look the part, the FHA swiped left.[6]

Even Jane Jacobs "lauded Radburn as a classic of residential planning" in her early days as a writer for *Architectural Forum.* During this time, she also "praised urban renewal and its large-scale demolitions and superblocks."[7]

But where were the traffic engineers?

Traffic engineers were happy to sit on the street network sidelines and focus on things like roadway classification. For instance, AASHO[c] was off working on its own set of pamphlets that had seemingly little to say about street network design. Its first pamphlet, the 1938 "A Policy on Highway Classification," focused entirely on classifying streets based on things like design speed and traffic volume.

The traffic engineering approach to roadway classification, however, played right into their tree-like hands.

Before we get to that, let's first ask, why do traffic engineers classify streets?

In theory, classifying streets helps give us a common terminology. When I describe something as a local street, collector, or arterial, another traffic engineer should know what I mean.

In theory, it helps with design consistency. We should be able to use similar design standards—such as design speed, design vehicle, and cross-section design—for all arterials.

In practice, there is little design consistency within functional classes. Without design consistency, our terminology—and words like *arterial*—can become meaningless.

---

[b] The FHA didn't issue mortgages itself. Rather, the government would repay loans in full if the homeowner defaults. For lenders, it's as risk-free as it gets.

[c] Recall that AASHO is the American Association of State Highway Officials. It later became the American Association of State Highway and Transportation Officials (AASHTO).

In practice, classifying streets helps determine how we distribute federal transportation dollars. And money drops a monkey wrench into any and all theories.

When we give more money to more miles of arterials, we incentivize classifying more streets as arterials.[d] Look at downtown Denver, where *every single street* is classified as an arterial.

But what is an arterial? Based on what traffic engineers call the ***functional classification*** system, an arterial is a road with a relatively high level of ***mobility*** and a relatively low level of ***access*** to abutting land uses. And when we say mobility, we historically mean car mobility.

That wouldn't be my definition of what I'm looking for in downtown Denver.

Anyway, a 1961 *Traffic Quarterly* paper gives a nice, simple overview of functional classification, saying, "Basically, a street or highway has two functions: to move traffic, and to provide access to abutting property. By properly combining these functions, it is possible to classify a street system."[8]

In other words, functional classification is based on an almost mutually exclusive scale of mobility and access. We have streets with relatively high mobility and low access like arterials. We also have streets with relatively low mobility and high access like local roads. Collector roads have a little of both mobility and access.

The big idea is that we can't have high levels of both mobility and access. But like in downtown Denver—with its high levels of land access on arterials—that big idea doesn't hold up in practice. It also doesn't quite hold up on arterials populated by fast-food joints, gas stations, and mattress stores. Keep in mind that it's a chicken-and-egg issue. A lot of businesses—and from their vantage point, rightfully so—look to locate themselves right along roads with lots of cars. The arterials we build are perfect for that. The functional classification disconnect comes later when we allow these land uses to come in and then try to facilitate access.[e] A 1963 *Traffic Quarterly*

---

[d] Don't get me started on how we also distribute federal transportation dollars based on vehicle miles traveled. The more you drive, the more money your state gets. This approach doesn't give state departments of transportation much reason to disincentivize driving, nor is it a safety neutral funding policy.

[e] That's how we end up with what Chuck Marohn and Strong Towns would call a stroad. A stroad isn't quite a street and isn't quite a road. By trying to accomplish

paper complains about the street "improving" that traffic engineers undertake in an attempt to "carry traffic," which is inevitably followed by "strip commercial development and inadequate curb cut control frustrating that purpose."[9]

But this chapter isn't about access management. It's about how something as fundamental to transportation engineering as the functional classification system influenced street network design and, in turn, road safety.

AASHTO describes the functional classification system using terms like "hierarchy of movement," which means that we want the "main movement" on big roads like freeways. This traffic then transitions onto arterials that distribute the traffic to collector roads. The collected traffic leads to local roads and land access.

Like it or not, this approach aims traffic engineers toward a certain target, and that target is a tree-like street network.

Don't get me wrong; we still apply the functional classification system to gridded street networks. But like downtown Denver, most gridded street networks predate the functional classification system. What happens when the street network comes after functional classification? The result often resembles a branchy deciduous tree with streets named after the trees we cut down to build it.

We can't blame a century of tree-like street network design on the functional classification, but it's not helping the cause.

Functional classification isn't all about mobility versus access. There is a second dimension. Some call it the physical setting. Some call it location. Some call it land use. Some call it context. Whatever the words, you might imagine a spectrum of contexts akin to the mobility versus access scale.

Nope. It's binary.

The functional classification system recognizes urban or rural.[f] That's it. Let's say you have a street that starts in a central business district and heads out to suburbia. Traffic engineers would classify that entire stretch as urban. To simplify design and construction,

---

both mobility and access, stroads do neither well (like how a futon is neither a good couch nor a good bed). It's probably not surprising that stroads form the basis of most high-injury networks.

[f] According to AASHTO, urban land use in the United States is defined as a contiguous area of census block groups with at least 1,000 residents per square mile and a minimum population of 5,000. Everything else is rural.

we might even design a single cross-section for that entire stretch. What's the big deal? Reid Ewing says it well in his 2002 *Transportation Quarterly* paper: "The most insidious respect in which reliance on typical sections may undermine context sensitivity is the tendency to adopt a single, typical section for an entire stretch of road, even when conditions change along the road. Having a typical section is convenient for the design engineer and construction crew. But it is not good policy."[10]

Traffic engineers don't think about the functional classification system as having much to do with road safety. But how could it not impact safety? Why wouldn't we want the street classification system to consider how people use streets other than in a car? Why would we want to classify the entirety of the United States as only urban or rural, with nothing in between?

Keep that in mind the next time you see a city street that doesn't seem to have a clue that it's in a city. It's the de facto design based on the functional classification of another place and possibly another time. Streets like that—backed by a faulty functional classification system—are killing us.[g]

---

[g] I was hoping to end this chapter with a "You're killing me, Smalls," but alas.

57 Bigger and Badder

___

So how did the *functional classification* system come to be?

The basics date back to the engineering pamphlets that AASHO put together in the 1940s. At the time, they were optional guidelines. They remained optional even after AASHO combined its seven pamphlets into the first *Policies on Geometric Highway Design* in 1950.[1]

But remember the Automotive Safety Foundation (ASF)? The one founded by the president of Studebaker, Paul Hoffman? They were the "primary organizer and funder" behind the "large, well-coordinated effort" to legislate functional classification during the 1950s and 1960s. Why did they bother? L. J. Aurbach found an old ASF pamphlet stating "that the purpose of functional classification was to funnel traffic and funding to the most socially important, high-speed, high-capacity freeways and arterials."[2]

These efforts started to pay off in 1968 when the newly formed US Department of Transportation (USDOT) released a report asking Congress for $292 billion.[3] The supplement to that report recommended a national functional classification system.[4] ASF then got everyone (AAA, AASHO, the Institute of Traffic Engineers, the National Association of Counties, the National League of Cities, the US Conference of Mayors, and so on) to testify before Congress about the dire need for a standardized version of functional classification being mandated nationwide.

Congress then funded a study about functional classification in its 1968 infrastructure bill. If you look at the 1968 manual that accompanied this study, you'll see that it's darn close to the functional classification system we use today. This test run of functional classification also found that urban arterials constituted 19 percent of road mileage but carried 75 percent of vehicle miles traveled. Numbers like that freak people out. So instead of $292 billion, the USDOT requested $592 billion from Congress simply to attain "minimum tolerable conditions" by 1990.

This "study" and lobbying worked. In 1973, Congress mandated functional classification. Since then, if you want a piece of the federal transportation funding pie, you must use the functional classification system.

But this mandate wasn't just about funding. The ASF pamphlet "predicted that functional classification would be a powerful shaper of future land use patterns."[5] Even back then, ASF wasn't alone in realizing that transportation and land use go hand in hand. The year after functional classification became the law of the land, the Eno Foundation held a conference on transportation safety. In the conference report, they worry about the impact of transportation policies—such as the functional classification system—on land use and ask, "How can there be national transportation policies without national land-use policies?"[6]

To my surprise, they act like this was something everyone should've already known: "Conventional wisdom has it that there is a two-way relationship between transportation and land use. In the short term, patterns of land use and economic activity largely determine travel demand. In the longer term, the accessibility potential offered by transportation facilities largely determines changes in land-use patterns."[7]

What sort of national land use policy should we have? It suggests one with "a density that can support sidewalks and a bus stop."[8]

That sounds nice, but a national land use policy? Really?!? That is almost laughable.[a] No wonder we didn't heed the alarms being rung by the Eno Foundation. Instead, we let the functional classification system rearrange the bones of our cities and towns. Heck, we *required* the functional classification system to do so. The result, as Aurbach concludes, is that "functional classification helped make American built environments—especially the thousands of square miles of new suburbs—firmly and almost permanently oriented to automobiles."[9]

It's been helping us build big, bad arterials since Richard Nixon was president.

But when you forecast traffic decades into the future, even the biggest and baddest arterials aren't going to be enough. If we really want to break the bones of US cities, we need the biggest and baddest in vehicular mobility. Freeways.

The USDOT knew that, and if you dig through that 1968 "needs"

---

[a] It shouldn't be.

report, it wants most of those billions spent on freeways. Not just on any freeways, but on freeways through cities.

Traffic engineers didn't hesitate to jump on board. Remember AASHO's 1957 manual specific to urban areas, *A Policy on Arterial Highways in Urban Areas*?[10] AASHO updated it in 1973 and called the new manual *A Policy on Design of Urban Highways and Arterial Streets*.[11] The titles seem similar, but the insides were not.

In 1957, AASHO only wanted to supersize our existing streets. By 1973, big arterials seem like an afterthought. The manual's chapter on arterial design doesn't come until well past page 600. Almost everything before that is about forecasting traffic demand decades into the future and how to accommodate that predicted demand by building freeways through cities. There is a fascinating chapter on how these urban freeways can "even serve as a catalyst to environmental improvement."[12] But the question of whether our cities need freeways cutting through them? That issue no longer seemed up for debate.

So we tore up our old street networks and cities in the name of solving traffic congestion. And for better safety. You could even say we did so to save our cities. But to understand the logic behind it all, we need to go further back in time.

58   Between Isn't Through

—

Historically, the federal government wouldn't allow its transportation dollars to be spent in cities.

Thankfully, people like good ol' Paul Hoffman from the Automotive Safety Foundation set out to change history. In a 1939 issue of the *Saturday Evening Post*, Hoffman shares a story about a conversation with his brother-in-law, Bill.

Brother-in-law Bill says that "what we need in this country is good highways around cities, so you can by-pass em." Paul spends the rest of the article setting his brother-in-law straight.[a]

Similar to the arguments made with Radburn a decade earlier, Paul says that "many of our cities are almost as antiquated, traffic-wise, as if they had medieval walls, moats, drawbridges. We crawl through some city streets at a mule's pace just as we used to crawl through road mud."

So, what should we do Paul?

"If we are to have full use of automobiles, cities must be re-made."

Re-made how?

"No city can be said to be equipped for the motor age unless all of its express highways are some type of limited-way facility."

Can't we just embiggen the arterials a bit more?

"The man on the street thinks we have to go in on present right of ways, the streets that grew from cow paths. We think, however, that this would only prolong the agony…. We are also thinking in terms of rights of ways 500 feet wide, actual highways 70 or 100 feet wide, with adequate parking alongside."

Wow. How do we know when it's time to start building this amazing future?

"There is no exact capacity formula, but … when the traffic load gets in the neighborhood of 2,000 cars per day, it is time to think of converting it into a multi-lane divided highway."

---

[a] Paul sounds like a fun hang.

And where are we going to put it all?

"The blighted areas offer the one opportunity we have to develop at reasonable cost the type of highway and parking facilities needed to meet the requirements of motorized communities."

What about the people that live in these so-called blighted areas?

"What real need is there for slum residents to stay where they are? Theoretically, the ideal situation for those who need low-cost housing is a five-acre plot in the country where they can be partially self-sustaining by raising their own vegetables, perhaps keeping a cow and chickens."[1]

In addition to wanting to ship the "slum residents" off to a farm, Paul Hoffman also isn't shy about reminding us that he's basically a car salesman.

"The greatest automobile market today, the greatest untapped field of potential customers, is the large number of city people who refuse to own cars, or use the cars they have very little, because it's such a nuisance to take them out. A waiting industry that will do wonders for prosperity will spring up when we revamp our cities and make it safe, convenient, pleasant and easy to use a car on city streets."[2]

Translation? The best way to sell cars is to make it unsafe, inconvenient, unpleasant, and difficult to do anything *but* use a car in our cities.

Hoffman then references Thomas MacDonald, chief of the Bureau of Public Roads,[b] whose nickname was the Chief,[c] saying that "Chief MacDonald insists that we must dream of gashing our way rather ruthlessly through built-up sections of overcrowded cities, in order to create traffic ways capable of carrying the traffic with safety, facility, reasonable speed."[3]

What's holding us back from "gashing our way rather ruthlessly through built-up sections of overcrowded cities"?

---

[b] Today's Federal Highway Administration used to be called the Public Roads Administration and before that the Bureau of Public Roads. And before the Bureau of Public Roads was the Office of Public Roads and Rural Engineering … and the Office of Public Roads … and the Office of Public Road Inquiries … and, dating back to 1893, the Office of Road Inquiry.

[c] It's hard to call anyone Chief if they aren't Sitting Bull or Robert Parish of the Boston Celtics (who, when I saw him at the airport when I was 10 years old, remains the one person I've ever asked for an autograph).

You might assume that Hoffman asks for money at this point. Nope. Money isn't the issue. In fact, all the government needs to do is buy up the "adjacent property" next to these new freeways. Once the freeway gets built, we can sell these former slums and "recover practically the entire cost" of the freeway.

The real problem? All the stubborn people who won't leave their homes and move to that farm upstate. Hoffman starts by saying we need "a new land policy." I was hopeful, but all he really wanted was to make eminent domain easier to use. That way, we won't have to wait "ten years to get" started.

Hoffman then says that having to wait to kick all these stubborn people out is literally killing us, as "certain fatal accidents are traced to the delay" of not getting that new freeway done as soon as possible.

How exactly will these freeways make us safer? It's all about getting people off those narrow and congested city streets and onto urban freeways, as Hoffman says in his 1939 book: "The effect of removing that much traffic from accident-prone streets would obviously be a large gain for safety."[4]

Obviously.

When Marty McFly goes back in time to 1955 in *Back to the Future*, Hill Valley is the idyllic "before" picture. By the time the Highway Trust Fund was established in 1956, we were fully on board with these freeways "gashing" right through our downtowns. The 90/10 split on funding—where the federal government picks up 90 percent of the tab—didn't help. A 1959 *Traffic Quarterly* paper questions whether cities will survive but isn't surprised by the funding priorities of Congress: "Obviously, cars are more powerful than cities. They are better organized, better financed, and far more persuasive."[5]

Even today, we still officially define freeways as "divided highways that carry longer distance major traffic flows *between* important activity centers and have fully controlled access at no at-grade intersections."[6]

Unofficially, the "between" in this definition became "through," and the results weren't always pretty. Remember the dystopian Biff universe in *Back to the Future Part II*? In many US cities, those scenes would be a good stand-in for the "after" picture.

59 One Shining Moment

—

Paul Hoffman wasn't the only one on this crusade. The Chief—Thomas MacDonald—also played a big role, both in giving Hoffman ideas and when the Bureau of Public Roads published *Toll Roads and Free Roads* in 1939.[1] According to MacDonald, putting these "express highways" in cities would solve congestion, save lives, and "stop the decay of city business districts."

In 1941, President Franklin D. Roosevelt commissioned a committee to investigate the need for a national highway system. When the committee submitted its "Interregional Highways" report to Congress in 1944, they acknowledged "that urban expressways generally had negative impacts" but hoped that "proper design could fix those problems."[2] The committee envisioned landscaped borders that would "insulate adjacent residential and business properties, churches, and schools from the noise, dust, and fumes of traffic."[3] The goal was for urban freeways to become "not the unsightly and obstructive gashes feared by some—but rather elongated parks bringing to the inner city a welcome addition of beauty, grace, and open green space."[4]

Slowly but surely, scare tactics helped shift this idea of urban freeways from being a giant question mark to a commonsense inevitability.

Safety was initially a crowd pleaser. Remember Colonel Robert Goetz, the editor of *Traffic Quarterly*? In his 1958 editorial, he says that flipping people from city streets onto freeways will save 4,000 lives each year.[5] By 1961, the Colonel was admonishing the cities that weren't on board: "It is unfortunate that with all its favor and the advantages of automotive transportation, some cities have appeared reluctant to make the most of it…. Many cities appear to ignore this important advantage by their failure to provide appropriate means to accommodate the automobile."[6]

It wasn't long until the ante was upped. In 1965, a *Traffic Quarterly* paper declares, "Accident prevention is a major motif … the lives saved in travel should exceed 8,000 annually."[7]

Time and time again, they argued that it would cost us more *not* to build freeways through cities. "Dollar value cannot be placed upon a human life lost nor on national freedom…. Yet, it is costing too much not to buy adequate and safe highways."[8] And it will cost us even more if we wait. "The cost of not having proper highways will increase as a geometric ratio."[9]

It seems odd to see "national freedom" thrown in, but safety isn't just road safety. These freeways will also save us from an atomic attack. In a 1956 *Traffic Quarterly* paper titled "Plans for Evacuating a Large City in Case of Atomic Attack," the basic premise is that the Russians will kill us if we don't have freeways in cities. "Millions of Americans can be killed…. And the best way to evacuate millions of people in America's target cities is by motor vehicles—on our highway."[10]

If atomic bombs aren't scary enough, we always have traffic congestion. One 1962 *Traffic Quarterly* paper called traffic congestion "the measure of inadequacy"[11] of a city.[a] Another claimed that "the most serious problem usually centers around providing adequate street and parking facilities at a rate equal to the demand. Too many cities have fallen behind in recognizing the problem and in developing programs to correct the deficiencies."[12]

Some traffic engineers pushed back. One 1959 *Traffic Quarterly* paper wasn't beating around the bush: "The family car is destroying the American city as a desirable place in which to live and work and play. The struggle is an entirely unequal one…. More than $6 billion of interstate money will have been spent for highways in urban areas. All this to make cities more comfortable for the family car to pass in and out of (or through)."[13]

Another 1963 *Traffic Quarterly* paper also sees some problems: "A freeway … can irrevocably damage the basic urban pattern and amenities. It can be a needlessly unattractive and offensive neighbor to other land uses."[14]

Even President Dwight D. Eisenhower had some questions. Here is an amazing quote from a memorandum of a 1960 Oval Office meeting:

> The President referred to a previous conversation with General Bragdon. He went on to say that the matter of running

---

[a] The opposite seems to be true.

Interstate routes though the congested parts of the cities was entirely against his original concept and wishes; that he never anticipated that the program would turn out this way. He pointed out that … he had studied it carefully, and that he was certainly not aware of any concept of using the program to build up an extensive intra-city route network as part of the program he sponsored. He added that those who had … steered the program in such a direction, had not followed his wishes.[15]

Yikes.

Did the president of the United States get his way? Nope. They told Eisenhower that ship had sailed. And even if he wasn't too late, they also told him he was wrong.

Here is a 1964 *Traffic Quarterly* paper, written by a man who works for AAA and had been working for AAA since 1924.[16] Like Adam Savage and Jamie Hyneman on *Mythbusters*—but without presenting actual evidence—Russell Singer sets out to bust some myths about problematic urban freeways. Here are the 10 myths he purportedly busts:

Myth No. 1: The automobile is strangling our cities.

Myth No. 2: Building freeways and providing parking space is self-defeating because such facilities merely attract more traffic, causing greater congestion and leaving the city worse off.

Myth No. 3: Freeways will hasten decay of downtown by drawing people and businesses to the suburbs.

Myth No. 4: Cities would be benefited by severely limiting the use of private automobiles downtown.

Myth No. 5: Rail transit is more efficient than highway transportation.

Myth No. 6: Rapid rail transit will eliminate the need for building many urban freeways not projected.

Myth No. 7: Freeways gobble up valuable land space, taking many acres off city tax rolls.

Myth No. 8: Mass transit should be subsidized so it can compete on even terms with highway transportation, which is heavily subsidized by the federal government.

Myth No. 9: Automobile travel is expensive.
Myth No. 10: Urban freeways are tremendously expensive, costing millions of dollars per mile, and represent a serious drain on public treasuries.[17]

It was hard to read this article without laughing. Singer's basic premise? Don't worry about it! Freeways won't hurt downtowns. In fact, as long as we give cars "adequate facilities and freedom of movement," they will contribute "to a brighter future and a rejuvenation of the downtown area."[18] He then gives the example of Midtown Plaza in downtown Rochester, New York. Built in 1962, "it is at the heart of Rochester's committed plan for freeways."

Papers like this helped quell concerns about urban freeways. It reached the point at which tearing our cities up for freeways wasn't even a question anymore. A 1963 *Traffic Quarterly* paper casually says that the benefits of urban freeways are undeniable: "While one could not and would not deny the importance and positive value of freeways to urban areas."[19]

Before we move on, here's a quick update on Rochester. Even though Singer assures us that "Rochester has become dynamic again," Midtown Plaza didn't stand the test of time. It shuttered in 2008. Six years later, this exact location won the Golden Crater award that Streetsblog hands out each March with its Parking Madness tournament: "An asphalt scar in Rochester, New York, has triumphed over 15 of the world's worst parking craters to become the Parking Madness 2014 champion. It was a surprising run. Who would have guessed a couple of weeks ago that this scrappy upstart would prevail over some of the sprawliest, most highway-marred urban spaces in North America?"[20]

What an incredible Cinderella story for Rochester. This unknown comes out of nowhere to lead the pack.

60 Doing Our Jobs?

—

Have you seen the PBS documentary *Divided Highways*?[1] It came out in 1997, released around the same time as Tom Lewis's book of the same name. Some parts of the show are dated, but it still holds up. It tells the story of how the interstate system both connected—and divided—us at the same time.

When I first watched *Divided Highways*, I recognized one of the older traffic engineers interviewed, Frank Griggs. He was the director of quality assurance at Clough, Harbour & Associates during my first few years working there. Frank earned his PhD in civil engineering back in the 1960s and worked for decades in the field. You can't overstate the amount of experience and respect he has as a traffic engineer.

In the documentary, Frank explains the traffic engineer's mind-set on where to build freeways:

> We were supposed to go through the least expensive land because land was a very high part of that cost. So we would go through swamps, the least expensive land. If we were in an urban environment, we would go through the slums because that was the least expensive land.... So we were trained, maybe not told specifically to do it, but we were trained to build the least expensive road to service the most people.

Frank was also quick to defend us traffic engineers:

> We were given this job to do. We were given guidelines to do it, and we did it. And suddenly people were unhappy. And we knew that we had done a good job. We had followed all the guidelines, all the criteria. We knew we could get people from A to B much more efficiently and much safer than we had done in the past. I think it was a tough time to be a

highway engineer because you felt you had done nothing wrong, and yet, you were being dumped on by the public.

When I first considered what went wrong, I figured it was a matter of traffic engineers trying to answer the wrong question. Traffic engineers were asked how to move cars fast and do so as inexpensively as possible. We ended up with problems because traffic engineers never bothered asking what the social, environmental, or long-term economic consequences might be.

The more I dug, the more I realized that Frank wasn't quite giving us the whole story. For decades, Thomas MacDonald—the Chief—liked to respond to critics of urban freeways by saying that "most of the buildings are of the type that should be torn down in any case, to rid the city of its slums."[2] In other words, we traffic engineers are doing you a favor by replacing your slums with a freeway.

The 1957 AASHO manual specific to urban areas includes a diagram intended to help traffic engineers locate where the big roads need to be built.[3] The graphic illustrates five possible locations: through undeveloped land or parks,[a] along railroad tracks or rivers,[b] or in "decadent" areas. I thought words like *decadent* generally referred to over-the-top extravagance, but decadent can apparently also mean "characterized by or reflecting a state of moral or cultural decline."

That same AASHO manual goes on to say, "Most cities have blighted areas slated for redevelopment. Where they are near general desire lines of travel, arterial routes might be located through them in coordination with slum clearance and redevelopment programs."[4]

That was the tip of the iceberg.

A 1953 *Traffic Quarterly* paper describes new urban highways as a good way to remove or cordon off the slums: "In each case, the highway locator has an opportunity to assist the community in the redevelopment of depressed areas and to open up for good development other areas not otherwise progressing.... Removal of low tax paying properties and development of new high tax paying properties have helped measurably to solve difficult city revenue problems."[5]

---

[a] The plot in one of my favorite Christmas movies, *Yogi's First Christmas*, revolves around traffic engineers wanting to tear down Jellystone Lodge to build a highway through Jellystone Park. Spoiler alert: Yogi and the gang save Jellystone.

[b] Big arterials, freeways, and parking lots along waterfronts drive me crazy!

The paper also makes the point that finding these slums takes more than looking at a tax map and identifying the least expensive land. The person hired as the "highway locator" needs to go out and do some "visual inspection." That is the only way to find that "fringe area between the present commercial and retail areas and the good residential areas which has been on the border of becoming slums.[6] The success story he gives? The 710 that runs through Long Beach, California.

In 1959, Wilbur Smith, one of the grandfathers of traffic engineering, wrote a *Traffic Quarterly* paper about how we need to rebuild our cities—and replace the "blighted areas"—for the sake of our cities: "These threats to the health of the downtown district have come about in the form of blighted areas, relative reductions in tax income and a general lowering of its relative attractiveness as a place to work, shop, set up a new business, or expand an existing one. The city as a whole has suffered."[7]

In a 1964 *Traffic Quarterly* paper, Boston traffic engineer William McGrath—the namesake of a big, bad arterial that runs through Somerville, Massachusetts—casually refers to the work that we do as slum clearance.[8] Another 1964 *Traffic Quarterly* paper then argues that we need to evolve past the "limited concept of slum clearance" and start looking for "blight in non-residential structures."[9] How do we measure nonresidential blight? According to the American Public Health Association, traffic engineers should be on the lookout for a "lack of off-street parking."[10] A 1972 *Traffic Quarterly* paper even weaponizes our work: "Modern transportation ... has provided an important weapon for attacking slums."[11]

Yup.

The best—or worst—examples were the cities they told us we should envy.

In the 1939 report *Toll Roads and Free Roads*, MacDonald and the Bureau of Public Roads called "the whole interior of" Baltimore a city that is "ripe" for these changes.[12] Years later, a 1967 *Traffic Quarterly* paper brags about the plan to put 20 miles of freeways right through Baltimore's slums. "Good news! An enormous stride has been taken toward creating heaven on earth."[13]

That's a bold statement Baltimore. Tell me more. "An examination of the present condition of the areas under consideration for a right-of-way shows that they are in a deplorable mess,... host to more thriving slums and crime than many areas now classified under urban renewal projects."[14]

And how did things turn out in Baltimore? I'll just assume that it was all smooth sailing in creating heaven on earth … and that *The Wire* is set in an alternate universe.

Detroit is another example. A 1959 *Traffic Quarterly* paper about improving traffic via urban freeways and street widenings boasts that Detroit's efforts will not only "eliminate slums" but also "prevent the formation of future ones." Slums weren't the only problem in Detroit; it was also the street network: "The street layout established in the urban renewal plan will consist of fewer, but wider, streets—because the project will provide a large amount of public and private off-street parking space—will be used primarily for moving traffic."[15]

What's the solution, Detroit? "Many projects planned for residential redevelopment will wipe out obsolete gridiron patterns and replace them with curvilinear systems such as are found in the best of suburban subdivisions."[16] But as a 1963 *Traffic Quarterly* paper tells us, obsolete, gridded street networks can be a blessing in disguise: "It is obsolescence, however, that can be a blessing, an opportunity to convert whole areas to the needs of man in the large modern city."[17]

Only the best for Detroit. Are there any specific projects you want to mention? "Detroit's Gratiot project will convert a slum area of about thirty small blocks into a ninety-acre superblock penetrated at intervals by cul-de-sac streets."[18] What about the people who live there? Not long before, "the customary acquisition procedure was to make an award of only one dollar at the time of the taking."[19] But in 1959? They will be getting 100 whole bucks for their trouble. Now take that 100 bucks and get lost.

The last example I'll give here is Nashville, Tennessee. It comes from a postretirement piece written by Thomas Deen[c] after four decades as a traffic engineer.[20] Deen started his career in Nashville in 1956, the same year Congress passed the Highway Act. Based on this act, "Nashville was to be the intersection of six legs of the Interstate, and was to include both an inner and outer loop to connect the branches."[21]

How was that done? Deen says "we really only had two criteria:" engineering and economic.

---

[c] Since 1992, one of the most prestigious career achievement awards given out by the Transportation Research Board is the Thomas B. Deen Distinguished Lectureship. Thomas Deen didn't win his own award until 2003.

Engineering means it needs to meet design "standards" while accommodating the 20-year traffic projections. Also, "from a mobility point of view, we were going to cut travel times in half." Economic means "we tried to fit it into the urban landscape as inexpensively as possible."[22]

That's what Frank Griggs was telling us. In fact, it's exactly what most traffic engineers would tell us.

But Deen goes on to explain what it really means: "As transportation planners we genuinely believe that one of the benefits of building the Interstate would be the chance to remove areas of decayed housing and replace them with new public housing. Finding such slum areas to build the roads appealed to our desire to build them as cheaply as possible."[23]

It seems simple enough, but Deen was a little taken aback when "we quickly began to get opposition from residents and representatives of poverty groups, who valued these neighborhoods and had no intention of moving."[24]

It didn't stop Deen—or Nashville—from building all those freeways.

How did things turn out? Well, Nashville isn't exactly a bastion for road safety. I wouldn't say Denver is either, but it's a useful comparison because the number of people who live in these cities is about the same. On average, Nashville's streets kill about 100 people each year, while Denver's kill around 60.[d] Over the course of a decade, that's an extra 400 lives lost in Nashville as compared to Denver.[e]

I've never seen the TV show *Nashville*, but the one thing I do know—spoiler alert—is that a car crash killed Connie Britton's character.[f] She leaves her office at Highway 65 Records in downtown Nashville and never makes it home to her mansion on the outskirts of the city.

She's not real, but the hundred or so people who die in Nashville every year are very, very real.

---

[d] Which is still 60 too many.

[e] Oslo, Norway, is also about the same size as Nashville. In 2019, it had one road fatality. That's it. And it wasn't even a pedestrian or bicyclist. Looking just at kids between the ages of 6 and 15, Oslo has had zero road deaths for at least the last 20 years.

[f] It's hard not to refer to Connie Britton as Tami Taylor of *Friday Night Lights*. "Clear eyes, full hearts, can't lose!"

61 Ain't That America

—

When it comes to urban freeways, I'm not talking about the safety of those *on* the freeways. I'm not even talking about the Tami Taylors of the world. Before we get to what I am talking about, let me step back and explain one crucial aspect of how road safety research works.

If I'm trying to publish a research paper that analyzes crash outcomes, I'm compelled to account for so-called *control* variables such as income and race. Why? Well, the existing research is clear that lower-income and minority populations in the United States are more likely to die or get injured on US streets than other Americans.[1] So if I want to study the relative safety of, well, anything, I need to include variables such as income, race, or even education to make things "fair." Otherwise, it's hard to make it through the peer review process.

Here's an example. Let's say I'm trying to understand the relative safety of different street network designs. We expect more crashes and more deaths in whichever neighborhood is poorer or whatever neighborhood has more people that identify as Black, Hispanic, or Native American—even if the street networks are exactly the same. Instead of asking **why** that's the case, we simply "control" for this difference with income and/or race as background variables in the statistical analysis. Once we do, we can then study street network design or whatever the real variable of interest might be.

Yet we've seen that traffic engineers systematically, intentionally, and repeatedly place big arterials and freeways directly through lower-income and minority neighborhoods. It's no wonder traffic engineers are disproportionately killing some people more than others.

It would be nice to say traffic engineers were oblivious to all that. After all, we were told to focus on engineering and economics. But that isn't really what happened.

A 1947 *Traffic Quarterly* paper by Nathan Smith—director of Public Works in Baltimore—questions us trying to "remedy slum conditions" with big roads and highways, even saying that adding more "traffic" may be "an important influence on the causes of urban blight and must be given careful attention in attempts to remedy slum conditions."[2]

He then suggests funding parks in these neighborhoods—instead of big roads—but laments that it's much harder to find funding for a park than for a big road.

But we didn't listen, and the next thing you know, we started putting freeways right through our existing and historic parks. Here is a 1977 *Traffic Quarterly* paper that doesn't like how things went down: "There are many examples of this. In Memphis, Tennessee, for instance, a 6-lane interstate was run through the middle of a 340-acre park; in Philadelphia, Pennsylvania, expressways go through and bound Fairmont Park, one of the largest city parks in the world ... and so on."[3]

If you've read Robert Caro's book *The Power Broker*[a] about New York City's Parks commissioner, Robert Moses, you learned that he was also chairman of the city's Slum Clearance Committee.[4] Moses liked to say that he located his freeways using sophisticated equations that kept getting more and more sophisticated each year. But math didn't have much to do with why the freeways ended up where they did.

Caro gives the example of the Triborough Bridge—now called the Robert F. Kennedy Bridge—which clumsily enters Manhattan at 125th Street instead of closer to 90th Street, which would make more sense given the location of the parkway in Queens. The reason for this location has little to do with engineering or economics. It came about because former congressman and newspaper magnate William Randolph Hearst owned a block of tenements near 125th Street that was losing money.[b]

Over the years, Moses "evicted tens of thousands of poor people in his way, whom, in the words of one official, he 'hounded out like cattle.'" He also dismantled the street networks of these places by forcing more than 627 miles of "expressways and parkways" through New York City and its surrounding areas. By doing do, Moses ripped out "the centers of a score of neighborhoods, many of them friendly, vibrant communities that had made the city a home to its people."[5]

---

[a] Congratulations, because it's 1,300-plus pages long.

[b] I guess it was about economics after all.

When building the Cross-Bronx Expressway during the 1950s, Moses "demolished a solid mile of six- and seven-story apartment houses—fifty-four of them—thereby destroying the homes of several thousand families" in the thriving East Tremont neighborhood.[6] After the expressway went in, East Tremont became a "vast slum,"[7] which is exactly what Moses was supposedly the chairman of getting rid of.

Did this improve road safety—or any kind of safety—in these neighborhoods, like Moses claimed it would? Of course not. Look at these neighborhoods now, and you'll also find high rates of asthma, lung disease, heart disease, and cancer.

If we dig into New York City's history of big roads to the time when Robert Moses was still in college, you'll come across Madison Grant. Grant was not a traffic engineer, but he was behind the construction of the Bronx River Parkway. The parkway was sold to the public as an "environmental improvement," but soon thereafter, Grant published a book called *The Passing of the Great Race* that became one of Adolf Hitler's favorite books. What happened next? As L. J. Aurbach explains, "Grant became an internationally known leader of the white-supremacist and anti-immigration movements, and one of the most prominent racists in America." Unfortunately, Aurbach continues, "Grant's fellow parkway commissioners echoed his bigoted sentiments," and they took pride in their parkways cutting through low-income, immigrant neighborhoods.[8] They also put us on a path toward generations of inequitable road safety outcomes.

In the words of John Cougar Mellencamp, "ain't that America."

62 Well, That Didn't Work

—

So we busted up the bones of our cities—and kicked out anyone who stood in the way—to build these urban freeways, and it didn't even work. It didn't improve road safety. It didn't solve the traffic congestion problem. Here is a 1975 *Traffic Quarterly* paper regretting how things ended up: "It is now argued that the Highway Act of 1956 can be blamed for this traffic mess in urban areas. It has created a highway-biased approach to transportation problems and thus resulted in economic inefficiency and inequity in urban transportation."[1]

The public grew disillusioned. A 1973 *Traffic Quarterly* paper[a] reminds us that "the public used to be enamored of highways," fawning over anything highway-related, starting with the General Motors Futurama exhibit at the 1939 World's Fair. But now, a "favourable public response" is more likely to await the "attacks on highways and highway administrators." In fact, it should be obvious that "in the field of transportation, it is plain that motorways no longer work magic."[2]

This paper continues, "Yet the roadbuilders are slow to recognize this fact."[3]

Of course they are.

Cities also couldn't help themselves because "the enormous 'bargain' of the 90-10 money makes it politically impossible to do anything but take the money as fast as possible." That quote is from a 1960 article—"New Roads and Urban Chaos"—by future US Senator Daniel Patrick Moynihan. He goes on to say that "it has been impossible for the cities to resist the offer of unprecedented amounts of money, however futile they might know it will be to spend it on highways alone.... The demand for 90-10 highway funds is so great that there is almost nothing, however sensible, that local governments would not do to get their share."[4]

There are many reasons a freeway might not be in the best

---

[a] Written by David Witheford, the former technical director of the Eno Foundation.

interest of a city, not the least of which is land consumption. One bizarre 1959 *Traffic Quarterly* paper, titled "Freeways Alone Are Not Enough," estimates that urban freeways consume about 30 acres of land per mile and 80 acres of land per interchange. The author says that Los Angeles will need to add 1,200 miles of freeways by 1980 and who knows how many interchanges. Even after all that, he frets that Los Angeles "may well have worse traffic conditions than exist today."[5] He wasn't wrong.[b]

Future senator Moynihan scoffs at the notion that we didn't know what would happen to these cities: "It is not true, as is sometimes alleged, that the sponsors of the interstate program ignored the consequences it would have in the cities. Nor did they simply acquiesce in them. They exulted in them."[6]

The 1956 Clay Commission Report[c] that President Dwight D. Eisenhower forwarded to Congress praises urban freeways for helping "to create a revolution in living habits." Because of these urban freeways, "our cities have spread into suburbs, dependent on the automobile for their existence."[7]

You would think cities might care about their bottom line. After all, when we turn privately owned land into freeways, "the land disappears from the tax rolls."[8] Why don't traffic engineers include this factor in their cost-benefit analyses? A 1971 *Traffic Quarterly* paper also wonders about how "the government has evaded paying the 'social costs'" of freeways.[9] What are the social costs? Health and safety topped this paper's list in 1971.

Here is a *Traffic Quarterly* paper from a couple of years later making a similar point:

> Even now, the tendency prevails among many highway planners to rely on cost as the principal, if not the exclusive

---

[b] Los Angeles County dedicates upward of 90,000 acres of land to its freeways and streets. On the other hand, Paris, France, has well over two million people living on just under 26,000 acres of land.

[c] Eisenhower pulled together what became known as the Clay Commission to advise him on the interstate program. Lucius Clay was chairman of a canning company at the time (not sure how that qualifies him for this job). His nickname was "The Great Uncompromiser." The committee also included Steve Bechtel of the Bechtel Corporation (which remains the largest construction company in the United States).

route determinant.... As a result, freeways are deliberately driven through areas of low-cost residential development. Displacement of poverty-stricken tenants to higher cost areas or to inferior overcrowded accommodations is happily accepted by everyone except the hapless evacuees themselves. Similarly, other social consequences, such as disruption of communities by intruding roadways or loss of communal amenities, do not find their way into the misnamed cost-benefit analyses complacently employed by planners and engineers. This attitude which regards the construction of a freeway or any other road as an end in itself is no longer tenable.[10]

Moynihan wrote in his 1960 article that "to undertake a vast program of urban highway construction with no thought for other forms of transportation seems lunatic." For whatever reason, we left it up to the traffic engineers even though: "Nothing in the training or education of most civil engineers prepares them to do anything more than build sound highways cheaply."[d] Moynihan then quotes John T. Howard, an urban planning professor from MIT, as saying, "It does not belittle them to say that, just as war is too important to leave to the generals, so highways are too important to leave to the highway engineers."[11]

Traffic engineers tried—and failed—to solve supposed traffic congestion problems some 20 or 30 years into the future. In doing so, we tore apart the street networks of our cities and created a whole new set of problems. Road safety is one of those problems.

To paraphrase Winston Churchill, we shape our cities, afterwards they shape us.

In some cases, they kill us.

---

[d] That's one reason I'm working toward changing how we educate transportation engineers.

Part 7

OK Data, Don't Mess This One Up

———

63 Statistically Significant Nonsense

___

"In God We Trust, All Others Bring Data."

This quote hung on the wall in NASA's mission evaluation room at the Johnson Space Center in Houston, Texas.[a] It's also something I've heard attributed to Michael Bloomberg during his time as mayor of New York City.[b] His thinking is simple. As he says, "You can't manage what you can't measure."[1]

When it comes to road safety, the push is always toward a more "data-driven" approach. Good luck trying to find a road safety plan that doesn't say that. Being data-driven will always be one of our key road safety tenets.

It would help the cause if we had some better data to work with.

How does crash data come to be? It usually starts with a police report. The police, however, are not collecting this data for the sake of improving road safety. Why do the police collect it? A 1959 *Traffic Quarterly* paper by Bruce Greenshields lets us in on a not-so-secret secret: "The standard police accident report is primarily designed and used to determine who is to blame and who should pay."[2]

In other words, the police help insurance companies figure out who to blame.

The underlying belief is that if everyone followed the rules of the road, we'd have no crashes. So when police officers show up, they do so presuming somebody must've done something wrong.[c] Was the driver speeding? Did the driver run a run light? Was the pedestrian jaywalking? Were any of them drunk? Or high?

What the traffic engineer might've done wrong is the furthest thing from their mind.

---

[a] As best as I can tell, something to this effect was first said by statistician (and electrical engineer and mathematical physicist) W. Edwards Deming.

[b] Bloomberg's version was slightly different: "In God we trust. Everyone else: bring data."

[c] And since tickets and court fees help pay the bills, the police might as well cite or arrest these rule breakers while they are there.

Our crash data reflects that mind-set.

"As a result"—in the words of safety researcher Ezra Hauer— "causes extracted from police data will tend to focus on the wrongs of road users."[3] It's therefore no surprise that traffic engineers look at this data and focus on the same so-called problems as the police.

Let's stop and think about a typical factor collected by police in crash reports: driver distraction. When police officers show up at a crash scene, how do they know whether the driver was distracted? A 2018 study interviewed police officers about this issue. Here is one takeaway: "Officers added that the potential lack of honesty from drivers was one of the most frustrating details of a crash investigation…. One officer explained that he knows when a driver must have been inattentive, although if the driver does not admit it, then it is nearly impossible to ensure it is correctly captured in the report."[4]

So the driver who ran over the pedestrian isn't being completely honest about what they were doing at the time of the crash? Shocking. But without a warrant or an eyewitness, what can the police do? It's not like there's an instant replay. The interviewed officer goes on to explain that "most officers don't even put anything [in the distraction section] because they feel it is just not true."[5]

In their 1963 manual, the Automotive Safety Foundation (ASF) acknowledges that "studies of accident occurrence are plagued by inaccurate and incomplete reporting of accidents."[6] As they went on to say, "This introduces errors which cannot be accounted for and place many research findings in question."

Missing variables in our crash data—such as with distracted driving—are a problem, but at least we have an obvious problem. What happens when the variable is right there in front of me but maybe unreliable? That's a far more insidious problem.

Let's look at speeding. It's a variable we expect the police to give us. It's also one constantly subjected to analysis.

After fatal crashes, investigators might measure skid marks and estimate impact speed based on that measurement. With a pedestrian death, they might measure how far the pedestrian flies after impact and estimate impact speed based on that.

In some cases—like that 18-year-old from Florida who was going 116 mph in his Tesla—they might be able to obtain the data from the car itself. In other cases, the police rely on eyewitnesses (who probably don't carry a radar gun with them like I do). Or maybe the only source of information comes from the drivers themselves. But again, the police don't often get to throw up the challenge flag and see an instant replay of what happened.

So, what do the police investigators do? They take the above into account and make their best guess.

How accurate are they? That's a hard question to answer.

I think a better question might be, how accurate can they be?

Before getting into that, let me give a quick overview of what is regarded as the best of the best when it comes to a national police-reported crash data in the United States. It is called FARS—short for the Fatality Analysis Reporting System—and it is run by the National Highway Traffic Safety Administration (NHTSA).

Here's how it works. Whenever there's a motor vehicle crash on a public road that results in a fatality, the state collects and sends more than 125 pieces of information about that crash to NHTSA. Turning police reports into what NHTSA needs is a complicated process, so the states all have analysts on staff trained in the FARS protocols. NHTSA not only collects this data, but it also works with the states to minimize errors and make sure the information is as complete as it can be. NHTSA usually releases its preliminary estimates every spring for the previous year. About eight months after that, we get to see the final data.

What's good about FARS?

It's the only source of nationwide crash data, which makes it useful for larger studies. It's got a whole lot of variables, which makes it great for identifying trends and running statistical models. Like me, it's been around since 1975, which makes this database great for longitudinal analyses.

What's not so good about FARS?

It only includes fatal motor vehicle crashes. If somebody ends up in a wheelchair or a coma for the rest of their life—instead of dying—that crash doesn't make the cut. Even if they ended up dying—but do so more than 30 days after the crash happened—that wouldn't count either. Plus, when FARS says "motor vehicle" crash data, it means it. A motor vehicle needs to be involved in the crash for it to matter. Not only that, but the motor vehicle needs to be moving. If a driver parks, opens the car door, and kills a bicyclist, that crash isn't in FARS. The same goes for the crash needing to be on a public roadway; if it happens in a mall parking lot, it doesn't exist for the sake of this database. Also, FARS has only been around since 1975. That isn't nearly long enough given that most of what traffic engineers do today dates to before 1975.[d]

So, when it comes to the speed variable, FARS has seven options.

1. Exceeded speed limit
2. Too fast for conditions
3. Racing
4. Speeding but specifics unknown
5. Unknown if it is speeding-related
6. No driver present
7. Not speeding related

In most states, the police don't select from this same list. Instead, it's the job of each state's FARS analyst to pick one based on their interpretation of the police report.

What is the difference between "exceeded speed limit" and "too fast for conditions"? It's hard to say, but I can tell you that "exceeded speed limit" gets the nod in about 42 percent of speeding-related, passenger-vehicle fatal crashes nationwide.[7] I can also tell you that "exceeded speed limit" gets checked off in 85 percent of such crashes in Massachusetts. Yet inexplicably, only 7 percent of similar crashes in Arkansas included the "exceeded speed limit" designation.[8]

For the sake of an example, let's say we combine "exceeded speed limit" and "too fast for conditions" into a single speeding-related variable. Over a recent six-year period, 28,642 pedestrians lost their lives on streets with a posted speed limit.[9] How many of the drivers who killed these 28,642 pedestrians were speeding? According to FARS, the answer is 2,015.

Go look at almost any road. Does it seem within the realm of possibility that the speeding problem is limited to 7 percent of drivers? Of course not, but somehow we are supposed to put faith in data that tells us that drivers were speeding in only 7 percent of pedestrian deaths.[e]

---

[d] It wouldn't be unfair to call them the dark ages. I'm hoping that today's crash data will eventually be considered the dark ages as well.

[e] Apparently, only 8 percent of drivers speed in bicyclist deaths (347 speeding drivers out of 4,335 bicyclist deaths).

That may not be realistic, but what do we expect? Asking police to collect precrash speed data is—unsurprisingly—an inexact science. I'd be hard-pressed to say that "science" is the right word. Yet we take this data and try to answer research questions with it.

FARS has also been collecting race/ethnicity data—via death certificates—since 1999. At least that variable is usually right, right? You would think so, but a nationwide study on the validity of demographic characteristics on death certificates found that we do pretty well for White and Black populations (about 98 percent accurate) but not great for Asian (about 83 percent) or Native American (about 57 percent) populations.[10] I also dug into the FARS data for a paper and found that the race data is included in "only 85.9% of fatalities."[11]

Having good fatal crash data is important. But when it comes to road safety research, it's just as important to study seemingly similar crashes that end up with less severe injury outcomes. For this analysis, we need nonfatal crash data. For that, we usually need to go to the individual states.[f]

Some states do this quite well. Other states do not. Suffice it to say that it's like the Wild West out there. To access California's data, I used to reach out to Ponch and Jon via the California Highway Patrol website. Now that data is collected as part of the Transportation Injury Mapping System developed by safety researchers at the University of California, Berkeley. Many states put their Department of Revenue in charge of aggregating the police data. Why the Department of Revenue? That again goes back to why we collect this data in the first place (and it's not for people like me looking to improve road safety).

As for the nonfatal data itself? In terms of key variables like injury severity, we use it and hope for the best. At the same time, you don't have to dig deep to see the cracks in the foundation.

Many states use the KABCO scale, which is considered the industry standard:

---

[f] For national nonfatal crash data, the other option is the extrapolated data in the General Estimates System. From 1988 to 2015, the National Automotive Sampling System collected nonfatal crash data from 60 locations and then extrapolated those crashes into an estimated dataset for the rest of the country. NHTSA now runs the Crash Report Sampling System, which essentially does the same thing, taking a sampling of crashes from the same 60 locations and turning them into the six million plus crashes we probably see each year.

- K = killed
- A = incapacitating injury
- B = nonincapacitating injury
- C = possible injury
- O = uninjured

The National Transportation Safety Board, however, admits that KABCO is overrated because it "does not effectively capture injury severity or actual injury outcomes."[12] Accordingly, some states use the Abbreviated Injury Scale, or AIS, which is somewhat similar to KABCO but with a couple additional severity levels:

- 0 = No injury
- 1 = Minor injury
- 2 = Moderate injury
- 3 = Serious injury
- 4 = Severe injury
- 5 = Critical injury
- 6 = Unsurvivable

Other states—such as California—use neither scale and instead roll out even fewer crash severity levels.

So for the sake of apples-to-apples comparisons, it would be nice if everybody could get on the same page. At the same time, how much would that really help the cause?

In terms of selecting the injury severity level, we generally leave it up to the responding officers. Neil Arason makes it clear that they don't do a great job of it, as police-reported data is particularly problematic with concussions and "grossly under-reports serious injury and usually defines it as a person admitted to the hospital."[13]

In other words, it's pretty darn arbitrary.

My former PhD student Nick Ferenchak and his coauthor Robin Osofsky compared the injury severity of pedestrians estimated by the police against the level of severity estimated at the hospital. They found that more than a fourth of the pedestrians the police estimated to be severely injured turned out to have only minor injuries when evaluated at a hospital. On the other hand, more than half of the pedestrians the police thought only had minor injuries were deemed to be severely injured by a hospital.[14]

An obvious upgrade would be to link our police-reported crash data to the corresponding hospital data whenever possible. The problem is that although the police data tends to be easily accessible, the hospital data is not. When I asked Nick about it, he told me that getting access to the hospital data took nearly two years.

So instead of asking medical professionals to determine injury severity, traffic engineers are content to rely on police officers. We're left with crash data that differs from place to place and from officer to officer.

Good luck making data-driven decisions with that.

64 Who Lives, Who Dies, Who Tells Your Story?

___

I happened to hear a story about wildfires on Colorado Public Radio one morning. The premise was that humans cause most of the wildfires in Colorado, but a lack of investigative resources—and data—means that few get held responsible. The reporter then laments that "after any serious traffic accident in Colorado, a report is filled out noting how it occurred and what factors contributed. Each of those reports, from every state, are pored over by experts, and patterns are identified that can result in changes as simple as new traffic lights or as complex as automotive recalls."[1]

Well, that version of the multiverse sure sounds nice. Unfortunately, that's not how things really work.

Let's start with the "what factors contributed" idea.

The US national Fatality Analysis Reporting System (FARS) says that 28 percent of road fatalities in the United States involve alcohol and 26 percent of them involve speeding.[a] Yet when we look at pedestrian and bicyclist deaths, drivers are inebriated in 17 percent of crashes and speeding in only 7 percent.

So if it's not drunk driving or speeding, what factors contributed to all those pedestrian and bicyclist deaths?

Lucky for us, the National Highway Traffic Safety Administration (NHTSA) publishes lists every year. Read through those lists and you might conclude that these pedestrians and bicyclists had it coming. The most common factor—on the part of both pedestrians and bicyclists—is their "failure to yield right of way" to the drivers.[2] Another big one is the "improper crossing of roadway or intersection," which ranks second for pedestrians and fourth for bicyclists. The converse is "not visible," which ranks second for bicyclists and fourth for pedestrians. Of course, we also have those that were "darting or running into road" or "under the influence of alcohol, drugs, or medication" themselves.

___

[a] Speeding has historically been an issue in about one-third of fatal crashes but inexplicably dropped to just one-fourth of crashes as of late.

What other factors contributed to our pedestrian and bicyclist deaths?

Some were "inattentive," too busy "talking" or "eating" to worry about getting run over. Others were using "portable electronics." Some were deemed too "emotional" to walk or bike down the street, while others had a "physical impairment" that apparently contributed to them getting run down.

NHTSA orders these lists by the relative prevalence of the contributing factor. But, if you look way down that the bottom, you'll see some big numbers next to "none reported" and "unknown." Why are those categories placed way down at the bottom? Who knows, but another NHTSA report looked at six years of FARS data and found that "no improper action" ranks fifth for pedestrian fatalities and second for bicyclist fatalities.[3] That information isn't obvious when we look at most lists of "what factors contributed," so our big takeaways almost always focus on what the pedestrians and bicyclists did wrong.

The Colorado Department of Transportation tweeted that drivers were only "at fault 1/3 of the time in pedestrian-related crashes" and reminded us to #CrossSafely.[4]

When I saw that tweet, my first thought was that 100 percent of dead pedestrians can't tell their side of the story. How surprised should we be that so few drivers were doing anything wrong?

If you ever want to get knocked off your high horse, look into the research that estimates how far pedestrians fly after getting hit by cars at various impact speeds. Researchers use math and physics to estimate the overall "throw distance"—that is, a combination of the initial impact speed, the amount of time the pedestrian is in the air, and how far the pedestrian tumbles and slides after landing. It's a dispassionate take on what's killed hundreds of thousands of real people over the years.

But the numbers themselves? They border on shocking. One paper estimated that if a driver hits an average pedestrian at 30 mph, the pedestrian would end up more than 60 feet away.[5] At 40 mph, it's more like 120 feet. At 50 mph, we are talking about more than 180 feet of throw distance. That is almost a full city block in Portland, Oregon.

Think about these numbers for second and then think about what police officers see when they show up at a crash scene. Was the driver speeding? Was the pedestrian in the crosswalk when they got hit?

Good luck figuring any of that out, especially when the driver and their passengers are the only ones alive to say what happened. Again, why should we be surprised that so few drivers did anything wrong?

Either way, good crash data has always been considered the starting point for improving safety. Here is our good friend Paul Hoffman in his 1939 book, *Seven Roads to Safety*: "Getting the facts on accidents (when, where, how, and why they occur) has been the starting point for improvement wherever improvement has been made."[6] In 1958, AAA agreed and said that we need "full accounts of the what, when, where, who, and why of accidents."[7]

I can't disagree with the sentiment.

But it's never that easy, as James Reason—the human error guru—suggests: "The primary sources of data are accident reports. Not only are these mostly concerned with attributing blame, they also tell a story that may be inaccurate or incomplete, even when the reports are prepared by experienced and relatively open-minded investigators."[8]

James Reason isn't talking about road safety with this quote, but it applies just the same.

In one of my papers comparing data for nonfatal crashes involving a bicycle and a car across eight different communities, we found few consistently coded variables.[9] Yes, we knew when these crashes took place and if the bicyclist was wearing a helmet. But everything else—from crash severity to crash type—does not give us anything resembling an apples-to-apples comparison. Even when looking at a single community, many variables lacked consistency from year to year.

The best—or worst—example from that paper comes from San Diego, which, of course, in German, means ... something. We found that 291 of San Diego's reported bike-car crashes were categorized as an "overturned" vehicle. Does that mean 291 cars flipped upside down onto 291 bikes? Of course not, which is why we think the police officers in San Diego figured that anything but an upright bike counts as an overturned vehicle. How a postcrash bicycle would remain upright is beyond me, but what actually happened in these 291 crashes? Good luck figuring that out. How can we use such nonsensical data to make our streets safer? It's hard to say. Instead, I'll just say this: you stay classy, San Diego.

This example may be egregious, but problems like these are pervasive. The related problem is the lack of consistency in what the problems even are. Unless we dig deep, how would we know what San Diego means by an "overturned" vehicle? Or that prior to 2020, the New York City police classified electric bicycle riders as "motor vehicle occupants" and didn't include their injuries among those of the acoustic bicyclists?[10] The problems with our crash data are all over the map, both literally and figuratively.

It might make sense to blame the police for the bad data, but it's not that easy. In that same piece on Colorado wildfires, the reporter goes on to say that "Colorado has averaged about 4,000 wildfires a year over the last five years, and the reports are often initially filled out by volunteer firefighters who have no investigative training and little time to examine why a field burned in the absence of lightning."[11]

A similar thing can be said for police officers. We don't give them investigative training on how to document pedestrian or bicyclist crashes. And even if we gave them such training, the crash report forms we give them to fill out tend to be so car-focused that it's impossible to document the specifics of a pedestrian or bicyclist crash properly.[b] Here is AAA again: "Pedestrian accident report forms must be improved. Existing forms strongly accent vehicle and driver rather than pedestrian factors."[12]

Even AAA knows we'll end up with biased crash data: "Police training has so emphasized driver factors that accident statistics, analyses, and conclusions naturally tend to relate much more to driver actions."[13]

Both of these AAA quotes come from 1958. I wish I could say that things were better today.

Despite the issues with our motor vehicle crash data, it's still amazing compared to what we normally collect for pedestrian or bicyclist crashes.

If we even collect that data at all.

---

[b] According to a survey of all 50 states by Kathryn Woei-A-Sack of Georgia State University, only eight states have bicycle-specific crash reporting requirements. She also found that whether or not the bicyclist wore a helmet was the only bicycle-related factor consistently recorded for all 50 states. Kathryn Woei-A-Sack, "A 50-State Survey of Bicycle Crash Reporting Policies," Georgia State University (2016).

65 We Don't Know What We're Missing

___

In Jeff Garlin's—of *Curb Your Enthusiasm* and *The Goldbergs* fame—stand-up comedy special, Garlin tells the story of accidentally breaking somebody's car window during a parking lot dispute in Studio City, California. Here he is explaining what the police told him: "They told me that if it was a Toyota Corolla from say 1998, I wouldn't be arrested. But because it's a new Mercedes S Series, I am being arrested for felony vandalism."[1]

What's the difference? In Los Angeles, property damage exceeding $400 is a felony.

But what does that have to do with crash data?

If nobody is injured, the police have a property damage dollar amount they use to determine whether or not to file a crash report.

In most places, that threshold is $1,000.[a] Two Tesla bump bumpers? Crash report. I total my first car—a 1983 Chrysler LeBaron that I paid $800 for—but walk away unscathed? No crash report.

Put another way, a car crash involving rich people? Report. A car crash involving not-so-well-off people? It's like a tree falling in the woods with nobody around to hear it.

One of my former students, Krista Nordback, was bicycling in Boulder, Colorado, when a driver hit her. The driver was looking one way for a gap in vehicle traffic and didn't notice Krista, and several other bicyclists, coming from the other way. The driver, failing to realize that she had hit Krista and had Krista's bicycle's front wheel under her car, was about to proceed. This would have crushed Krista. But another bicyclist pounded on the car's window and kept the situation from becoming a tragedy. Krista wasn't hurt, but her bike got mangled.

___

[a] In Oregon, police don't bother with noninjury crashes. Instead, such crashes are supposed to be self-reported. Not surprisingly, Oregon ends up with about half the number of total crashes as other states. It doesn't mean Oregon is safer; it's just that a lot of these crashes never get reported.

Krista was in the midst of her dissertation work on bicyclist safety and wanted to see this crash end up in our database. The Boulder police came to the scene but refused to write a crash report. Why? For one, Krista only had a few bruises. She wasn't injured to the level needed for a crash report. Another reason is that the property damage didn't meet the crash report threshold either. If Krista had been riding a more expensive bike, the crash would've made the cut.

So Krista is involved in a bike crash where she comes close to dying, but we'll never see this crash in any crash database.[b]

It's not all a function of an injury or dollar threshold. A lot of people—for good reason—might not be keen on dealing with police, or with hospitals, unless it's absolutely necessary. Maybe they don't know the crash is supposed to be reported or how to do so. Or maybe they'd prefer their insurance company never find out.

Given all that, you might ask yourself, who cares? Why bother collecting data on such crashes. Do these crashes really matter?

Let me answer that with an example. For streets with both on-street parking and a bike lane, we usually put the bike lane right where people open their car doors. We call this space the door zone. A "dooring" crash is one when the driver of a parked car opens their car door just in time to hit a bicyclist.

The research tells us that a bike lane is safer for bicyclists than no bike lane. At the same time, the research doesn't account for dooring crashes because we don't collect that data. Why? The car wasn't moving. The crash doesn't qualify.[c]

So what's safer? It's hard to say, but now picture that same street as a blank canvas. Where would you least want to ride your bike? There are a few options, but you'd probably want to stay away from wherever we put the door zone of the parked cars. Yet that is exactly where we put most bike lanes and exactly where we tell bicyclists it's safest to ride. Even worse, we don't know if it's safe because we knowingly undercount bicycle crashes.

Sharrows—more professionally known as shared-lane markings—developed under a similar deficiency of crash data. Invented

---

[b] Krista is now a fantastic bicycling safety researcher with the University of North Carolina.

[c] Although if another driver drives into the door zone and tears off the door of a parked car, that crash would indeed show up in a crash database.

by Denver traffic engineer James Mackay in the early 1990s, they hastily became a staple of the traffic engineering toolbox. The reason? They were a cheap and easy way to add so-called bicycling infrastructure. With a little paint, your city's scattershot bike map would suddenly look like a connected network.

Proponents of sharrows said they improve bicyclist safety, particularly by helping them avoid dooring crashes. Did any of the research account for dooring crashes? Of course not. In fact, most of the so-called safety research didn't use any crash data whatsoever. Instead, it looked at bicyclist lane positioning on streets with sharrows versus streets without. Since bicyclists seemed to shift away from parked cars by a couple of inches, traffic engineers assumed sharrows were safer. Only one question remained: what exactly should our painted sharrows look like?

Back in 2005 when I was a graduate student at the University of Connecticut, the city of Cambridge, Massachusetts, asked us to test where bicyclists rode on different versions of a sharrow. So I headed to Cambridge, laid out inconspicuous black duct tape on the ground, and filmed for hours and hours. I later reviewed the videos and measured where the bicyclists were with respect to the duct tape and how much space the moving vehicles gave them. When Cambridge changed the sharrow marking, the plan was to go back out and do the same thing.

Cambridge never changed the sharrow marking, so we never finished that study. And even without actual crash outcomes—much less dooring crashes—the Federal Highway Administration added shared-lane markings to the *Manual on Uniform Traffic Control Devices* in 2009.

The safety research on sharrows remains sparse, but so far, the consensus seems to be that sharrows are *worse* than nothing. Without accounting for dooring crashes, a 2013 study from Toronto found that streets with sharrows had a higher risk of injury for bicyclists than streets without.[2] After Chicago started collecting dooring crash data, I did a before-and-after study with my former student Nick Ferenchak.[3] Once a neighborhood put in sharrows, the rate of traditional police-reported bike crashes nearly doubled, while dooring crashes more than tripled. Across the rest of the city, the rate of dooring crashes dropped, including by more than 60 percent in neighborhoods that added real bike infrastructure instead of sharrows.

When talking about what prompted his invention of the sharrow, James Mackay says that as Denver's traffic engineer responsible for bicycle infrastructure, "I was always under pressure to do less."[4]

That is exactly what sharrows are: less.

We can do better, but we need better crash data—particularly for bicyclist and pedestrian crashes—because the level of underreporting is staggering. Take, for example, a 1999 Federal Highway Administration study, which found "that anywhere from 40 to 60 percent of the bicycle-motor vehicle cases were not reported in official State files." For pedestrians, the results "showed that 35 to 55 percent of these events may go unreported."[5]

Combine those numbers with the somewhat dubious facts embedded within our current bicyclist and pedestrian crash data and what do we have? Not much. In other words, we don't know nearly as much as we think we do.

66 Better Data, Better Insights

—

Elaine Herzberg, as covered earlier, had the unfortunate distinction of becoming the world's first pedestrian killed by an autonomous car. The public debate surrounding the crash focused on who to blame. Was it the pedestrian, Elaine Herzberg, for jaywalking and doing so at night? Was it the operator, Rafaela Vasquez, for watching *The Voice* on her cell phone at the time of the crash? Or was it the autonomous vehicle company, Uber, for programming the car to "see" objects such as Elaine Herzberg as false positives to ensure a smoother ride?

But what fascinated me about the crash was that we had multiple videos of the entire incident—showing what was happening both inside and outside of the car.

Imagine if we had videos like that for every crash.

From a technology standpoint, it wouldn't be difficult. Politically or socially? That's a different story.

Either way, it's hard to think of anything that would revolutionize crash data more than video showing the outside and inside of every car for every crash. Not only would we now have a puncher's chance of deciphering what went down, but it might also influence what goes down. We wouldn't be capturing what normally happens. We'd instead be capturing what people behave like when the cameras are rolling.

Cameras could also help give us a sense of the street itself and the surrounding context at the time of the crash. At best, most crash investigations look for anything out of the ordinary when it comes to the underlying transportation system. But as you now know, the "ordinary" may be the problem.

Let's say a driver hits a pedestrian who is crossing a street in a location without curb extensions. Or maybe the driver hit a bicyclist in an unprotected bike lane. Since not having curb extensions or bike-lane protection is normal, crash investigators never bother noting their nonexistence. Out of mind, out of sight.[a] Here is safety

---

[a] The order of this was intentional.

researcher Ezra Hauer talking about this issue in a 2020 paper: "If a pedestrian crossing the road midblock between signalized inter-sections is hit and injured, the absence of a median, of a pedestrian refuge, or of a protected pedestrian crossing will not be noted by the investigating officer."[1]

Maybe we could see these things with cameras, but my point isn't about whether or not we use cameras to collect crash data. The more fundamental issue is that we need better crash data. That's been an issue for as long as we've had crash data. And I don't just want to know what the officer thought happened in the instant right before the crash happened. Truth be told, I want a much more complete timeline.

Figuring out the bigger story behind what happened could be a game changer.

Let me give you an example. Back in 2010, Raquel Nelson—a 30-year-old mother of three from Marietta, Georgia—wanted to do something fun for her kid's birthday. Raquel took her three kids—ages 3, 4, and 9—out for pizza and to Walmart. Marietta, a suburb of Atlanta, isn't exactly known for its incredible transit service. So, when they missed the bus, Raquel and her three kids had to wait an hour for the next one. At this point, it's getting late. Having three kids myself, I'd guess that all four of them were tired and wanted to get home.

Eventually, the bus dropped them off directly across the street from their apartment complex. The nearest crosswalk is about a third of a mile up Austell Road, a five-lane arterial with a speed limit of 45 mph. Suffice it to say that it's not a crosswalk I would ever use as an example of a "good" crosswalk. The same goes for the sidewalks between the bus stop and the crosswalk.

One option was to walk a third of a mile up the street with three tired kids on iffy sidewalks next to a five-lane arterial with scarce lighting to cross at a questionable crosswalk and then walk a third of a mile back on similarly iffy sidewalks. Another option was to "jaywalk." So Raquel did what a lot of reasonable people would do—including others on the same bus—and tried to cross right between where we placed the bus stop and the apartment complex.

Raquel and the kids got halfway across the street and were stand-ing on the median. Her 4-year-old son, AJ, saw one of the other bus riders run across the rest of the street. AJ got loose of his mom's grip and tried to do the same thing. He was killed before he got to the other side.

The driver who hit AJ, Jerry Guy, was partially blind, had been drinking earlier that day, and was on pain medication. Thirteen years earlier, he had been convicted of a hit-and-run crash on the same road where he killed AJ. Even crazier, Guy had been convicted of a second hit-and-run crash that had happened on that very same day in 1997.[b] Guy also fled the scene after killing AJ with his van.

Guy was originally charged with hit and run, first-degree homicide by vehicle, and cruelty to children. For whatever reason, the prosecutors dropped all charges except for the hit and run. Guy pleaded guilty to that charge and was sentenced to six months in jail.

Raquel Nelson does not own a car, but prosecutors still charged her with vehicular homicide in the death of her son AJ. She was convicted and sentenced to 36 months in jail.

Wait, what?

Once this case made national news, the judge threw out Raquel's prison sentence and instead gave her 12 months' probation and a $200 fine for jaywalking.

Take a look at the crash data, and we have a simple case of a pedestrian fatality where the pedestrian failed to yield right-of-way.

The crash data does not mention where AJ was coming from, who he was with, or where he was going. It does not mention where the nearest crosswalk was located or if that crosswalk was any good. Nor does it mention what the sidewalks were like in between. Was there decent lighting along these sidewalks? The crash data doesn't know.

The crash data also doesn't give us any information about where Jerry Guy was coming from or where he was going. We don't really know how fast he was driving or where his attention was at the time of the crash. How long has he been awake? When was the last time he had a vision test? The crash data doesn't know any of that.

My bigger issue with the crash data is that it doesn't give us any indication that traffic engineers need to do better. Yet, I would say that we failed AJ. We failed Raquel. We failed their family and friends. We continue to fail the millions of people who live, work, or go to school near terrible roads like this one.

---

[b] Guy pleaded guilty to these 1997 hit-and-runs and was sentenced to two years in jail but ended up serving less than a year.

To do better, we need better data. We need a timeline of events to help us understand people's mind-set as the situation is unfolding. *Why* did it make sense to them to do what they did?

As is, just about everyone breaks the law just about every time they use the transportation system. That makes it easy for the crash data to blame anything and everything on so-called human error. It also makes it hard for traffic engineers to learn from our mistakes. It's difficult for traffic engineers to even get to the point where we might see something we did as a mistake.

In 2016, Volvo pledged that starting with model year 2020, nobody who isn't "a suicidal maniac or a total idiot" would be killed or seriously injured in a new Volvo.[2] Is that a marketing plot? Maybe, but Volvo is putting in the effort to collect some real data. Every time a Volvo is involved in a serious crash in Sweden, Volvo sends its own investigation team to the crash scene, no matter the time of day or day of week.

In a perfect world, I'd have access to a comprehensive crash database that includes hospital data, coroner's reports, information from the department of motor vehicles, and facts about the transportation and the surrounding built environment. Names should be redacted, but the database would still include an overview of those involved and what they were up to for much more than the few seconds right before the crash.

Why can't we have data like that? To start with, it would be a pain in the neck to collect. But the bigger hurdle is liability. For fear of discipline or prosecution, people don't have much incentive to tell their true stories. Maybe we need a confidential way for people to do so outside the police report, but as is, the system encourages secrecy to the point where basic solutions like cameras are nonstarters in most places.

On one level, we end up with crash data where it's hard to look beyond the so-called human error problem. On another, it makes it hard for traffic engineers to see themselves as part of the problem— or the solution—when it comes to our real road safety problems.

Wouldn't it be nice if we collected crash data for the sake of improving road safety instead of for the sake of liability?

Part 8

The Blame Game

_____

67 The Liability Boogeyman

The movie *John Wick* doesn't really kick off—spoiler alert—until the guy who played Theon Greyjoy kills John Wick's dog. Not long after, his father—the Russian crime boss Viggo Tarasov—explains who they are dealing with.

> Viggo Tarasov: They call him Baba Yaga.
> Iosef Tarasov: The boogeyman?
> Viggo Tarasov: Well, John wasn't exactly the boogeyman. He was the one you sent to kill the f$%&ing boogeyman.
> Iosef Tarasov: Oh.

In transportation, our boogeyman is liability. It terrifies drivers and pushes our crash data into the fringes of fiction. It worries traffic engineers and restricts us to whatever we find in our manuals. Why? When it comes to liability, here is our thinking: "If you don't meet the standards, you could be sued. That's why we blindly follow the standards no matter what."

These words were said by a Denver traffic engineer who was a guest lecturer in my road safety class on the day before I wrote this chapter. Reid Ewing says a similar thing about the AASHTO Green Book in a 2002 *Transportation Quarterly* paper: "State and county DOTs blindly follow this bible of geometric design. They do so for fear of tort[a] liability."[1]

We'll soon get into how to overcome this boogeyman, but most traffic engineers just live with it. And "most traffic engineers" have a valid point when it comes to liability because "courts only ask whether

---

[a] A tort is a legal term describing a violation where one person or entity causes damage, injury, or harm to another person or entity. Basically, the government—and its traffic engineers—have the legal duty to exercise ordinary care for the safety of road users who are exercising ordinary care themselves. If the government breaches this duty and somebody gets hurt, it could end up with a tort liability claim for government negligence.

the applicable standard was adhered to; they do not ask whether the standard was based on evidence about its road safety consequences, nor whether adherence to it makes the road reasonably safe in use."[2]

So wrote road safety researcher Ezra Hauer in his 2019 paper about the lack of a link between road safety research and design. It's hidden way down in the middle of the 21st footnote in his paper. What is he saying here? He's saying that actual safety outcomes are irrelevant as long as we meet the so-called standards.[b] So for traffic engineers, safety is defined by whether or not we conform to standard practice. As for whether or not we improve real safety outcomes? That's beside the point.

What's crazier is that the development of new standard practice by no means requires the consideration of safety outcomes. Hauer goes on to say that "when writing a standard, there is no need for the 'State' to find out what the safety consequences of adherence to it will be. On the contrary, knowing what road safety consequences to expect may limit the 'State's' freedom in coining it."[3]

In other words, it's easier to pretend we don't know anything about the safety implications. We only care if we met whatever standard design we made up, not if the standard design improves safety outcomes or not.

Imagine that the kind of toaster you bought on Amazon killed 40,000 Americans last year, including somebody you loved. Instead of recalling the toaster, they tell you that everything is fine because the toaster met the industry standard. Who decides what's the industry standard? Well, the toaster industry. They then go on to suggest that your loved one was probably just using it wrong. If only they had read and followed the instructions that came with the toaster.

You are welcome to sue, but it's going to be an uphill battle if they followed the industry standard.

Instead of a toaster, imagine that it's a street. Instead of toaster instructions, we have rules, regulations, and signage to tell the street customers what to do.

Traffic engineers built this street according to industry standards. It has a high design speed, a big clear zone, and the ability to carry

---

[b] Using "so-called" here suggests that even though I'm calling them standards, they aren't. Most are just guidelines that can be overruled with engineering judgment and a rational approach. Both of which I'll explain in more detail soon,

an awful lot of traffic that we don't expect to see for another 20 years. Despite the speed limit "instructions" posted, drivers go fast. Still, your loved one attempts to walk across this street and, unfortunately, is killed by one of those fast-moving drivers.

Let's put on our lawyer hats and figure out who can we blame.[c]

If your loved one didn't use a crosswalk, it's easy. It's their own fault.

If your loved one *did* properly use a crosswalk, the answer isn't as clear-cut as you might think.

In a great *Freakonomics* podcast episode called "The Perfect Crime," Stephen Dubner kicks off this way: "Let me warn you: What you're about to hear is a sick idea. Let's say, I want to kill someone. But I also don't want to go to prison. In fact, I don't want to be punished at all. So what do I do? Well, there's this one idea I have."[4]

After checking in with a couple of experts, Dubner lays out the best way to murder someone and get away with it. He says that he'd "wait 'til they were outside, walking down the street, maybe crossing at the light, and then I'd run them over in my car. Now, I'd have to make sure that no one knew I was trying to run them over. But they'd be dead and I, especially in New York City, would in all likelihood go scot-free."[5]

Why would he go "scot-free"? One study delved into the police reports for 880 pedestrian fatalities in New York City over a four-year period and found that drivers were "largely or strictly culpable" in 651 of these fatalities and "partly culpable" in another 141.[6] Add those up and drivers were at least partially to blame in 90 percent of pedestrian deaths. Yet the podcast tells us that "only about 5 percent of the drivers who kill a pedestrian in New York are arrested."[7] Why so few? In New York, they have the "rule of two," which means "you need two significant violations of traffic laws in order to bring a charge, including some incredibly reckless or criminally negligent act. Otherwise, it's just an accident."[d]

Back to the question at hand: who can we blame if your loved one *did* properly use a crosswalk?

Assuming that the police can prove that your loved one was in the crosswalk, we might be able to get the driver for manslaughter. In a

---

[c] I feel like more lawyers should wear hats.

[d] They've been trying, unsuccessfully thus far, to get rid of the "rule of two."

place with the rule of two, it would still be hard. Driver was speeding? No dice. Driver was texting? Maybe, but probably not enough. Driver was speeding, and speeding significantly, while also texting? We might have a case.

A lawyer who used to prosecute organized crime wrote a 1968 *Traffic Quarterly* paper complaining that there is less justice in transportation than organized crime. He says that "speed and right-of-way violations lead the list of those offenses which have a causal relationship to accidents resulting in personal injury. Yet most of the persons cited for such offenses never see the inside of a courtroom."[8]

The author is flummoxed as to how "someone can run a red light, hit a pedestrian in the crosswalk, totally disable the victim, and never even have to make an appearance in court. The offender simply writes out his check for $10 which he posts as bail and then forfeits."

He goes on to say that "of course if the unfortunate pedestrian dies, the motorist will be charged with vehicular manslaughter, but it takes a homicide to get you into court unless the underlying moving violation is an extremely serious one."[9]

But in Maryland, the consequences for the driver have nothing to do with the outcome for the pedestrian. According to a 1992 *Transportation Quarterly* paper, "The motorist who goes through a red light and kills someone pays the same fine as if no injury had occurred."[10] It then gives an example of a 1987 case where "a driver who killed a blind couple and their guide dog" paid a $40 fine and got three points added to their license.

But what about the traffic engineer? What if they provided an uncontrolled crosswalk that research tells us is dangerous? Or what if they didn't provide a crosswalk at all? As long as they followed industry standards, good luck blaming them.

What if industry standards led to a dangerous situation? Doesn't matter. How about if traffic engineers developed these industry standards absent real crash outcomes? Also irrelevant. What if the traffic engineer designed this street in 1955 with now antiquated 1955 industry standards? Not good enough.

Once upon a time, property owners were held "strictly liable for all the damages people suffered on, or because of, his property."[11] But rules like that would hold us back from making the most of the Industrial Revolution. So lawyers came up with the idea of negligence theory. With negligence theory, it was fine to maim and kill as long as you could show that you were using "reasonable care." What's reasonable care? Well, it means that defendants could now argue for reasonable care if they were "following industry custom and were at least no worse than most other people who were doing what they did at the time they did it."[12]

For the transportation system that was to come, this is a big deal. As you can probably guess, it's hard to argue that any given dangerous street is worse than all the other dangerous streets you see everywhere you look. It's even harder when the street dates back to yesteryear. If I come up with a new, safer design today that becomes an industry standard, the older streets do not automatically become negligent. In other words, good luck trying to establish that something that was originally not created negligently, over time, became negligent.

Given how many of our streets predate any and all industry standards—as well as traffic engineers altogether—you might think the lack of standardization left us with streets that were dangerous when built and that have only become more deadly over time. But the opposite is usually true. Streets built before the advent of traffic engineers are some of our safest, often with far fewer fatalities and severe injuries than the new-and-improved, fully traffic-engineered versions.

Find yourself an example of an old, narrow street in your favorite part of any older city. Watch how people use this street. Take some measurements. Then ask a traffic engineer if they'd design a "substandard"—yet safe—street with the same dimensions.

Most traffic engineers wouldn't go anywhere near this because "tort liability arises when substandard designs results in accidents."[13]

So because of the liability boogeyman, empirical road safety outcomes—and common sense—take a back seat to whatever the so-called standards tell us to do.

We may need Baba Yaga to pay the boogeyman a visit.

68　The Guidelines Won't Save Us

There was a crash in New York City, where, according to court records, "a twelve-year-old child rode his bicycle across a four lane highway at night. A car traveling at nearly twice the speed limit struck the child, causing severe injuries."[1]

The kid was Anthony Turturro, and he was bicycling along Gerritsen Avenue in Brooklyn when Louis Pascarella hit him. Anthony was fortunate to survive, but his injuries were extensive, to the point where he needs lifelong assistance for daily activities.

In most cases like this, we blame the driver for speeding or the kid for not bicycling where he should or for wearing dark clothes.[a] In some cases, we blame nobody and chalk it up as a so-called unfortunate accident.

In this case, Anthony's parents sued the City of New York and won. The city was deemed to be 40 percent at fault. The driver received 50 percent of the blame, and Anthony himself got 10 percent. The jury awarded Anthony tens of millions of dollars in damages.

The first appeal agreed with the ruling but reduced the damages to $10 million. Then, the state Court of Appeals affirmed the liability findings. Here is an excerpt from that decision:[b] "Although the driver of the automobile eventually pleaded guilty to a felony charge of reckless assault, a jury determined—in a verdict affirmed by the majority on this appeal—that the City of New York was liable for the accident. Why? Because the city failed to take adequate measures to prevent the speeding motorist from breaking the law."[2]

Gerritsen Avenue in Brooklyn was an unsafe street known for speeding, and the city got lots of complaints. Still, the avenue seemed to meet the engineering guidelines.

---

[a] Unless it's a motorcycle crash because, for whatever reason, nobody would ever suggest that motorcyclists wear high-visibility clothing (let alone a helmet), even though most drivers who hit motorcyclists say they looked but didn't see the motorcyclist.

[b] To make these court opinions easier to read, I removed references to other cases.

The dissenting opinion in the state Court of Appeals case reads as follows:

> I do not disagree with the majority that the government, act-
> ing as a landowner, has a proprietary duty to maintain road-
> ways in a reasonably safe condition. That duty requires the
> government, among other things, to install adequate traffic
> control signals, erect and maintain proper barriers, provide
> adequate warnings of unsafe road conditions, remove dan-
> gerous snowbanks, remove standing water, fill potholes, and
> generally clear roadways of debris and other hazards.[3]

The dissenting judge goes on to say that "the proprietary duty to maintain safe roadways, however, extends no further than making the road safe for individuals who follow the rules of the road."[4]

Most traffic engineers would wholeheartedly agree. If people don't follow the rules of the road, don't blame us. In fact, the city's lawyers tried to blame Anthony's injuries on the police due to a lack of **enforcement**.

But the majority opinion ruled that "the specific act or omission by the City claimed to have caused Anthony's injuries was the City's failure to adequately study or implement roadway design changes intended to reduce speeding in response to repeated complaints."[5]

So the city was held liable for failing to redesign a street that met our guidelines but also had a known history of speeding drivers and bad safety. In other words, the city lost this case *because* it blindly followed the guidelines and ignored the empirical evidence.

What should New York City have done differently? The majority opinion says that "there was a rational process by which the jury could have concluded that traffic calming measures deter drivers such as Pascarella from speeding, and that the City's failure to conduct a traffic calming study and to implement traffic calming measures was a substantial factor in causing the accident."[6]

This "rational process" is pretty simple. But it's also powerful because it allows traffic engineers to circumvent the liability fears of not following the so-called guidelines that can lead to unsafe streets.[c] The basic idea is that we need to apply a logical approach to

---

[c] Yes, we need to fix the guidelines as well. We'll talk about that in the next part.

solving a known problem. In a 2003 *Transportation Quarterly* paper about traffic calming, Reid Ewing lays out the four steps.[7]

The first step is to document your problem. It may start with public complaints, but it needs to be supplemented with crash data, by measuring speeds, or maybe by counting near hits.

The next step is to consider your options. Here is where you need to do your due diligence because the intent is to show that you gave fair consideration to the gamut of viable solutions to the problem. We then want to highlight the valid reasons and research behind selecting the preferred alternative over the other options.

The third step is to pilot and test the intervention. That doesn't necessarily mean putting in a permanent change right away; it could mean testing a low-cost, temporary version of the approach. The key is to test the intervention by making sure we measure something that lets us assess the original problem.

The last step is to follow up and evaluate the intervention. We hope it's working as intended and improving the problem we set out to fix. If so, great. Now you can go ahead and make the intervention permanent. If not, try to figure out why. Improve it and test it again ... and perhaps again.[d]

Why does this strategy work? How could a process so simple give traffic engineers the wherewithal to build safer streets regardless of what the guidelines say?

Fundamentally, the government cannot exercise power—which includes its power over the streets—in ways that are considered "arbitrary, capricious, or unreasonable." By documenting a problem and working through a logical progression of steps to solve that problem, we've put ourselves in a position to jettison the possibility of our design being considered "arbitrary, capricious, or unreasonable."

It also falls under the ***engineering judgment*** umbrella. Almost every one of our guidebooks—probably to protect the guidebooks themselves from liability—says traffic engineers should always use their engineering judgment above almost anything the guidebook says. So, when I said this rational planning process "allows traffic engineers to circumnavigate the so-called guidelines," the intent isn't to pull the wool over anyone's eyes. The reality is that our ***current*** guidelines aren't what is stopping traffic engineers from building

---

[d] This process sure sounds an awful lot like tactical urbanism.

safer streets. There is already plenty of flexibility built into the guide-lines—that is, if traffic engineers are willing to use it.

The amount of flexibility in our design guidelines shouldn't be news to traffic engineers. In the late 1990s and early 2000s, the con-cept of context-sensitive design started taking hold. The idea was that we could better integrate transportation infrastructure with the interests of the community during the design process. In a 2001 *Transportation Quarterly* paper on this, Nik Stamatiadis explains that context-sensitive design helps us "recognize the flexibility that exists in the current design guidelines."[8] The problems with our current designs? While the guidelines might not help the cause, we were the ones who decided to "focus on providing high levels of mobility" instead of common sense.[9]

Still, traffic engineers don't mind letting everyone believe our hands are tied when it comes to following the guidelines. In reality, traffic engineers have much more design flexibility than we want to admit. Traffic engineers may have more design flexibility than they even realize.

If you think about Anthony Turturro, it's easy to blame the driver. But traffic engineers failed Anthony as well, and the state of New York's highest court agreed. New York City ended up with a liability problem *because* its traffic engineers let the guidelines outweigh what was actually happening in the street.

That hasn't always been the case. For instance, in the 1992 case of *Manna v. State of New Jersey*, the courts found that "immunity is not lost even if new knowledge demonstrates the dangerousness of the design, or the design presents a dangerous condition in light of a new context."[10]

After discussing this 1992 case, Ewing's 2002 *Transportation Quarterly* paper says that "if the right body or official approves a design (or the standards upon which a design is based), even if flawed, the decision is immune from tort liability."[11]

But if the Anthony Turturro ruling becomes the norm, our guidelines would become just that: guidelines. Treating them like standards would no longer be a viable approach to protecting against liability.

To the risk-averse traffic engineer, it may be time to rethink your relationship with liability. The so-called standards might not save you.

69   Hard to Say I'm Sorry

—

In the second season of the TV show *Parks and Recreation*, Amy Poehler's character, Leslie Knope, tries to forgo the governmental red tape and fill in the infamous pit using the old ***beg forgiveness instead of ask permission*** ploy. She forgot that Chris Pratt's character, Andy Dwyer, was living in the pit. So when the backhoe starts dumping fill right on top of him, Andy ends up in the hospital.

As Leslie rushes to the hospital with balloons and a stuffed pig, the city attorney stops her in the hallway. Here is their exchange.

> LESLIE KNOPE: What I'm going to do is go in and say we're so sorry. It's entirely our fault.
> CITY ATTORNEY: No, no, no. You can't say any of that. It admits liability. You can't say I'm sorry, or I apologize. It implies guilt.
> LESLIE KNOPE: That's insane. I have to apologize. Andy was the victim …
> CITY ATTORNEY: You can't say victim.
> LESLIE KNOPE: … of an extremely unfortunate situation.
> CITY ATTORNEY: You can't say unfortunate, and you can't say situation.
> LESLIE KNOPE: I can't say the word situation?
> CITY ATTORNEY: No. It implies there was a situation.
> LESLIE KNOPE: Can I give him the pig?
> CITY ATTORNEY: Yeah, pig's fine.

When I first moved to Denver in 2009, I showed up expecting to teach a site engineering course. On the Friday before the first day of classes, the head of my department canceled that class and said I'd be teaching an online GIS, or geographic information systems, class instead.[a]

---

[a] That was a fun weekend!

Most people think about GIS as something we use to make maps. That's true, but GIS is more of a spatial analysis tool that can help reveal spatial relationships that might otherwise go unnoticed. To prepare for teaching the GIS course, I searched for cool and interesting GIS layers that might be locally available. At the time, Denver had a surprisingly good GIS database with far more data layers than most cities. One of these was a sidewalks layer.

Even today, decent sidewalk data is hard to come by in most cities. Back in 2009, Denver's sidewalk data was a shocking find. But the sidewalk data had not been updated since it was first collected by the city in 2003.

I asked somebody with the city why this information wasn't being updated regularly like everything else. The reply? It's not allowed. The city's lawyer said not to.

Wait, what?

The lawyer's thinking? If the city knew there was a problem, it would be liable. If the city stuck its head in the sand, it could feign ignorance. How can it possibly fix a problem it doesn't know it has? So the city looked the other way on sidewalks while averaging nearly 250 pedestrian injuries each year.

Unfortunately for the city, the old ostrich trick doesn't work. Nor does trying to pawn sidewalk maintenance onto the adjacent property owners.[b] Like many cities with similar sidewalk policies, Denver was sued under the Americans with Disabilities Act. The city settled the case, promising to install at least 1,500 new sidewalk ramps each year for as long as it takes.

From 2009 until 2016, when the regional metropolitan planning organization started collecting sidewalk data, at least 100 pedestrians died in Denver. Did Denver send out a team of investigators to every fatal crash location to figure out what went wrong and how things could be safer?

---

[b] An audit of Denver's so-called sidewalk repair program says that "the responsibility for sidewalk repair should remain with Denver property owners to protect the city from liability." Denver also published a "Homeowner's Do-It Yourself Guide for Hazardous Sidewalks" that suggests renting a "Masonry Rotary Grinder" from Home Depot if your sidewalk has a tripping hazard as big as 1.5 inches. Why this recommendation isn't also a liability concern is beyond me. Timothy O'Brien, "Audit Report: Neighborhood Sidewalk Repair Program, Department of Transportation & Infrastructure," City and County of Denver (2020).

Of course not. Making a change might imply that we did something wrong in the first place.

It doesn't have to be that way. New York City started sending out the Severe Accident Forensic Evaluation, or SAFE, team to fatal crash sites in 1988. Once notified of a fatal crash, the city's Department of Transportation unleashes this group of specially trained investigators, technicians, and traffic engineers. Their job is to "expedite priority regulatory repairs and recommend other corrective measures that may prevent future incidents."[1] Some sort of "corrective measures" get implemented in almost half their cases.[2]

Denver recently followed suit by initiating the Vision Zero Rapid Response Program. The team reviews all fatal crashes, as well as severe injury crashes involving pedestrians, bicyclists, or motorcycle riders.

Yet most cities treat road fatalities like we treat sidewalk data. Traffic engineers remain so worried about liability that instead of looking for ways to make streets safer, we feign ignorance about anything other than a human error problem.

How real of a concern is this liability problem? As Reid Ewing says in his 2003 *Transportation Quarterly* paper, "Lawsuits and damage claims are not nearly the problem they are made out to be."[3]

Some states, such as Delaware, still have sovereign immunity, which basically shields the government and all government agencies—including transportation departments—from lawsuits. Sovereign immunity was the norm for traffic engineers nationwide until Congress passed the 1946 Federal Tort Claims Act. Since then, most states have enacted tort claims acts or replaced sovereign immunity with limited discretionary immunity. For the many states that retain "discretionary" immunity, usually this immunity includes the planning and designing of public infrastructure.

Ewing overviewed traffic calming interventions across 50 cities, including nearly all major US cities. At the time, few of these interventions appeared in any of our guidebooks. Yet Ewing found that "only two lawsuits against traffic calming programs have been successful, and one of those was overturned on appeal." He goes on to say that "the legal maneuvering has more often involved city attorneys, concerned about potential liability, than private attorneys, claiming actual damages."[4]

According to a 1987 *Transportation Quarterly* paper, 85 percent of engineering-related lawsuits do *not* involve injuries or deaths.[5] For traffic engineers, most claims are property-damage-only problems. In the medical field, nearly 100 percent involve injury or death.

Given these numbers, you might think that the medical profession would be more anxious about liability concerns than traffic engineers. Maybe they are in some respects, but I couldn't help but notice the burgeoning *sorry movement* in medicine.

Sorry Works! is a nonprofit organization started by Doug Wojcieszak in 2005. Wojcieszak's 39-year-old brother Jim died in 1998 when the hospital mixed up his file with his father's file. The Wojcieszak family wanted somebody to talk to and some assurances that this kind of mix-up wouldn't happen again. But nobody at the hospital would talk to them.

The hospital's reasoning for not talking is simple. As Alina Tugend says in her 2011 book, *Better by Mistake*, "Lawyers often advise their clients, whether someone involved in a car accident or a multinational corporation, not to apologize. They know that saying 'I'm sorry' could be interpreted as 'I'm guilty.'"[6]

But research suggests, at least in the medical world, that saying "I'm sorry"—and doing it well—makes people much less likely to sue. One study looked at more than 1,100 medical cases and found that saying sorry—or implementation of what the study called the "disclosure program"—resulted in significantly fewer lawsuits, faster resolutions, and lower costs.[7] Another study—titled "Risk Management: Extreme Honesty May Be the Best Policy"—looked in-depth at a single hospital that was reeling from a couple huge malpractice judgments.[8] After changing to a "proactive full disclosure" policy, seven years of data suggested "that an honest and forthright risk management policy that puts the patient's interests first may be relatively inexpensive because it allows avoidance of lawsuit preparation, litigation, court judgments, and settlements at trial."[9]

Why do they think saying you're sorry and being honest worked? They found that it "diminishes the anger and desire for revenge that often motivate … litigation."[10]

Plus, most states have laws stating that apologies can't be used as an admission of guilt in a court of law. In Massachusetts, 16-year-old Claire Saltonstall was killed in 1974 while out riding her bike.[c]

---

[c] The Claire Saltonstall Bikeway starts in Boston along the Charles River and runs 135 miles to Provincetown at the tip of Cape Cod. Since 1980, the Pan-Mass

Her father, William, was appalled that the 19-year-old driver never expressed any contrition for veering off the road and killing his daughter. But William was also a state senator, so he introduced a bill—later passed—giving "safe harbor" to apologies. The specifics of such laws vary from state to state. Colorado, for instance, grants immunity for the apology but not if you admit guilt. You can say you're sorry, but you need to make sure the apology is framed correctly.

What does saying sorry have to do with traffic engineers? As is, traffic engineers feel unable to speak up for fear of liability. But silence no longer works. James Reason, our sage of human error, concludes his last book by saying, "One thing is certain: the traditional 'deny and defend' strategy is unacceptable in the 21st century."[11]

My takeaway?

Traffic engineers can say sorry when problems arise. Traffic engineers can be proactive about collecting data on problems and work toward fixing them. But traffic engineers first need to get our heads out of the sand when it comes to liability.

Challenge has used this route for an annual fundraising ride to benefit cancer research and the Jimmy Fund. Apparently, "no other single athletic event raises or contributes more money to charity" than the Pan-Mass Challenge. Pan-Mass Challenge, "Pan-Mass Challenge Organizer Optimistic about Exceeding $31 Million Goal," press release (2010).

70   If Only

—

Find a news article about a recent traffic fatality. Chances are the article will mention something about what one of the road users did wrong.

If only they weren't speeding…. If only they saw the stop sign….

It's worse when a pedestrian or bicyclist is involved. What the pedestrian or bicyclist did "wrong" doesn't even have to be illegal.

If only they were wearing a helmet…. If only they weren't wearing dark clothes….

Also take a close look at how the news media constructs their sentences. The clearest way to explain things would be to say that A hit B. Instead, they flip things into the passive voice and say that B was hit. In other words, instead of a driver hitting a pedestrian, it reads that a pedestrian was hit.

Starting in 2018, one of my master's students, Molly North, dug into 288 news articles about 91 fatal child pedestrian crashes for her thesis.[a] She found that 80 percent of headlines said the pedestrian "was hit" instead of saying that a driver or car did the hitting.

Not only that, but the media habitually depersonalizes the driver. Instead of a pedestrian getting hit by a driver, they'll say that a pedestrian was hit by a car. Molly found that 88 percent of headlines made no mention of a driver. Within the articles themselves, 82 percent start by focusing on the car instead of the driver, and 40 percent of articles neglect to mention a driver existing at all.

Unless the car is autonomous, or unless the pedestrian walked into the side of a stationary car—which wouldn't result in a crash report—saying that a driver hit a pedestrian with a car shouldn't be up for debate. Imagine that I drove my car into a brick wall. The headline would never be "Brick Wall Hit by Car" because that's absurd. That is, unless it was a superfamous brick wall with a Banksy mural that I hit.

---

[a] Thanks to Ryan Archibald for collecting all these articles.

Here's a real example. In 2015, 24-year-old Mallory Weisbrod was walking on the sidewalk along Second Avenue in Manhattan. Around 4:30 in the afternoon, 64-year-old Dimas Debrito lost control of his Mercedes, drove up on the sidewalk, hit Weisbrod along with two others, and killed her.

Here's the initial *New York Post* headline: "Woman Suffers Grisly Leg Injury after Car Jumps Curb."[1] The article itself—with the *New York Post* being careful not to blame the driver before the "facts" came out—made no mention of a driver whatsoever.

Nine days later, the *New York Post* updates the story. Here's the new headline: "Woman Dies from Injuries after Car Jumps Curb." The article explains the crash by saying that "Weisbrod was hit by a Mercedes-Benz after it mounted a curb near Second Avenue and 49th Street." It then tells us that "the car stayed at the scene, no arrests have been made in the incident, cops said."[2]

The car stayed at the scene? Really?!? More than a week after the crash, and there was still no mention of this car having a human being inside, let alone what this person was doing.

That seems to be the end of the *New York Post* reporting on this story. If you wanted to know that Dimas Debrito was driving more than twice the speed limit when he killed Mallory Weisbrod, pleaded guilty to two misdemeanors, but only lost his license for six months, you'd have to look elsewhere.

In another great paper, a research team led by Tara Goddard asked 999 people to read different versions of the same article.[3] What they found was that the sort of reporting Molly quantified in her thesis (pedestrian was hit by car) led respondents to place more blame on the pedestrian. Flip that around (driver hit pedestrian), and people have a very different take on what happened and how we should fix it.

Let's say you are driving and you hit a cow instead of hitting a pedestrian. For whatever reason, the media does a better job with those stories.

Here are a few recent headlines:

- "Driver Crashes into Cow on Utah Highway, Officials Say" from the *Sacramento Bee*
- "Driver Hits Cow on Route 38" from Butler Radio in Pennsylvania

- "Naked Driver Arrested after Hitting Vehicles, Dog, Cow" from the *Methow Valley News* in Twisp, Washington

We can ignore that last one. Still, it is worth pointing out that we give more attention—and rights—to cows in the street than we do pedestrians. Here is another headline that caught my eye: "Woman Hits Wandering Cow; Virginia Law Says She Responsible for Damage, Cow Costs" from NBC 12 in Richmond.

Why is that the case?

Many states have open range laws, which means that ranchers don't have to keep their cattle off the road. Not only can't animal owners be held liable if somebody drives into their cows, some states—including Virginia—force the driver to pay the rancher damages.

If I'm driving and hit a cow, I'm at fault. If I hit a pedestrian, the pedestrian is.

The impact of media bias or bias in the legal system is an issue worthy of discussion. But notice how we aren't talking about traffic engineers anymore? We get so wrapped up in who did what wrong that we miss the systemic problems.

Both the media and the police can do better. But you can't blame them for trying to point out where people went wrong, because almost everybody breaks the law almost every time they use the transportation system.

When we put out the scofflaw survey back in 2015, we got nearly 18,000 responses from all over the world in just a couple months. Every single one of those 18,000 respondents admitted to breaking the law in the transportation system on a regular basis.

Maybe they drove a little too fast. Or rolled through a stop sign. Or walked against the pedestrian signal when no cars were around. Whatever the misdeed, these *scofflaws* did not consider themselves to be criminals. Society doesn't treat seemingly minor infractions as shameful or reprehensible either. If we did, we're all criminals.

We don't pay these misdeeds much mind because we all do them and because most don't lead to crashes. James Reason tells us in his 1990 human error book that "very few unsafe acts result in actual damage or injury."[4] Yet I like how Sidney Dekker describes it in his 2017 human error book: "Murphy's law is wrong. What can go wrong usually goes right."[5] However we explain things, the lack of bad results help cement scofflaw behaviors as the norm.

But think about what that means when a crash does happen. If everyone is breaking the law nearly every time they use our transportation system, every crash becomes a debate over which road user was the bigger culprit.

We also found that the overwhelming majority of bicyclists are not the reckless bike messenger types we imagine, rebelling against society and its laws. Instead, they are mostly rational individuals simply trying to get where they are going safely and efficiently, even if it means doing so illegally.[b]

This fact speaks to my intentional use of the word *scofflaw* in this research. I had recently watched the Ken Burns documentary film series about Prohibition. In 1924, the *Boston Herald* held a contest where it asked readers to come up with a new word for people who broke Prohibition laws but weren't thought of as criminals because it was socially acceptable to do so. The word *scofflaw* emerged as the winner and, nearly 100 years later, a useful way to describe this research. Most pedestrians and bicyclists who break the law are just trying to survive a trip in the transportation system we've given them but not designed for them.

Let's say a bicyclist is sitting at a red light on a street with no bike lane and is waiting next to a line of cars. If there is no cross traffic, your options as a bicyclist are to either (1) wait for the light to turn green and compete with the adjacent car traffic for street space or (2) run the red light and get out ahead of the cars to establish your space in the street before the cars get to the next block.

The second option is illegal in most places. But who's to say it isn't safer?

Yet red-light-running bicyclists seem to anger drivers more than

---

[b] I recently found some 1970s research that I hadn't uncovered when I wrote my literature review for that scofflaw paper. The researchers all complain about how often bicyclists break the law but weirdly put failing to signal with their arm when turning and improperly carrying a package as infractions on par with running a red light (see Sommer and Lott, 1971; Henszey, 1977; Drury, 1978). Drury (1978) notes that we should "question the applicability of particular laws to bicyclists" since "most of the time, a bicyclist can break the law with impunity—no accident ensues, no legal sanctions are taken, and the bicyclist has saved a little time and energy." Robert Sommer and Dale Lott, "Bikeways in Action: The Davis Experience," Bicycle Institute of America (1971); Benjamin Henszey, "Bicycles: A Need for Comprehensive Regulation," *Traffic Quarterly* 31, no. 1 (1977); Colin Drury, "The Law and Bicycle Safety." *Traffic Quarterly* 32, no. 4 (1978).

anything else.[6] At the same time, the world doesn't blink when Audi puts out a Black Edition that includes "Stealth Mode." What is Stealth Mode, you ask? Audi's ad suggests that will it help you get away with breaking the law because "stealth, whether it's on a fighter jet, a battleship or even a car, is all about avoiding detection for sneaking under the radar."[c]

Still, whenever road safety becomes an issue at city hall, the police start by cracking down on all the scofflaw pedestrians and bicyclists.

But here's the thing. We know that drivers running red lights is a massive factor in crashes that result in an injury or fatality.[7] Waymo—or the artist formerly known as the Google Self-Driving Car Project—even had to program its autonomous fleet to pause for a second or two after a light turns green because actually going when the light turns green turned out to be the biggest crash risk. But the connection between safety outcomes and scofflaw pedestrians/bicyclists? It's dubious at best. Here is Kenneth Todd in a 1992 *Transportation Quarterly* paper about pedestrian safety: "The notion that compliance with the law saves lives is a myth."[8]

Other researchers agree with him that "the commonly held notion that lawful behavior equals safe behavior is not borne out by statistics."[9]

I'm now imagining the guy from the 1989 Weird Al Yankovic movie *UHF* saying, "Statistics? We don't need no stinking statistics. We just need pedestrians and bicyclists to follow the rules." But, as Todd goes on to say, "No statistical analysis is needed to demonstrate that crossing is safe in the absence of moving vehicles and unsafe in their presence, regardless of the control device installed at the location. Safety is not assured by looking at a signal display and relying on someone else to stop."[10]

Going back to the ol' "if only" argument about everyone following the rules of the road, we should acknowledge that comparing drivers to pedestrians or bicyclists is not apples to apples. Here is AAA, in a 1958 report, talking about pedestrian-car crashes: "There is no justice in such a situation. Consider the pedestrian's chances in traffic conflicts. A collision between a motor car and a pedestrian, whatever the circumstances, is a grossly uneven affair. The heavier,

---

[c] The ad goes to describe the "darkened windows and ... blackened grills," as well as "special-ops styling" that will "make your Audi even more extraordinary."

sturdier, faster-moving car may suffer no more than scratches; the pedestrian, on the other hand, is almost certain to be either painfully injured or killed."[11]

We've long known that cars constitute a useful tool but a deadly weapon. When a driver runs a red light, more than half of those killed weren't the ones who ran the red light. But we've also long made the same excuses for cars that we do for guns. You might remember the movie *Happy Gilmore* where Mr. Larson (played by 7-foot-2 Bond villain Richard Kiel) wears a "Guns Don't Kill People. I Kill People" T-shirt. Here is a 1966 *Traffic Quarterly* paper making a similar point: "An automobile is not by nature dangerous in the same sense that an atomic bomb, or strychnine, or a mad dog is dangerous. It is the automobile plus the human element that adds up to a dangerous potential."[12]

Sure, but compared to drivers, pedestrians and bicyclists are holding water guns.

Let me be clear. I'm not looking for a lawless free-for-all for pedestrians and bicyclists. But if we really want a data-driven approach to road safety, we should acknowledge that pedestrians and bicyclists rarely harm anyone but themselves. As that same 1958 AAA report reminds us, we should give "special attention to violations which are linked with serious accidents."[13] This focus may shift as we see more e-bikes that are faster and heavier than the analog versions. But at this point, there are more people who die by vending machine each year than there are pedestrians who are killed by bicyclists.[d]

The other thing we need to remember is that people respond to the transportation system we put in front of them. Todd says in his 1992 paper that most "pedestrians do not obey traffic lights when they find it safer to disobey them."[14]

The same goes for drivers, because most also behave rationally given the transportation system we put in front of them. Another 1992 *Transportation Quarterly* paper studied how often drivers run stop signs and found that it was four times worse on streets with three lanes instead of two.[15] These researchers also found that the

---

[d] The best data I could find suggests two per year killed by vending machines and 1.4 by bicyclists.

violation rate decreases as land-use intensity increases, particularly with the "most hazardous" offenses.

If true, our scofflaw behaviors should be considered, at least to some extent, a design issue. What do we expect when we combine high design speeds with low speed limits? Or don't include a crosswalk anywhere near where people want to cross? The result is a world where, as a 1959 *Traffic Quarterly* paper says, "while perhaps well-intentioned ... the letter of the law makes everyone a violator."[16]

Well-intentioned rules where everyone is a criminal? At least the rules are based on safety, right? Right?!?

There's a 1957 *Traffic Quarterly* paper that says our "rules are not based on any study, or on any knowledge of whether they are basically sound or not."[17]

Well, that's disappointing. The other issue is that, as an Uber employee once bragged, "The law isn't what is written. It's what is enforced."[18]

That's true, except when it comes to understanding crash liability. Once a crash happens, all of a sudden the unenforced, written rules matter again.

In the words of Ben Stiller as White Goodman in *Dodgeball: A True Underdog Story*, "That is pure poppycock."

No matter our systemic design problems, the liability reverts back to the road users who weren't following the rules of the road that we laid out for them. This lets us explain away everything while also explaining nothing.

Most people, James Reason reminds us, "involved in serious accidents are neither stupid nor reckless."[19] In other words, most are rational people who are behaving reasonably given the transportation system. Acknowledging where they went wrong is only one piece of the puzzle. We also need to acknowledge that traffic engineers built a system where only those who use it are at fault.

Tara Goddard's crash article experiment included a third version where the researchers add context about previous road safety issues on the street. With that information, people were less likely to blame those involved and more likely to want infrastructure changes. This type of information is what's needed to make things better.

When it comes to understanding crash liability, emphasizing the rules of the road benefits traffic engineers but harms everyone else.

71   Safer Designs Please

The Netherlands made it legal for pedestrians to jaywalk in 1997. Back then, more than 130 pedestrians lost their lives each year. The population of the Netherlands has since grown by more than 10 percent, but now the Netherlands sees fewer than 50 pedestrian deaths annually.[1]

Transportation infrastructure in the United States isn't quite the same as what you might find in the Netherlands. But another interesting difference is how the Netherlands handles assigning blame in a crash, as John Pucher and Lewis Dijkstra explain in their 2000 *Transportation Quarterly* paper: "Even in cases where an accident results from illegal moves by pedestrians and cyclists, the motorist is almost always found to be at least partly at fault." And when the pedestrian involved in the crash is a kid or an older person, "the motorist is usually found to be entirely at fault."[2]

Pucher and Dijkstra go on to say that "in almost every case, the police and the courts find that motorists should anticipate unsafe and illegal walking and cycling. Having the right of way by law does not excuse motorists from hitting pedestrians and cyclists."[3]

In terms of liability, how do insurance companies in the Netherlands handle crashes between cars and pedestrians or bicyclists? "The insurance company for the motorized vehicle automatically pays the damages, regardless of guilt. The only exceptions are cases where the pedestrian or cyclist can be proved to have deliberately caused the accident."[4]

It's not about giving pedestrians free rein of the streets. Rather, it's about setting up a system where those using bigger, heavier, and more dangerous modes are responsible for the safety and well-being of those using smaller, lighter, more sustainable, and more vulnerable modes. If you are a driver in the Netherlands, the best way to protect yourself from liability is to do your darndest to make sure that you don't hit any pedestrians or bicyclists.

Compare this policy to that in China, where only at-fault drivers have to pay damages in a crash with a pedestrian or bicyclist. China's liability rules seem more logical than how the Netherlands does business. But since it costs less to pay for a funeral than a lifetime of medical bills, many Chinese drivers live by the very real mantra that "if I hit someone, I'll hit him again and make sure he's dead."[5]

The result is cases where drivers protect themselves from liability by doing just that, driving back and forth over their victims to make sure they're dead. A 2015 *Slate* article tells some insanely sad stories, one of which focuses on a BMW driver backing out of a parking spot and running over a 3-year-old boy.[6] The driver then drives forward and hits the boy again. He gets out of the car and looks around before backing up yet again, hitting the boy a third time. After that, he drives away. The incident was captured on video, but the driver claimed that he thought the boy was a bag of trash.[a] The court later declared the driver not guilty of intentional homicide. I used this *Slate* article to help spark a discussion in my transportation system safety graduate class a few years ago. One of the Chinese students in that class, Yu Liu, told us that stories like this one were all too true and all too common.

If autonomous car technology ever catches up with the hype, I'd like to think it would help shift liability into something more like the Netherlands. Or better yet, something more like Disney's monorail. Imagine the outrage—and lawsuits—if Disney's monorail killed or maimed even a fraction of what we see happen in the real transportation system?

Now imagine a world where we put the onus on car manufacturers to protect themselves from liability for any and all crashes, particularly those with pedestrians and bicyclists. Instead of headlights aimed low so not to annoy opposing drivers, they'd be aimed toward where we might expect pedestrians and bicyclists. Instead of bull bars, we might have soft, fuzzy cars with exterior airbags.[b] Still want a bull bar? Sure, but it'll cost you more because the liability risk is higher.

But given what we're seen so far, I'm dubious that we'll see such a seismic shift in how we treat liability.

We've talked about Elaine Herzberg a few times now. She was the

---

[a] A bag of trash that, for some reason, needed to be run over multiple times?!?

[b] Google was granted a patent for external airbags in 2015.

world's first pedestrian killed by an autonomous car. We know that Uber's autonomous car LiDAR technology "saw" the pedestrian with more than enough time to stop but treated her as a false positive. We know that too many false positives make for a bumpy ride, so Uber's programmers disabled the emergency braking system and prioritized a smoother ride over safety. Yet blame still fell on Elaine Herzberg for walking where she shouldn't and the operator for not paying attention to what the autonomous car was doing.[c]

It's not just about so-called full autonomy. Car companies continue to add safety technology such as automatic emergency braking. They sell these features to prospective buyers but don't mention that the chance of them working may only be slightly better than a coin flip.[7] Nor do they mention that the odds drop at night or for pedestrians with darker skin.[8] Despite the inequities and ineffectiveness, liability from a pedestrian or bicyclist death isn't likely to land on the car companies.[9]

Nor will it likely land with the traffic engineers.

Traffic engineers protect themselves by overestimating their design liability risk. We rightly worry about a design mistake costing someone their life, but we discount the idea that better design could save many lives. We say that we encourage creativity and innovation, but that only applies to technology as a panacea.

On the other hand, traffic engineers may *underestimate* liability risk on another front. Instead of treating design as a "rational process," we allow it to become a ***squeaky wheel*** issue.

If your neighborhood has the wherewithal to set up petitions and complain to the right people, you now have a shot at seeing one of our safer designs. If your neighborhood doesn't have that sort of time or political clout, tough luck.

The reverse is also true—and to a greater extent. Sometimes, all it takes is a few vocal constituents to squash a safer design. Given some of the atrocities perpetuated by traffic engineers in the past, taking community input seriously makes sense. But safer streets shouldn't be subject to a popularity contest.

Opting against a safer design after a few drivers grumbled about it forcing them to drive slower? Canceling a protected bikeway after

---

[c] Uber settled a civil case with Herzberg's family but was never criminally charged for killing her.

a few neighbors moaned about losing some on-street parking or the perceived aesthetics of the street?

Such decisions are common for traffic engineers, and they should invite a liability problem. Even an unfrozen caveman lawyer could argue that basing what gets built on the whims of a few neighbors—instead of a "rational process"—could fall under the umbrella of "arbitrary" or "capricious."

So instead of getting at the root of the problem with better design, it's easiest for traffic engineers to protect themselves by throwing up a warning sign.

Every so often, you'll see a sign that says something like, "High Collision Area." Traffic engineers post such signs for fear of liability, but the underlying reason speaks more to which lawsuits win. In his 2003 *Transportation Quarterly* paper, Reid Ewing found that the only lawsuits that seem to win are those that fail to add a warning sign.[10]

Yes, it's frustrating that the legal system incentivizes adding signs over safety outcomes. But now flip this thinking. As long as traffic engineers add a warning sign, designs that don't match the guidelines are—as Ewing found with the early nonstandard, traffic calming treatments—"unlikely to succumb to a legal challenge."[11]

Roadside memorial signs don't fall under this liability umbrella, but you see them almost everywhere, to the point where most people rarely notice them. Some are ad hoc, placed by the family and friends of those who died. Others are real street signs, placed by the powers that be.

The Colorado Department of Transportation (CDOT) has an official program for these memorials. Once you die on a Colorado street, your loved ones can fill out the appropriate form. If approved, CDOT will take their $100 and fabricate, install, and maintain the sign for six years. The bottom half of the sign will say "IN MEMORY OF," followed by your name. The top half of the sign varies. Based on the DOT form, your loved ones will have four options. With a driving under the influence conviction or a toxicology report, they could opt for the "DON'T DRINK AND DRIVE" sign. If you weren't wearing a seat belt, the "PLEASE BUCKLE UP" sign becomes an option. If you were riding a bicycle or motorcycle, your family could select a "PLEASE RIDE SAFELY" sign. For any other fatality, or for any of the above, the "PLEASE DRIVE SAFELY" sign is the default.

One of these signs used to sit at the corner of Leetsdale Drive and Monaco Parkway in Denver. It commemorated Amelia Bates, a 15-year-old high school sophomore who was killed there in 2008. George Washington High School sits near one corner of this intersection. On the opposite corner is a grocery store. Amelia was a writer for the school's newspaper and had walked over to the grocery store to buy a birthday cake for her editor. On her way back, Raj Sundaram hit and killed Amelia in the crosswalk as he made a right turn.

The day after Amelia died, the *Denver Post* wrote an article about Amelia and the crash. While I don't love the title suggesting the absence of a driver ("Classmates Remember Girl Hit, Killed by Car"), I did appreciate that they provided at least a little context: "The intersection where she was hit crosses five lanes on Leetsdale, a major east-west artery that hums with quick-moving traffic."[12]

Unsurprisingly, Leetsdale is on Denver's high-injury network. This designation refers to the 5 percent of streets where we see about 50 percent of the fatalities and serious injuries. The speed limit is 35 mph, and despite being a designated school zone, 50 mph speeds are not uncommon.[d] The same goes for the intersecting road, Monaco Parkway, a six-lane arterial. Want to take a right turn on red? Go right ahead. Heck, we'll make it easier for you by adding a slip lane with a huge turning radii designed for high speeds. Since the slip lane immediately merges with through traffic, you'll have to focus on cars to your left. Oh, but also be sure to simultaneously look right for pedestrians because we've thrown in an unsignalized crosswalk. You may only see a pedestrian 1 out of every 100 times you make this turn, but past performance is no guarantee of future results.

That is where Raj Sundaram hit Amelia Bates.

For reasons unbeknown to everyone, the Denver police spokeswoman said that it "was just a very tragic accident."[13]

But since Raj failed to give Amelia the right-of-way, the crash data faults him.[e]

The sign seems to agree and reminds us to "PLEASE DRIVE

---

[d] Two of my former students, Jenny Godwin and Lily Lizarraga, collected speed data in this area for my class tactical urbanism project.

[e] Raj Sundaram was charged with failing to exercise due care and careless driving resulting in death. I'm guessing the case was settled or the charges dropped because an hour of internet sleuthing didn't help me figure out what happened.

SAFELY." Sure enough, *"IN MEMORY OF AMELIA BATES"* follows.

How did traffic engineers improve this tragic intersection? They added a yellow pedestrian crossing sign with a little arrow pointing toward the still uncontrolled crosswalk.

That sign might help protect them from liability issues the next time something bad happens. But fundamentally, the intersection remains the same dangerous place it was when Amelia was killed.

Sorry Amelia. We need to do better. Until that happens, your memorial sign should read "SAFER DESIGNS PLEASE."

Part 9

Standard Issue

___

72   The Pirates' Code

—

In the first *Pirates of the Caribbean* movie, Keira Knightley's character, Elizabeth Swann, negotiates a deal with Geoffrey Rush's character, Captain Barbossa. When Barbossa gets what he wants[a] and immediately reneges on the deal, Elizabeth reminds him of the Pirates' Code.[b] Here's the exchange.

> ELIZABETH SWANN: Wait! You have to take me to shore. According to the Code of the Order of the Brethren …
>
> CAPTAIN BARBOSSA: First, your return to shore was not part of our negotiations, nor our agreement, so I must do nothing. And secondly, you must be a pirate for the Pirates' Code to apply, and you're not. And thirdly, the code is more what you'd call guidelines than actual rules.
>
> Welcome aboard the Black Pearl!

Traffic engineering **standards** aren't that different from the Pirates' Code. Most of our standards are more what you'd call **guidelines** than actual standards.

Actual design standards in traffic engineering are few and far between. We can't install green stop signs, so it's fair to refer to the rules surrounding traffic signs, signals, and markings as published in the *Manual on Uniform Traffic Control Devices* (MUTCD) as something close to standards. The ADA Standards for Accessible Design (built on the Americans with Disabilities Act of 1990) are also a passenger on the standards boat.[c]

All our other 1,000-page manuals? They're more what you'd call

---

[a] A cursed Aztec medallion.

[b] Officially, the Pirates' Code is the Code of the Order of the Brethren.

[c] The ADA wasn't passed until 1990, but the 1972 Federal-Aid Highway Act required curb ramps on all federal projects (unfortunately, that doesn't mean they always exist).

guidelines. Some traffic engineers like to refer to them as standards and wave them around as if they were handed down to Moses on Mount Sinai. This gives these manuals an air of authority and shuts up the nonbelievers. And since safety is supposedly baked right into the so-called standards, we can "design without explicitly analyzing the safety consequences of design decisions," as Ezra Hauer says in his 1989 book chapter.[1]

The most infamous of these manuals has long been the American Association of State Highway and Transportation Officials (AASHTO) Green Book. We've talked about AASHTO and its manuals a lot in this book, but it is worth digging into their origins to give you a sense of what traffic engineers are working with.

Over the course of six years in the late 1930s and early 1940s, the group formerly known as the American Association of State Highway Officials (AASHO)[d] published seven policy brochures. In 1950, AASHO stapled them all together and called the volume *Policies on Geometric Highway Design*.

AASHO ditched that book and separately published the rural Blue Book[e] in 1954 and the urban Red Book[f] in 1957.

After a couple of iterations of each, AASHO combined the Blue rural and Red urban books in 1984 with *A Policy on Geometric Design of Highways and Streets*. Neglecting that blue and red make purple— as well as the popularity of Prince in 1984—they put a green cover on it.[g]

Although 70 years in the making, the Federal Highway Administration admitted that even the 2011 version "updated the earlier policies but was, itself, a work in progress."[2]

"A work in progress" is a good way to put it, but that isn't how traffic engineers view and use these manuals. In a 1998 paper about the history of AASHO's design manuals, Jerome Hall and Daniel Turner tell us they "now represent national guidelines in the eyes of

---

[d] Created in 1914, Babe Ruth's rookie year, AASHO added "and Transportation" to its name in 1973 and became AASHTO.

[e] *A Policy on Geometric Design of Rural Highways*.

[f] *A Policy on Arterial Highways in Urban Areas*.

[g] Glad released its "Green Seam" plastic bag design the next year, in 1985. The Tom Bosley commercial that I remember so well—with kindergartners showing us how easy the bag is to seal—didn't come out until 1987. Yellow and blue make green. It's sealed!

most users." But, they also say "that has not always been the case."[3] For much of AASHO's history, "the information in the AASHO documents was rarely adopted in full" by states.[4]

Designing a street using today's AASHTO manual feels so scientific. We routinely use equations that come across as deeply steeped in the physical sciences.

The original intent of **design speed**, for example, was to mathematically link design speed with curve radius, superelevation,[h] and the side friction factor. In fact, the title of Barnett's 1936 paper in which design speed was first introduced is "Safe Side Friction Factors and Superelevation Design."[5] Design speed was not the main attraction. Instead, Barnett's goal was a "balanced design" where all these factors would work in harmony to counteract centrifugal forces and allow drivers to take corners without slowing down.

The same basic superelevation equations we use today date to 1920.[6] At the time, "there was considerable public hostility to providing any superelevation due to the opinion that it encouraged high speeds and reckless driving on curves."[7]

Despite the lack of consensus, AASHO made up its mind in 1930 that superelevation on curves was their official recommendation. The next step was to figure out the **side friction factor**, meant to represent the coefficient of friction between the vehicle's tires and the road.

In 1935, the Bureau of Public Works called on volunteers to participate in a study. "Several hundred drivers" were asked to take their own cars and ride along curves of known superelevation and radius.[8] These drivers were then asked to report the speed at which they felt the car begin to "pitch outward." As you might expect from such a study design, the results were all over the map. Nevertheless, the researchers averaged whatever drivers told them—combining both wet and dry conditions—and ended up with a value of 0.16. AASHTO uses almost the exact same value today.

Today's Green Book tells us that this equation "governs vehicle operation on a curve" and is based on the "laws of mechanics."[9] It certainly looks the part, with variables representing vehicle speed, superelevation, and curve radius being placed in an elegant

---

[h] Superelevation refers to degree of banking of a curved road. The more the curve is banked, the faster you can go.

mathematical equation with a gravitational constant and the side friction factor. But this equation doesn't tell us anything about the point at which cars will start skidding off the road. It seems like it does, which is why traffic engineers treat design values like speed, curve radii, and superelevation as **absolute minimums**. But all it really tells us is how comfortable drivers are in taking corners at certain speeds. We assume that this value equates to safety.

In the early days of traffic engineering, skidding was a huge safety issue.[i] One of the lasting legacies of the early skidding research is **perception-reaction time**, or the time it takes a driver to perceive and react to an impending hazard. If a driver sees something in the middle of the road, we want them to have enough time to see that object, figure out what they want to do, and then put that decision into effect. But how long should we give them to do so?

The first **reaction time** study dates back to 1925, when F. A. Moss and H. H. Allen placed a gun loaded with red paint under the car. They sat alongside the driver and fired the gun into the pavement. Upon hearing the gunshot, the driver would brake as quickly as possible. The application of the brakes caused another gun to automatically shoot red paint onto the pavement. Knowing the speed, they measured the distance between the two red spots and calculated the reaction time. They tried this for 57 drivers and found an average reaction time of just over half a second (ranging between 0.3 and 1 second).[10]

We didn't add **perception time** (the time it takes a driver to see and register the hazard) to our reaction time until a Massachusetts Institute of Technology study in 1934.[11] This time, researchers put drivers in racing cars and told them to follow a lead car and apply their brakes as soon as the lead car did.[j] They calculated an average perception-reaction time of 0.64 seconds (with a range between 0.2 and 1 second), but the researchers assumed that this wasn't enough time and suggested something between 2 and 3 seconds to give us a

---

[i] The other big research topic back then was making sure hills weren't too steep because cars at the time couldn't get up them (as covered in Frederick Cron's "Highway Design for Motor Vehicles—A Historical Review" in *Public Roads* [1975–1976]).

[j] There was no need for the driver to figure out what maneuver is appropriate. In this study, the driver knew that they should brake as soon as the car in front of them started braking.

"factor of safety." AASHO agreed and, in 1940, set the perception-reaction time as 2 seconds for 70 mph and 3 seconds for 30 mph (their thinking was that drivers would be extra alert at high speeds). For the 1954 Blue Book, AASHO simplified it and went with 2.5 seconds for all speeds. AASHTO still uses 2.5 seconds today (1.5 seconds for perception time and 1.0 second for reaction time).

We use 2.5 seconds despite knowing that true perception-reaction time very much depends on the situation. It depends on street lighting and background clutter. It depends on the color of the object. Or, if it's a pedestrian, it depends on the color of their clothes, or, sadly, maybe their skin. It depends on driver expectations, which depend on location. Drivers may perceive and react faster to a moose in Bangor, Maine, than in San Diego, California. Drivers may also perceive and react faster to a pedestrian in the street in downtown Boston than one in rural Wyoming.

You might think that perception-reaction time would also depend on the driver's age. Fortunately for traffic engineering ease of design, a 1986 study "proved" that age doesn't matter. Paul Olson and Michael Sivak had 49 younger drivers (under the age of 40) drive with them through rural Michigan in a 1980 station wagon for 10 minutes or so before directing them to a road with a yellow piece of foam stuck in the middle of the lane.[12] They noted when the driver released the accelerator and when they hit the brakes. They then asked the driver to show them where the car was when they first saw the yellow foam. The estimated time in between the driver's self-reported "seeing" of the object and when they released the accelerator was deemed the *perception time*. The time between that and when the driver hits the brakes was deemed the *reaction time*.

They did the same thing for 15 so-called older drivers (age 50 and above) and found that "there is no evidence that older drivers require substantially greater sight distances." Olson and Sivak readily admit that "the number of older subjects in the sample is relatively small," but traffic engineers still cite this paper as "proof" that age doesn't matter. Nor does it matter that "past research has consistently shown that reactive time increases with age" or that their definition of an "older" driver was 50 years old.[13] I'd also venture to guess that older drivers with vision or other such issues don't often sign up for research studies like this.

Olson and Sivak said their results solely apply "to the specific situation investigated," which was a piece of yellow foam in the road and a suspect methodology for figuring out when drivers first saw it.[14] These drivers were probably more engaged than most, having only driven for a short time and doing so with a researcher in the car with them. And like almost all other perception-reaction time research, it was done during daylight hours on a clear day with drivers who weren't distracted or drunk.

Reading the Olson and Sivak paper, it feels like they set out to disprove a 1983 Federal Highway Administration study suggesting that the 2.5 seconds "currently used ... is too short."[15] Why would they care? As Olson and Sivak say, "Overestimating PR time unnecessarily increases roadway construction costs."[16]

How so? When it comes to design, we use the perception-reaction time to calculate what we call the *stopping sight distance*. Let's say there's something in the middle of the road. The thinking is that we should give drivers enough time to apply the brakes in time to stop and not skid into whatever it is. Seems logical, right? We determine this time factor with another sciencey equation that includes vehicle speed, relative road steepness, and gravity. Combine that with 2.5 seconds of perception-reaction time, and we end up with a stopping sight distance. We then design the road to make sure drivers have a line of sight for at least that distance.

In the 1920s, AASHO assumed that the object in the road was 4.5 feet tall. German traffic engineers of that era assumed that the object in the road was 8 inches tall. Objects closer to the ground would require a longer *sight distance*, so this meant that Germany's minimum sight distance was "more than double the AASHO minimum."[17] AASHO thought "that the American rule for measuring sight distance was risky because it did not allow for small obstacles such as fallen rocks or small animals" but also didn't want construction costs to skyrocket with huge sight distances.[18] Eventually, safety concerns put enough pressure on AASHO for it to go with the "so-called dead cat rule" of 4 inches.[19]

All these somewhat arbitrary assumptions are embedded into our stopping sight distance equation. The equation seems completely scientific. After all, it's got nearly 100 years of research behind it. It also comes to us in a 1,000-page manual that dozens of people and committees holding countless meetings have come to a consensus on. As Frederick Cron says in his historical review of design criteria,

"Design practice generally evolves slowly over a long period, and crystallizes into formal standards only after a strong consensus has developed among informed engineers and administrators … standards in the United States are the distilled essence of the experience and judgment of hundreds of engineers and administrators—a consensus of many informed people."[20]

It's hard to argue the "distilled essence" of hundreds of engineers, especially when you are a young traffic engineer.

We take it for granted that those who came before us did their homework and found us a safe solution. But dig into many of the design criteria that we use today—like the side friction factor or perception-reaction time—and you'll see that we were flying by the seat of our pants.

In a 1989 book chapter, Ezra Hauer describes traffic engineers like the bumbling pirate crew: "In the absence of relevant factual knowledge, professionals have to resort to the waving of manuals."[21]

I've quoted Cron's 1976 historical review of design standards a few times here. Here is how he introduces his work: "The account which follows traces the principal threads of the national research program in the fields of traffic and geometric design up into the 1950s. Research is still in progress in all of these fields, but the fundamental principles were largely established by 1950."[22]

If we believe that "the fundamental principles were largely established by 1950," we've got some work to do. But that is why these manuals have become so powerful, and that is what we are working with.

73   Don't Blame the Manuals

A couple of years ago, I assigned a "charming streets" case study project to my sustainable transportation class. The idea was for the students to find a street they found "charming" or "beautiful" or whatever synonym you like. They next found a more conventionally designed nearby street (of the same functional classification) with a not-so-great road safety record. Once they selected their two streets, they dug into our traffic engineering manuals and our municipal rules and regulations to see what hurdles they'd face in trying to redesign the unsafe street to be more like the charming street.

Every time, my students found that their biggest hurdle was *not* our traffic engineering manuals. If they knew where to look—and what to ignore—there was usually enough flexibility on that front.[a]

For nearly all design criteria in the Green Book, AASHTO includes a range of values that run from what it considers the absolute ***minimum*** to something it deems more ***desirable***. Their values tend to be "conservative to a fault," especially "when it comes to lane and shoulder widths, vertical and horizontal curve radii, lateral clearances and offsets at the street edge, and other geometric features," as Reid Ewing says in a 2002 *Transportation Quarterly* paper.[1]

Given where we started, it's easy to understand why we err on the side of bigger. Take street width, for example. Back "in 1915, 16 feet was considered adequate width" because it was "wide enough for two vehicles to meet without the need for either of them to run onto the shoulder."[2] But drainage was also a top-of-the-list concern in those early days. To get rainwater off the street as quickly as possible, we'd put the highest point—called the crown—smack dab in the middle of

---

[a] I also make it clear to my students that they should ignore the perception that manuals like the AASHTO Green Book represent standards. Although the Green Book slips in phrases like "standard minimum radius" or "standard turning area," it admits that it only gives us "guidance based on established practices." It's a policy book that, as AASHTO says, "is not intended to be a prescriptive design manual that supersedes engineering judgment."

the street.[b] We even did it on curves, despite it banking the outer lane in the wrong direction. Speeds were slow enough that it didn't matter much at first. But as speeds increased, drivers started "cutting the curve to utilize the banking afforded by the crown in the opposing lane."[3]

It isn't ideal for safety when drivers take the inside lane on curves directly into oncoming traffic, so wider streets seemed like the obvious solution. But even back then, people worried that wider streets might lead to increased speeds and safety problems. Several early research papers quickly quelled those fears.[4] These papers found "no correlation … between average speed and pavement widths."[5]

This conclusion simplified things for AASHO. Bigger is better. Give them a "***minimum***," toss out some "***desirable***" values, and we're good to go.

Decades pass, and we forget that the widest streets these early authors studied were narrower than most of today's narrowest streets. In fact, the widest street in one 1934 study was 20 feet wide[6] and 22 feet wide in another.[7] In the 1945 study, it was 24 feet wide.[8] What endures is the simplistic thinking that "drivers drive as fast on narrow pavements as on wide ones."[9]

That may be true when comparing narrow streets to other narrow streets. But is it true as streets get wider and wider and wider?

We didn't know. But unless you go back and read the original papers, it sure seems like we did. And that thinking gets "copy and pasted" again and again. But like Michael Keaton in the movie *Multiplicity*, we lose a little bit of clarity when we make a copy of a copy. But instead of fading, the loss of clarity makes the finding stronger to the point where, eventually, it seems like a given instead of as a research question that hasn't been answered yet. No wonder we still haven't wrapped our heads around the speed—and safety—implications of basic street dimensions.[c]

Local design criteria may be more restrictive than our traffic engineering manuals, especially when it's for an arterial controlled by a state Department of Transportation. For my students, there was enough design flexibility to re-create the basic dimensions and design characteristics of the charming street in almost every case.[d]

---

[b] Doing so minimized the distance the rainwater had to move to get off the street.

[c] It didn't help the cause that when researchers eventually studied safety, they focused on fender benders instead of injuries and fatalities.

So what was the biggest hurdle?

Sometimes it was the fire code. Fire marshals want to make sure they can get to emergencies quickly. They typically "mandate overly wide streets, requiring 20 feet of unobstructed path for new or significantly improved streets."[10] Good intentions aren't enough if these streets entice speeding and spawn more serious crashes that might not have occurred otherwise.[e]

Most times, as we'll see next, the biggest hurdle was *level of service*.

---

[d] It did help the cause when the Federal Highway Administration published a Guidance Memorandum in 2013 that supported "taking a flexible approach to bicycle and pedestrian facility design" and encouraged the use of more progressive guidebooks such as those published by the National Association of City Transportation Officials (NACTO) and the CNU-ITE *Designing Walkable Urban Thoroughfares: A Context Sensitive Approach*. One problem with these guidebooks is that they are too new. They don't yet have decades of time on their side to shroud their flaws.

[e] Over the last five years, the Denver Fire Department responded to three times more car crashes than fires. Denver Fire Department, "2022 Annual Report," (2023).

74   Level of Frickin' Service

Level of service (LOS) refers to the capacity-focused letter grade we calculate for streets and intersections. Like a school report card, it ranges from A to F. An A equates to a relatively empty facility with free-flowing traffic, while an F signifies a traffic jam. LOS originated in the 1965 *Highway Capacity Manual*.[a]

Like the AASHTO Green Book, we should remember that the *Highway Capacity Manual* is only a **guidebook**. Yet here is a 1966 *Traffic Quarterly* paper's take on the Highway Research Board's new manual:[b] "In conclusion, it should be reemphasized that the 1965 manual like its predecessor, the 1950 manual, is a guidebook rather than a policy document. However, putting together 11 chapters and four appendixes of significance under one cover, together with 130 drawings and 78 tables in 397 pages, makes for a guidebook that is difficult to ignore."[1]

In other words, we get the typical guidebook disclaimer, but it's followed by a reminder that they did such a great job in creating this guidebook that we'd be fools not to use it.

So we used it, and traffic engineers in the 1960s focused on systematically "eliminating inadequacies in the existing street network" so as to "meet minimum standards of right-of-way and roadway design" and "achieve a balanced capacity."[2]

It would be one thing if level of service is limited to a methodology in a guidebook. The real problem is that municipalities, agencies, and state governments codified LOS into law. Some did so directly with minimum LOS grades, but most embedded LOS into their **traffic impact analysis** requirements. If I'm trying to build a

---

[a] Voltaire wrote that "the Holy Roman Empire was neither holy, nor Roman, nor an empire." In a similar vein, the *Highway Capacity Manual* is neither focused on highways, nor does it actually measure capacity, nor is it really a manual. Voltaire, "Essay on the Manners and Spirit of Nations" (1756).

[b] The Highway Research Board became the Transportation Research Board in 1974.

new coffee shop, I'd have to estimate how much traffic that coffee shop will generate and add that traffic to the street network. If the additional traffic changes the LOS grade of the nearby intersection or street, we could have a so-called significant impact on our hands. If so, I'd have to mitigate that impact—and get the LOS back closer to where it started—for my coffee shop to be approved. As for mitigation, that usually means increasing the capacity of the adjacent intersection or street.

It doesn't sound like a big deal, right? But instead of a coffee shop, try proposing the sort of changes that might make for a more charming—or safer—street. Narrow the street. Constrict the travel lanes. Reallocate space to wider sidewalks, bike lanes, or bus rapid transit. Reduce the design speed. Minimize the turning radii. Remove a slip lane. Add lane-narrowing bulb-outs to slow turning cars and shrink the pedestrian crossing distance.

Do even a few of these things, and we'd likely end up with a safer street.

But it's not that easy. Why? Because we don't typically have the same flexibility in the LOS and traffic impact analysis regulations that we do in our street design guidelines. Anything that adds to driver delay or slows car speeds could make LOS worse. Let's say we want to add a midblock pedestrian crossing: "Pedestrian interference at midblock causes friction, which increases the travel time of vehicles traveling along the segment. As a consequence, the average travel speed and the level of service along the segment are lowered."[3]

Defining pedestrian activity as friction "conditions" and automobile "interference" seems problematic—especially given that the quote comes from a 2021 paper, not one from the 1960s. The bigger problem is that if these "friction conditions" cause driver delay or car speeds to dip beyond a certain threshold, we've got a significant impact that needs to be mitigated.[4] If that's the case, the path of least resistance means to keep doing nothing. Why bother trying to make our street more charming … or safer?

Where else might this come into play?

Let's say we are trying to figure out how much time we need to give pedestrians to cross the street. For decades, traffic engineers, when timing pedestrian crossings signals, assumed that pedestrians walk at 4.0 feet per second. We did so knowing full well that "about one-third of all pedestrians cross streets at a rate slower than 4.0 feet per second."[5] This quote comes from AAA's 1958 report on pedestrian safety, but it took the Federal Highway Administration until the 2009 version of the *Manual on Uniform Traffic Control Devices*[6]—the document that sets the standards for traffic signs, signals, and markings—to change the recommendation to 3.5 feet per second.[c] We still use this 3.5 feet per second assumption today even though we know it's still too fast for about 15 percent of the population.

Why would we do that? Why would we purposefully design crosswalks that don't give enough time to millions of people that might want to cross? It comes back to level of service.

Have you ever heard of the Barnes Dance? Instead of pedestrians having to compete with turning cars, the Barnes Dance gives them an all-pedestrian signal phase. Not only that, but it's timed so that pedestrians can cross diagonally in one fell swoop. It was used in Boston in the 1920s but was popularized by traffic engineer Henry Barnes in the early 1950s when he implemented it throughout downtown Denver.

The name Barnes Dance stuck when newspaper reporter John Buchanan said that "Barnes has made the people so happy they're dancing in the streets."[7]

As Henry Barnes says in his 1965 autobiography, his goal was safety:

> As things stood now, a downtown shopper needed a four-leaf clover, a voodoo charm, and a St. Christopher's medal to make it in one piece from one curbstone to the other. As far as I was concerned—a traffic engineer with Methodist leanings—I didn't think that the Almighty should be bothered with problems which we, ourselves, were capable of solving. Therefore, I was going to aid and abet prayers and benedictions with a practical scheme: henceforth, the pedestrian—as far as Denver was concerned—was going to be blessed with a complete interval in the traffic signal cycle all his own.[8]

It worked. Denver's Central Business District was averaging seven

---

[c] Given the role this manual is supposed to play, this recommendation seems outside its purview.

road fatalities each year, so Barnes converted 50 intersections. "The year after the installation, there were no fatal accidents in that area."[9] Sacramento followed suit, converting 10 intersections, where it "saw a 47% drop in ped accidents and a 15% drop in vehicle ones."[10] A 1980s study looked at 1,297 intersections across 15 US cities and found that the standard pedestrian signal—where pedestrians cross at the same time as cars moving perpendicularly—was no safer than having no pedestrian signal whatsoever.[11] But give pedestrians a Barnes Dance, and you end up with five times fewer turning crashes.[12] Later research suggests a 29 percent reduction in pedestrian crashes.[13]

Around 2012, I met with Denver's head traffic engineer, Brian Mitchell, at a downtown coffee shop. After we argued about some road safety issues in another part of the city, he said he wanted to remove the Barnes Dance traffic signal timing from all of downtown. The local transit agency was adding a fourth car to its downtown trains. This necessitated an overhaul of the downtown traffic signal timing because adding that fourth train dropped the level of service below what he decided was an acceptable level. Instead of taking green time away from cars, he took it away from the pedestrians. He then told me that getting rid of the Barnes Dance was going to happen sooner or later, so he might as well do so before he retires and make life easier on the next head traffic engineer.

Eliminating the Barnes Dance wasn't required by our traffic engineering manuals. It was a policy decision based on a protocol that—as we shall soon see—has nothing to do with safety. We could've gone in the other direction and given more time to pedestrians. AAA, in its 1958 report on pedestrian safety, recommends that the time allowed "should never be less than the time needed for waiting pedestrians to cross the roadway, plus at least five seconds to provide time during which pedestrians may begin to cross."[14]

Five extra seconds? On top of the time we already give to pedestrians? Imagine suggesting that to a traffic engineer today. You might get a response similar to the words of the president (played by Peter Sellers) in the 1964 Stanley Kubrick film *Dr. Strangelove*: "This is preposterous. I've never approved of anything like that."

75 Unfinished LOS Business

—

Before handing our design reins over to the reign of *level of service*, or LOS, you might assume that ... well, you might assume a lot of things.

One thing I didn't realize was that we had to be talked into it. According to a member of the original LOS committee, "The idea of measuring and estimating traffic flow capacities had to be sold and it took a long time. It did not begin to be generally accepted as a useful tool in highway design, operations and traffic control, regulation and management, until after World War II."[1]

That LOS committee was led by O. K. Normann, a man they called Mr. Capacity. Described as the kind of man who "became so focused on the discussion that he would light cigarette after cigarette, not realizing he already had several already burning," Normann studied capacity at a time when "congestion on multilane highways was virtually unseen."[2] The pesky lack of congestion didn't stop Mr. Capacity, so he did the best he could to estimate capacity by counting cars and how far apart they stay from one another at different speeds.

The underlying idea was to treat traffic flow like fluid in a pipe[a] and to maximize its kinetic energy. You don't need to be a physics professor to know that more kinetic energy would be bad for safety. Yet that same 1964 paper tells us that LOS was "based on maximizing the kinetic energy" of the "traffic stream."[3] Even today, higher vehicle speeds mean "better" LOS for urban streets.

The A to F letter grades that traffic engineers use to score LOS originated with a contentious vote by the committee in 1963. The idea passed, but "the Committee membership remained divided and uncertain about how this concept would be implemented."[4]

---

[a] According to the 1964 paper mentioned here, "The relationship between level of service and traffic volume (flow) is analogous to the relationship in classical hydrodynamics between energy and momentum."

To fill out the A to F range, the task force first transferred its existing capacity definitions. For instance, "Level of Service E was intended to replicate the notion of 'possible capacity' as defined in the 1950 *Highway Capacity Manual*."[5]

The next two levels—which ended up being LOS C and LOS B—were then added "to represent the notion of 'practical capacity' as defined in the 1950 *Highway Capacity Manual*" in the urban and rural context, respectively (LOS C was meant for urban and LOS B for rural). Practical capacity refers to design capacity, whereas possible capacity gives us something more like maximum traffic under prevailing conditions.

A guy named Karl[b] then said that California highways already handle a lot more traffic than what Mr. Capacity estimated. The task force agreed and stuck LOS D in the middle "to reflect the maximum sustainable" traffic.

Then a guy from New Jersey named Charles[c] said he wanted his new toll roads to be better than the best. So the task force added LOS A "to provide a standard of service higher than 'practical capacity.'"

The task force brought these five levels of service (A through E) back to the committee. The debate was described as "passionate," with one member shouting that "it appears the committee believes that, in the beginning, God created the Heavens, the Earth, and five Levels of Service!"[6]

Still, LOS passed, but not before the committee tacked on LOS F as an "afterthought" meant to be a "catch-all" for anything that doesn't fall into the other levels. The committee then gave O. K. Normann a gold watch engraved with the words "Mr. Capacity."

He died a few months later.

The committee carried on and published its new *Highway Capacity Manual*[d] the next year. The book kicks things off with a nice tribute to Mr. Capacity and soon thereafter defines LOS: "Level of Service is a qualitative measure of the effect of a number of factors, which include speed and travel time, traffic interruptions, freedom to maneuver, safety, driving comfort and convenience, and operating costs."[7]

---

[b] Karl Moskowitch, longtime Caltrans traffic engineer. (I'm using first names here to convey what it often feels like in such meetings.)

[c] Charles Noble, chief traffic engineer for the New Jersey Turnpike.

[d] Now with six levels of service!

Safety is listed right there in the middle as one of the factors that "should be incorporated in a level-of-service evaluation." But for now, "there are insufficient data to determine either the values or the relative weights of certain of the six factors listed." In other words, the 1965 *Highway Capacity Manual* admits that level of service is a work in progress and that safety should be included in future versions.

Did we ever include safety? Nope. In subsequent editions of the *Highway Capacity Manual*, we ghosted safety. As Wayne Kittelson says in a 2000 paper, "Safety and operating costs were dropped from the definition in the absence of any historic or current attempts to include specific criteria related to those items."[8]

Take a look at any edition of the *Highway Capacity Manual* and you won't find the word *safety* in the index.

While we never added safety, we did take what was originally defined as a **qualitative** concept and shove it down a **quantitative** hole. And we did so with a stepwise function where, for example, reducing delay from 55 seconds to 54 seconds would shift the level of service from E to D. But if we take the same intersection and increase delay from 55 seconds to 80 seconds, the LOS remains at E with both.

For most traffic engineers, LOS is a black box. That haziness was intentional. Here are the words of the original LOS committee: "Although definitive values are assigned to these zone limits …, no explanation is given as to how these values were obtained. This is in no way intended as a criticism since … the function of any manual is essentially that of a handbook and … should not include a method-ical discussion of the facts."[9]

So "in lieu of more detailed information," traffic engineers seem content with this haziness while allowing LOS to dictate design—and safety.

When I first learned about LOS, my professor told me that it represents driver perception. What's funny is that LOS doesn't even do that well. As Kittelson says, "Although LOS is intended to be a quality measure describing the user's operational experience on a facility section, [*Highway Capacity Manual*] methodologies have not been based on extensive user surveys."[10]

Studies suggest that LOS accounts for about 35 percent of driver perception.[11] At the same time, we know that isn't consistent from place to place or even time to time. Drivers in Wahoo, Nebraska, have different expectations than drivers in Boston. And drivers in Boston circa 1965 have different expectations than drivers in Boston today.

The early days of the COVID-19 pandemic instantly "improved" the level of service for every street and intersection. Most traffic engineers assumed that road deaths would plummet as well. But—surprise, surprise—the number of people dying on our streets climbed and climbed sharply. The National Highway Traffic Safety Administration blamed it, at least in part, on "speeding on less-crowded" streets.[12] Related to this thinking is that LOS—one of our primary design considerations—apparently has nothing to do with road safety.

Well, that isn't quite right. The truth is that we never finished the job. We never bothered to figure out what LOS has to do with road safety.

You might think that the *Highway Capacity Manual* would stick to its original LOS designations without giving us a value judgment as to what is good or bad. But the *Highway Capacity Manual* isn't shy. Here is the 2022 edition, which doesn't hesitate to tell us what's what: "Quality of service describes how well a transportation facility or service operates from the traveler's perspective. Level of service (LOS) is a quantitative stratification of a performance measure or measures representing quality of service. The LOS concept facilitates the presentation of results through the use of a familiar A (best) to F (worst) scale."[13]

Take a good look at these three sentences. Given what we just talked about, how much of that is true?

But don't blame Mr. Capacity. We're just using his creation in ways he never intended. It's our own fault for putting safety on the back burner and never taking it off.

There are no engineering standards saying we need six or eight lanes instead two or four. These decisions derive from us leaning into LOS as a design consideration. Why don't we analyze the safety consequences of the design alternatives? Well, it's easier not to. To be honest, most traffic engineers wouldn't even know where to start. Besides, the manual will defend us.

76　Blind Faith in the Normal

—

Pile up all our 1,000-page traffic engineering books and the only one that isn't merely a "***guidebook***" is the *Manual on Uniform Traffic Control Devices* (MUTCD). Dating back to the 1930s, the MUTCD regulates our traffic signs, signals, and markings. But, as Ezra Hauer explains in a 2019 paper, it doesn't give us much to go on in terms of safety: "Just like the Green Book, the MUTCD gives no information about the safety consequences of the choices and decisions the engineer has to make."[1]

But without any safety information, how does the MUTCD help traffic engineers decide what to do? It uses what we call ***warrants***. Warrants represent thresholds for deciding, for instance, when we have enough pedestrians to "warrant" a crosswalk.

The obvious problem with this approach is the chicken-or-egg issue. We build a street that is too dangerous to cross and then tell you that it doesn't "warrant" a crosswalk because not enough pedestrians risked their life to do so in the hour that we happened to count them.

The less obvious problem is what's *not* included in the MUCTD. It does not include references to any research or study that explains where these warrants came from. So why do we need at least 93 pedestrians[a] crossing a dangerous arterial in a single hour to warrant a crosswalk? Don't worry about it. We just do.

Despite MUTCD status as "more than a guideline," the truth is that we don't actually *need* to hit these thresholds. We have the ability to use ***engineering judgment*** to supersede whatever the warrants tell us. Heck, we have no obligation to even say why because the MUTCD states that "documentation of engineering judgment is not required."[2]

But it's easier to tell the general public that the warrant has our hands tied. It's also easier to let you think these warrants are steeped

---

[a] And yes, the number for one warrant is actually 93.

in 100 years of science. If we told you the truth, these warrants would seem like the back-of-the-envelope rules of thumb that they are. We can't have that.

The MUTCD does give us other factors—above and beyond the warrants—to take into consideration. Safety is not at the top of the list. Here's Ezra Hauer again: "The MUTCD lists five factors to be considered by the engineer in choosing the appropriate traffic control. Surprisingly, the expected safety consequences of the choice is not one of them."[3]

So where could we turn for those "safety consequences"?

Your first stop might be our only 1,000-page book completely dedicated to safety: the American Association of State Highway and Transportation Officials (AASHTO) *Highway Safety Manual*. What's immediately different about this manual is that, unlike all other books that date to the 1930s, '40s, and '50s, this one didn't come out until 2010.

On one hand, that lack of historic baggage is a good thing. At the same time, the *Highway Safety Manual* doesn't want to overstep its role as the new kid on the block. There are no **minimums** or **desirables**. There are no **standards** or **warrants**. At no point does the manual even begin to tell us what to do. All it does is give us the mathematical relationship between the number of crashes and the number of cars for different categories of roadways. That's it.

Compared with some of the other manuals we've talked about, that sounds pretty good, right? Well, give me a minute.

For the past handful of years, I've been lucky to have one of the foremost *Highway Safety Manual* experts, Jake Kononov, speak in my safety class. Jake's research is cited multiple times throughout the *Highway Safety Manual*. He's a huge proponent of the manual but readily admits that it "is so complex that it's almost unusable."

Simultaneously, the *Highway Safety Manual* is also shockingly simplistic. **Average daily traffic**—for cars—is the only variable the first and current edition considers. It doesn't account for pedestrians or bicyclists. It doesn't account for whether there is a school nearby or a bar. It doesn't account for the number of cars on the road at the time when the crash took place versus that over an average day. Here is a 1967 *Traffic Quarterly* paper, in a precursor to the *Highway Safety Manual* methods, making that case: "The volume occurring during the hour the accident occurred would have a greater bearing on the accident than the average daily volume. In fact, the volume or density at the instant the accident occurred would seem to have the most bearing but, at present, this is not obtainable."[4]

That volume data is more readily available now, but we choose to keep things simple and focus on *average daily traffic*.

Add it up and all the *Highway Safety Manual* lets us do is compare the number of crashes on our street segment or intersection to the norm for street segments or intersections with a similar number of cars per day.

Sure, this is useful, but what does it mean? Does being safer than the norm equate to safety?

It used to be normal for 1 out of every 20 women to die during childbirth. A death of 1 in 30 is better, but I wouldn't say 1 in 30 dying is good.

The norm may also be changing for the worse. For decades, a kid being in the 50th percentile for height and weight would be considered in the healthy range. Over the years, we've had to update those percentiles to account for a rise in obesity. At some point, the 50th percentile might equate to obesity.

What am I am getting at? Well, normal isn't necessarily healthy, and normal isn't necessarily safe.

When my class asked Jake about the *Highway Safety Manual* considering safety as relative, he agreed with it being a problem, saying, "It is perfectly normal to have 10 percent head-on collisions on a two-lane road with 9,000 cars per day, but it isn't normal if you are in that crash."

It's also perfectly normal to add lanes to a road like this in response to an LOS analysis that predicts traffic 20 years into the future, but as Jake said, "As the number of lanes increases, so do the degrees of freedom for something to go wrong."

You might see better safety at first, but the *Highway Safety Manual* tells us that it'll worsen over time. Signalized intersections are perfectly normal as well. But, Jake continued, "if we take a signalized intersection and replace it with a roundabout, we'd expect a 70 percent reduction in injuries, even in a rural environment."

Yet the manual would never tell us to use a roundabout or not to add extra lanes. It is careful to say that it's only meant as an analysis tool. It doesn't supersede other manuals. It doesn't require or mandate anything. It doesn't give us a legal standard of care. Nor does it create a public duty for traffic engineers to change what is normal. The *High-way Safety Manual* has its flaws, but what it could do well—like compare what is normal to what could be—gets lost in the shuffle.

The *Highway Safety Manual* isn't our only newish guidebook. Groups like the National Association of City Transportation Officials (NACTO) and the Congress for the New Urbanism (CNU) have been putting out pretty good ones for much of this century.[b] But despite approvals from the Federal Highway Administration, these newer guidebooks are too easily ignored. Many traffic engineers haven't heard of them.[c] Others like to conveniently forget they exist. Many consider them to be too new to bother with. They haven't yet stood the test of time like our other books.

Are they too new? Maybe yes, maybe no. Either way, I think we can safely say that the old guard guidebooks aren't what we say they are. Maybe we can at least acknowledge that the path the old manuals led us down has *not* stood the test of time?

So where should we look for answers? How about starting with something that has stood the test of time?

Part of the point of the "charming streets" assignment I gave my class was just that. Many cities have streets or entire neighborhoods built before any of these guidebooks—or traffic engineers for that matter—ever existed. Many of these places continue to work well to this day. Dig into the road safety data and you'll probably find they outperform their well-engineered counterparts.

What they represent are empirical examples of good design. And we don't need a guidebook to learn from them.

In the early 1970s, Toronto hired Alan Littlewood to help redevelop an industrial area into a neighborhood. Alan had worked at the same firm with Jane Jacobs's husband, so he reached out to Jane.[d] Alan started by showing off all the traffic engineering studies he was about to commission. Jane was appalled, saying that "they aren't the right studies. What you want to do is just replicate the parts of the

---

[b] CNU published its *Designing Walkable Urban Thoroughfares: A Context Sensitive Approach* in collaboration with the Institute of Transportation Engineers in 2010. In 2013, NACTO released its first *Urban Street Design Guide*.

[c] Many states still don't require any sort of continuing education to continue being a licensed Professional Engineer.

[d] Jane and her husband moved to Toronto in 1968 to protect their sons from the Vietnam War.

city that you know work well—take their characteristics and copy them into St. Lawrence. Start by creating a regular street system in the area and go from there."[5]

Alan was flummoxed. That advice was the antithesis of how traffic engineers do things. But he gave it a go and "studied blocks in various neighborhoods that he thought worked well, and measured their width and depth."[6] Alan realized that Jane was right, saying, "We have to replicate what is the best of existing communities in Toronto. We don't have to look elsewhere. We have wonderful neighborhoods right here in the city."[7]

So that is what Alan did. He treated the best parts of the existing city as his guidebook. According to John Sewell,[e] one of the city councilors at the time:

> Alan's … plan showed how things worked on streets like Brunswick and Howland, Palmerston, and Collier. It showed that successful blocks were about 200-feet wide, and 300-feet wide where there were significant buildings. It showed the distances between buildings, where front and back yards should be, how all buildings should have doors facing towards the street—just like on existing streets in the city. It talked about how wide the paved area of streets should be. In short, there was nothing experimental about what was planned; Alan was suggesting that the plan for St. Lawrence should be more of the Toronto that was already built and had worked well for scores of years.[8]

Today's traffic engineers could do the same thing, and we could do so in a systematic manner where we create evidence-based approaches to safer streets based on existing streets. It's not our so-called **standards** that prevent us from doing so. The problem is more about **standard practice**.

We could develop guidelines that compel safer streets. Instead, it's standard practice to use the existing guidelines to shield ourselves and our work from scrutiny.

We could prioritize safety over level of service. But it's standard practice to minimize driver delay whenever and wherever possible.

---

[e] John Sewell later became the mayor of Toronto.

Take the George Washington Memorial Parkway. It's a four-lane scenic road that follows the Potomac River in northern Virginia. After a series of fatal crashes, a state delegate complained that "it is too narrow and curvy, has no shoulders, very few median areas, and too many cars speed on it. It is almost impossible for bikes and pedestrians to cross."[9]

No traffic engineering standards required four lanes. That was a choice. It could have been two lanes with sidewalks, protected bike lanes, and a grassed median. It could still be, but we may need to change standard practice to make that happen.

So yes, maybe it's time to throw out our old guidebooks and start fresh. But that isn't going to be enough to solve our problems, because the guidebooks aren't the only problem. We also need to transform the discipline of traffic engineering. Here's a 1962 *Traffic Quarterly* paper that gives us some insight as to where to start:

> The younger student is quite used to learning, but lacks experience, and will accept almost without question ideas which to the older person will appear startlingly novel or bizarre; on the other hand … the older engineer … often requires drastic re-appraisal of concepts which he has held for years…. Such a change of thought can be disturbing.[10]

It's high time we disturb the traffic engineering discipline.

Part 10

Safety Edumacation

———

77 An Empty Silo

—

In the postscript to his last book, James Reason—our *Human Error* spiritual guide—summarized a report into the problematic culture of a British hospital. The inquiry found several issues with the doctors and hospital administrators that also could be levied against traffic engineers. Well, that is if you replace the word *patients* with *safety* in this list:

> "A lack of openness for criticism"
> "A lack of consideration for patients"
> "Defensiveness"
> "Looking inwards, not outwards"
> "Secrecy"
> "Misplaced assumptions about the judgements and actions of others"
> "An acceptance of poor standards"
> "A failure to put the patient first"[1]

Sure, some of these criticisms might be a little unfair to traffic engineers. But a lot of them hit the mark; it would be hard to argue otherwise. Traffic engineers aren't trying to do a bad job, though. In most cases, we're just doing what we were taught.

So, what do we teach them?

One common refrain I hear is that we have a silo problem. In other words, traffic engineers are way too specialized and too deeply focused on the traffic engineering world.

But, to be honest, we don't actually teach them all that much about traffic engineering.

Let's say you graduate from high school and get into an undergraduate civil engineering program. How many transportation courses will you take in your first year? The answer is zero. How about your second year? The answer, again, is zero. If things go well, you will probably take your first transportation engineering course when you are a junior in college.

When you get to your senior year, you have more choices. Do you want a second transportation course that typically focuses on highways and designing their vertical and horizontal curves? If not, feel free to take design courses in structural, hydrology, or geotechnical engineering. Add in a capstone course or a full-year thesis course like I had at the University of Virginia[a] and you graduate with a civil engineering degree.

In total, you will probably graduate with one or two transportation engineering classes under your belt.

Some students will graduate with a civil engineering degree having taken *zero* transportation courses. Shocking as it may seem, accredited civil engineering programs do not need to offer *any* transportation-specific courses whatsoever. As long as the school offers courses in at least four other "technical areas appropriate to Civil Engineering," it meets the accrediting criteria.[2]

The Bureau of Labor Statistics says that fewer than 25 percent of practicing civil engineers have a graduate degree. In other words, more than three-fourths of civil engineers start off with only what they learned as an undergraduate. Combine that undergraduate education with whatever they learn from our guidebooks or on the job from other engineers, and that's the extent of it for most traffic engineers.

Not having any transportation courses in your background won't preclude you from taking the transportation Professional Engineering exam. As long as you've gained some work experience under a licensed Professional Engineer, you are good to give that test a shot.[b] Should you pass, you are a licensed Professional Engineer.

Is it possible to get good transportation engineers with this kind of education system? Sure, but you still have to ask, what the heck

---

[a] My undergraduate thesis was supposed to be about the trenchless tunneling methods used for the Big Dig in Boston. Looking at it now, I wrote more about the generational opportunity for a major city to reclaim dozens of acres of downtown land from urban freeways than I wrote about trenchless tunneling technology.

[b] You usually need a civil engineering degree and to have passed the Fundamentals of Engineering (FE) exam (a precursor exam the Professional Engineering exam). These days, 10 percent of the questions on the FE are related to transportation.

are we doing? How do we expect to produce exceptional transportation engineers—or even competent ones—if we aren't teaching them about transportation engineering, let alone road safety? Why should we be surprised that traffic engineers hide behind our guidebooks and the status quo when we haven't given them the basic wherewithal to understand their own silo?

The author of that British hospital report mentioned above said that fixing things will require "openness, transparency and candour throughout the system." His thinking was that "duty of candour" should be a legal requirement akin to "duty of care."[3]

Traffic engineering needs much of the same.[c] Still, we've put ourselves in a tough position. We're the supposed experts, but our undergraduate education system isn't set up to give us that expertise. We hope we'll pick it up on the job, but most firms aren't looking to educate their young engineers beyond getting tasks done. Even if these firms were, are we sure we want the "experienced" engineers teaching it using the same old guidebooks? Either way, there isn't much time for traffic engineers to sit back and figure out what's behind the curtain of why we do what we do.

And traffic engineers don't want others peeking behind our curtain because we don't quite know what's back there ourselves.

So it's easier to be overconfident—and defensive—of both ourselves and our guidelines. We scoff at questions about the quality of our work. Of course we've done our job and done it well. Any suggestions—let alone data—to the contrary are an attack on our personal and professional pride, as well as on 100 years of science.

When you combine such an undergraduate education with these sorts of self-justifications, it makes sense why traffic engineers are unaware of the limits of our expertise and unaware of what we've done wrong.

A week after the 1961 Bay of Pigs fiasco, President John F. Kennedy made a speech. In his address, Kennedy said his administration "intends to be candid about its errors; for as a wise man once said: 'An error does not become a mistake until you refuse to correct it.'"[4]

Traffic engineers are well past that point. Our errors have become mistakes, and we continue to perpetuate those mistakes and kill millions of people on our roads.

---

[c] And that is one reason I wanted to write this book.

In the early 2000s, Frank Gross and Paul Jovanis reached out to every transportation engineering school in the United States, multiple times, to see which ones offered road safety courses.[5] Their "scan of U.S.-based university courses in safety identified relatively few current offerings within engineering programs."[6] Out of the 117 universities that responded, only 29 had a safety course.

They dug deeper and found that 10 of the 29 safety courses are not offered on a regular basis. They then compared each of these courses against a list of 37 learning objectives based on five road safety "core competencies" that had been developed over a multi-year period by "educators, researchers, public officials, and leaders of industry and professional associations" of a task force on safety and workforce development. The core competencies represented what the committee saw as the baseline knowledge and skills needed by every road safety professional. And when they say *baseline*, they mean *basement*: "The competencies do not represent all safety knowledge that a safety professional should know; they represent the core that they must know."[7] In other words, the core competencies "are intended to define the floor of knowledge."

Whether or not we differ on the core competencies needed for a good road safety professional,[d] their findings are still depressing: "A survey of the curricula and course content of university engineering … suggests that they are not playing a major role in instilling safety-related knowledge and skills in the road safety workforce. Safety-related content is spotty in both undergraduate and graduate programs, and no program provides a comprehensive basis for competency in road safety management."[8]

What's even more depressing is what most so-called college safety courses mean by safety. Here is how a National Academies report describes it: "An in-depth review of course materials revealed that many current safety courses … utilize engineering texts (both design- and operations-oriented) [that] represent the content as 'safety-oriented.'"[9]

You want a safety course? We've got a safety course. But what are we actually teaching you? "Although many transportation courses

---

[d] In the words of the ancient secret society known as the Stonecutters, to which Homer Simpson becomes the Chosen One, we do. That is, we do indeed have different thoughts on the core competencies of a road safety professional.

indicated a strong presence of safety content, the majority only incorporate aspects of safety with primary content in design and operations."[10]

In other words, we teach the **guidebooks**—which have a dubious connection to safety—and assume that doing so equates to safety. Safety, after all, is just what happens with we focus on things like capacity, right?

Here are Gross and Jovanis again: "Some have argued that traffic safety is not a field of legitimate scientific inquiry, but is somehow derived from 'good' planning, design and traffic operations. Proponents of these positions argue that adherence to existing professional guidelines (e.g. AASHTO Green Book and MUTCD) adequately addresses safety."[11] Six of the so-called safety classes they reviewed didn't even have the word *safety* in the course title.

Most traffic engineers get roped into thinking that safety is implicit to our business-as-usual approach to design. As Ezra Hauer says in his 2005 paper, many believe that adhering to guidebooks "will automatically ensure that a proper amount of safety is built into roads" but "such a belief, while honestly and passionately held, is without foundation."[12]

Ezra goes on to discuss his own teaching: "For 27 years I taught traffic engineering, highway design, and transportation planning to budding civil engineers…. True, they may take a course or two about traffic and highway design. In the traffic course most of the time will be devoted to capacity and delay; in the highway design course to geometric design standards. The road safety consequences of their engineering design decisions will not be mentioned." And toward the end of his paper, he asks a good question: "How can researchers be trained in road safety and in road safety research methods if no university offers such a program?"[13]

In total, Gross and Jovanis found only four universities with a second safety course. Still, a Transportation Research Board Special Report counts 100,000 workers with road safety responsibilities, and "nearly 10,000 of them spend all or most of their workday managing road safety."[14]

Given how "few universities offer a systems-level road safety curriculum, and there are few places where public agencies can recruit trained safety professionals,"[15] where did all these road safety professionals all come from?

The same Special Report says that "the building of the road safety profession has been a largely ad hoc and unstructured process," so "most appear to have relied on on-the-job learning after migrating into the road safety workforce from other disciplines."[16]

In another part of the report, an interesting caveat is added to the end of a similar thought: "Core safety knowledge and skills are often obtained on the job, if they are obtained at all."[17]

If they are obtained at all?!? It's no wonder things haven't turned out so well.

78 Cultivating Engineering Judgment

—

To sit for the Professional Engineering exam, most states require that you demonstrate, as one website claims, a "maturation of engineering judgment." And once you get enough *engineering judgment*, you get to tell the guidebooks what to do (instead of vice versa).[a] For instance, the AASHTO Green Book concedes that it "is not intended to be a prescriptive design manual that supersedes engineering judgment."[1] Even the *Manual on Uniform Traffic Control Devices* (MUTCD)—which provides standards for traffic signs, signals, and markings (and might be the book with the least leeway)—still mentions engineering judgment 167 times.[b]

We don't teach traffic engineers enough about road safety in school. Still, the education of a traffic engineer isn't complete until they get their engineering judgment.

So what is engineering judgment, and how do I get some?

Ezra Hauer's 2019 paper on the subject cites a 1977 paper called "Advice to a Young Engineer."[2] In that paper, renowned civil engineering professor Ralph Peck makes the point that engineering judgment is vital to engineering but hard to define: "Almost all people in the practice of engineering would agree that successful practice requires a high degree of engineering judgment, but few would agree on the meaning of the word judgment itself."

Most of our guidebooks talk about the importance of "sound" engineering judgment being "necessary," yet few of those same guidebooks define it for us. The MUTCD is the exception. In the 2023 edition, the MUTCD—which is published by the Federal Highway Administration—describes engineering judgment as "the evaluation of available pertinent information including, but not limited to, the safety and operational efficiency of all road users, and the application of appropriate principles, provisions, and practices as contained in

---

[a] Suffice it to say that few traffic engineers would do such a thing.

[b] I was looking at the 2023 edition of the MUTCD.

this Manual and other sources, for the purpose of deciding upon the design, use, installation, or operation of a traffic control device."[3]

MUTCD historically focused on traffic control devices with this definition and didn't include the words "safety" or "all roads users" until the 2023 edition. Still, engineering judgment more generally relates to all the planning and design work we do. In other words, the idea is that we should use whatever data and resources we find appropriate in making educated, engineering-related decisions.

Doing so could open the door for anyone to cultivate engineering judgment, but the MUTCD is quick to remind us that only engineers—and those who engineers supervise—could possibly possess engineering judgment: "Engineering judgment shall be exercised by a professional engineer ... with appropriate traffic engineering expertise, or by an individual working under the supervision of such an engineer, through the application of procedures and criteria established by the engineer."[4]

So where might a young engineer acquire said engineering judgment? One obvious answer might be via the write-ups of past engineers and their engineering projects. Wouldn't it be nice to read through their engineering decision-making processes and what went into those judgment decisions?

Of course it would, but do traffic engineers routinely jot down where we found our "pertinent information" or "other sources" before using our engineering judgment? Nope, because the most surprising part of the MUTCD definition is the last sentence. The MUTCD is clear that the "documentation of engineering judgment is not required."[5]

So if not from learned experiences of other traffic engineers on previous projects, where does engineering judgment come from? Ezra Hauer cites a 2018 paper about cognitive biases complaining about a lack of a "precise definition" but saying that "engineering judgement is intended to mean the formation of opinions or decisions based on experience and knowledge gained as an engineer."[6]

Acquiring engineering judgment based on your own experiences gained as an actual engineer sounds good. But how is that possible—especially when talking about road safety—without a feedback loop of the road safety outcomes?

As is, our system is set up for traffic engineers to perform repetitive tasks and forge ahead without checking back to see how our designs—and the people who use them—fare.

That's not the engineering judgment we want. That's us perpetuating what we've always done with a never-ending cycle of older traffic engineers—with shockingly little road safety education—handing down the same old beliefs based on the same old guidebooks.

But what if traffic engineers established a real feedback loop where not only do we keep track of *what* happened, but we also ask *why* it happened? In asking why, we don't just blame human error and go back to business as usual. Instead, we continually ask what role our work played in *why* these road users did what they did. If it means going back to the drawing board—and questioning our underlying design and design philosophies—so be it.

How can we create such a system? Maybe traffic engineers just need a growth mind-set.

The difference between a *growth mind-set* and a *fixed mind-set* is something that my kids' elementary school talked about a lot. Alina Tugend's book on mistakes also hits on this difference.[7] Children (and traffic engineers) with a fixed mind-set "usually see mistakes and failures as beyond their control."[c] They "aren't willing to hear feedback that contradicts their own views, because they see it as a personal attack." On the other hand, children (and traffic engineers) with a growth mind-set "may not embrace mistakes, but they tend to see them more as part of the process of learning than as a reflection of their intelligence or abilities."[8]

As is, many states don't have a continuing education requirement for those with a Professional Engineering license. It doesn't matter if you got your license prior to the first Apple Macintosh computer, whatever engineering judgment you amassed along the way must be good enough.

How does that make any sense without a feedback loop?

In talking about growth mind-set in schoolchildren, Alina Tugend reminds us that "adversity can lead to growth." The world certainly endures plenty of adversity on our streets. Let's now use that adversity to cultivate some useful engineering judgment.

Take the growth mind-set page out of an elementary school playbook, and we just might end up with some engineering judgment worth having.

---

[c] Perhaps they even blame their mistakes on the errors of others…

79 Generalists Are Special
—

What do our engineering coursework and *engineering judgment* teach us about the impact of a freeway on the people in the flanking neighborhood? Or its impact on climate change and rising sea levels? Or economic resiliency to rising gas prices? Or its impact on inequities in road safety outcomes?

How in the world can a traffic engineer acquire that sort of bigger picture understanding?

Years ago, I saw Andrés Duany, one of the founders of the Congress for the New Urbanism, give a talk about why cities need generalists more than specialists.

Andrés started by showing a drawing from Sir Raymond Unwin's 1909 book, *Town Planning in Practice.*[1] It's an image of a beautiful street with incredibly detailed houses, trees, and stone walls, all hand-drawn in three dimensions. Lined up underneath the perspective drawing is an engineering plan view of the street with dimensions for street widths, footpaths, and grassy buffers with street trees. Unwin was an engineer. And a planner. And an architect. Andrés explains, "That is what planners were like in the 19th century. They were all generalists. You cannot tell that Raymond Unwin was an architect because he was just as good with plants and with Civil Engineering."[2]

A 1996 *Transportation Quarterly* paper makes a similar point. Once upon a time, these disciplines were one in the same: "Civil engineers and architects have the same roots. About 200 years ago, there was no difference between both in many countries. Then a split happened into different professions. The architect became more and more of an artist, the engineer more and more a technician, the latter based on mathematics."[3]

Or, as Andrés goes on to say, "In the post-war period, people became specialized. The traffic engineers only know how to make roads.... The architects only know about architecture ... and the landscape architects are doing their own thing ... all quite independent from one another."[4]

What ends up happening when everyone is a specialist? Here is Andrés again: "The traffic engineers are putting the roads through quite independent of whether there are houses there, trees there, people there, or dogs there. They just know the traffic."[5]

He doesn't mince words: "What is destroying our cities is that everybody is a specialist."[6]

Traffic engineers have long acknowledged such specialization to be a problem. Here is quote from a 1963 special committee charged with reviewing the purpose and scope of the Institute of Traffic Engineers (ITE): "The traffic engineer can no longer be content with the geometries of design or the mechanics and hardware of traffic operations. He must look beyond the technical requirements of his work and give consideration to patterns of land use trends, to the desires and needs of people for transportation as well as the effects of all other means of transportation on street and highway traffic."[7]

Nevertheless, this same committee decided against flipping the Traffic for Transportation in the ITE name.[a]

The solution, as described in a 1980 *Traffic Quarterly* paper, seemed simple: "It was necessary to bring other disciplines, such as geography, mathematics, and sociology, in to deal with the factors of urban growth to which transportation facilities must be related."[8]

One problem is that we give final say to the most mathematical and seemingly technical of all the disciplines. In most cases, that is traffic engineering. And that's why traffic engineers need their specialization but also need to be a generalist.

How can we cultivate something like that?

When I consider my most successful graduate students, I'm struck by how many of them never attended a civil engineering undergraduate program. Both Rachael Bronson and Molly North showed up with undergraduate biology degrees, one with a minor in French literature and the other with French as a double major. Both were passionate about transportation. Both turned out to be great graduate students and great transportation professionals.

What's the problem? Why not recruit more Francophile biologists or other such students with atypical backgrounds? The answer: it's not that easy.

---

[a] That change didn't come until 1975.

To be eligible for most transportation engineering graduate programs, we kind of want you to have an undergraduate degree in civil engineering. Or something very close to that.

How did we make it happen then?

In the mid-1990s, well before I got to Colorado, the civil engineering department at the University of Colorado Denver wanted to offer a master's degree in geographic information systems. Most prospective students did not have engineering degrees. So instead of offering a master's of science degree, the university created the master's of engineering degree. What's the difference? A far different list of prerequisites. You didn't take three semesters of calculus? No problem. One will be enough, and it's unlikely you'll need much calculus anyway. Eventually, the university started offering a similar degree option—with fewer prerequisites—in the other civil engineering subdisciplines, including transportation.

As long as Rachael and Molly had a decent math and science background, they only needed to clear a few more prerequisite hurdles. How different were their paths than my own? I got a civil engineering undergraduate degree, but when I got to graduate school, my transcript only had one more transportation engineering class than Rachael or Molly.

I'm not saying the traditional path never works. But which path is more likely to lead to traffic engineers who can design a great street, neighborhood, or city while simultaneously considering the neighbors, the environment, and road safety inequities?

Most transportation engineering graduate programs still play gatekeeper, only letting in those with enough civil engineering credentials in their backgrounds. But we need more students with the chance to be a generalist as well as a specialist.

An out-of-place 1975 *Traffic Quarterly* paper touts the benefits of "crossing of disciplinary lines" because even though transportation systems are but "one subsystem in the urban environment," they "so obviously affect virtually every other facet of urban life."[9] To do so requires us to "see the transportation system as extending far beyond the narrow objectives of moving people and goods" because "the costs of transportation are reflected far beyond the transit fare box, traffic congestion, right-of-way acquisitions, and capital outlays; and that the benefits of transportation may be realized in the diversity and pleasantness of neighborhoods and landscapes, in the impact on political representation, and in the diminution of current social problems such as crime and racism."[10]

Take a hard look at the traffic engineering discipline. How exactly do current practices prepare students for these complexities? Toss in problems such as road safety inequities and I see a discipline desperately in need of including students from more diverse educational backgrounds. To have a chance at fixing these problems, we will also need more diversity in gender, sexual orientation, race, ethnicity, socioeconomics, and personality.

It feels like a messy recipe; I get it. Do you want to be a road safety professional? Instead of statics, fluid mechanics, and three semesters of calculus, how about classes in public health, behavioral science, sociology, human ecology, public policy, and statistics? Add some transportation engineering and road safety classes to the mix, and wouldn't that make for a much better—and less messy—foundation?

Or maybe we just need more messy recipes.

80 Transportation Is Made of People

___

Not far from where I work in Denver, 23-year-old Rogel Aguilera-Mederos was driving a semitrailer truck carrying lumber along Interstate 70 on a sunny April afternoon in 2019. He was driving too fast, his brakes failed, and he panicked.

In part due to the steep elevation changes, the I-70 corridor has more runaway trucks than any other interstate in the United States.[a] The runaway truck ramps are used so often that the Colorado Department of Transportation (CDOT) added variable message signs to let truck drivers know when the ramps are full.

But Aguilera-Mederos was rattled and missed the runaway truck ramp.[b] The sunny skies turned black with a massive plume of smoke after Aguilera-Mederos slammed into 12 cars and three trucks. He killed four people.[c]

When a big truck hits, well, anything, it's likely to be deadly. A 1983 *Transportation Quarterly* paper tells us that a crash between a large truck and a car has a fatality rate "as much as 80 times higher" than in a car-car crash.[1]

It's not like deaths are equally distributed either. Given all crashes that involve a large truck, the people in the large truck represent about 3 percent of deaths. Per usual, the driver of the truck, in this case Aguilera-Mederos, walked away from his fiery crash with minor injuries.[d]

In many calculations, traffic engineers like to simplify things. When accounting for large trucks, we use what we call ***passenger***

___

[a] At least that is what somebody from CDOT told me.

[b] CDOT subsequently added a second runaway truck ramp near the crash site.

[c] Miguel Lamas Arrellano, Doyle Harrison, William Bailey, and Stanley Politano tragically died that day.

[d] He didn't really walk away though. He was convicted and sentenced to 110 years in jail. After some public outcry, Colorado's governor Jared Polis reduced the sentence to 10 years. Aguilera-Mederos will be eligible for parole in five. During his vehicular homicide trial, Aguilera-Mederos cried on the witness stand, asking, "I ask God

car equivalents, or PCEs. The idea is to find "the number of passenger cars that will result in the same operational conditions as a
single heavy vehicle."[2] So, despite the abundant kinetic energy—and
safety risk—of heavy trucks, we typically assume one large truck to
be equivalent to two passenger cars. Or if the terrain is rolling like
the I-70 corridor west of Denver, one large truck would be three
passenger cars. As you might've guessed, PCEs focus on capacity,
not safety.[e]

But what about crash severity? What would the safety PCE of a
large truck be?

We could determine that. It sure seems like it would be worth
knowing and analyzing. But we don't. Safety PCEs only exist when
talking about contaminated drinking water. We might try to limit
large trucks due to noise pollution or pavement wear and tear,
but safety is rarely a good enough reason.[f] It's not even a good
enough reason to derive truck safety PCEs in the first place. Why?
Because nobody is asking us to measure the difference in safety.
All traffic engineers need to calculate are metrics like *level of service*. So we teach traffic engineers where to find the table that tells
them how many passenger cars equate to a large truck in terms of
capacity. We don't teach them how to determine this measurement
for safety.

This chapter isn't really about trucks or PCEs. It's also not about
runaway truck ramps or so-called human error. To have a chance
at fixing our road safety problems, the traffic engineering discipline
needs to start by being honest with itself. We need to reexamine
everything we do and admit to ourselves that, when it comes to road
safety, our system to train traffic engineers is lacking, oversimplified,
and often focused on the wrong things.

If you've read this far, you have a decent idea what we teach traffic engineers about road safety—and what we don't. As Ezra Hauer
says in his 2005 paper, "Civil engineers graduate from a 4 year

---

too many times why them and not me. Why did I survive that accident?" "Judge
Sentences Rogel Aguilera-Mederos to 110 Years in Prison for Fiery I-70 Crash,"
CBS News, December 13, 2021.

[e] PCE numbers also used to be a lot higher in older versions of the *Highway Capacity
Manual*.

[f] Some facilities do separate trucks from cars, such as what the New Jersey Turnpike
started doing in the late 1960s.

program without being taught about the link between the design decisions they will make and the crash frequency and severity that will follow."[3]

You'd think that graduate school would do the trick. Yes, graduate school should help, but it's not just an undergraduate problem. The Transportation Research Board Special Report about road safety education makes that clear: "Safety-related content is spotty in both undergraduate and graduate programs, and no program provides a comprehensive basis for competency in road safety management."[4]

Where do we go from here?

That same report gives us a nice, generic answer as to what we need: "There is little doubt that the scale and complexity of the road safety problem will require an increasingly rigorous and systematic approach to safety management that is carried out by a highly skilled, analytical, and multidisciplinary safety workforce."[5]

So our future road safety professionals need to be "skilled, analytical, and multidisciplinary" so that they can do "rigorous" and "systematic" road safety work. Sounds good, but it's hard to say what that really means.

Here's an example.

After COVID-19 hit, the University of Illinois enlisted two physics professors to model the various back-to-school scenarios the school was considering.[6] Given a scenario with required masks, mostly online classes, testing twice a week, contact tracing, and an exposure notification app, the professors modeled a worst-case scenario of 500 total COVID cases per semester.[7] The school tried it and ended up with thousands of cases, despite adding a lockdown along the way.

I followed this story more closely than I should have because of an article about these physics professors bragging that "for them, transitioning to epidemiology was easy."[8] It was so easy that this work "just doesn't provide the same intellectual thrill" as physics. Where did they go wrong? As their chancellor later said, the physicists failed "to focus on human dynamics and behavior."[9]

Things also tend to go badly when traffic engineers don't focus on human behavior.

That's one reason I left the structural side of civil engineering for transportation. With structural engineering, I felt like I could plug numbers into a computer and get an answer. Once I realized the computer didn't really need me, I knew structural engineering didn't really need me either.

In the outer-space book *Project Hail Mary* by Andy Weir, the main character—an unintentional astronaut named Ryland Grace—says, "Got to love computers. They do all the thinking for you so you don't have to."

Later in the book, Ryland commends the intelligence of his buddy Rocky. Rocky isn't as impressed with himself: "Math is not thinking. Memory is not thinking. Memory is storage. Thinking is thinking."[10]

Transportation is never that easy. Transportation requires thinking. Everything we do has a human element that can throw a monkey wrench into whatever our simplistic equations tell us. That's why transportation needed me ... and that's why transportation needs you.[g]

Together, we need to be able to evaluate the safety consequences of our design alternatives. As Ezra Hauer says, "Only scientific research can do so. This is well accepted in medicine, education, and most similar fields."[11]

He wasn't the first. Here is Frank McGlade in a 1962 *Traffic Quarterly* paper, complaining that a lack of personnel is the crux of the problem, with varied aspects:

> 1. There is a dearth of skilled researchers with special preparation in safety education and accident prevention techniques;
> 2. There is a lack of safety education and accident prevention specialists, and the few specialists in this area are inadequately prepared in related fields of knowledge; and
> 3. The multivarious nature of the traffic accident problem (the interaction of man, vehicle, and environment, with a host of intervening variables) necessitates the involvement of many scientific disciplines, if traffic accident research is to be productive.[12]

Ezra follows up his call to action with one of my favorite monologues in an academic paper: "And yet, for some unfathomable reason there exists a widespread ... notion that common sense and an engineering degree are sufficient to do road safety research. Civil

---

[g] Drop me a line if you want to come learn with me.

engineers typically do not receive any training in the kind of research method needed in road safety, nor do they graduate with much factual knowledge about road safety. Furthermore, nothing in routine engineering practice helps to relieve them of this innocence."[13]

We also need more of a reason to do that. You would think millions of people dying our streets would be enough. Apparently not.

At the beginning of this chapter, I mentioned how we don't have truck safety PCEs because nobody asks for them. The same goes for the crash frequency and severity impacts of different design alternatives. Nobody requires that information. Nobody asks for it. Here is Ezra Hauer again: "Today one can devise a long term transportation plan for a region, one can get approval for [a] road network in a new subdivision, one can implement a traffic signal coordination and timing plan for a metropolis, one can design a new highway, and in all this, never consider the future crash frequency and severity differences between options and alternatives."[14]

It's not that we couldn't. We could. We just don't need to. As is, most municipalities require a **traffic impact analysis** to make sure that our **level of service** is going to be OK. If it's not OK, we need to mitigate the dreaded traffic congestion.

How many municipalities require a **road safety impact analysis** that would do the same thing but for safety?

Zero.

And because nobody requires it, nobody teaches it. We need you to get out there and create a demand for this.

A 1962 *Traffic Quarterly* paper talks about the study of road safety as the study of human behavior. If we want safer streets, it "then behooves us to study the many facets of human behavior." And what is human behavior? Well, this paper also tells us that "unfortunately, human behavior is not easily defined or studied in its many contexts."[15]

A 1971 *Traffic Quarterly* paper says that most "engineers work routinely with mechanical and other physical systems with errors of about 1 percent," but traffic engineers deal "with humans" and "are fortunate if they can even roughly predict what will happen."[16] In other words, "understanding events, never mind predicting them, is difficult."

Got it. So, what else do we need from you?

Get out there and experience our streets. Personally, I like doing so on a bike. David Byrne of Talking Heads fame does a nice job of explaining why in his book, *Bicycle Diaries*:[h]

> "On a bike, being slightly above pedestrian and car eye level, one gets a perfect view of the goings-on in one's own town."
>
> "This point of view—faster than a walk, slower than a train, often slightly higher than a person—became my panoramic window over much of the world over the last thirty years—and it still is."
>
> "I felt more connected to the life on the streets than I would have inside a car or in some form of public transport."[17]

By experiencing our streets and intersections on my bike, or even by foot, I get a better understanding of how the transportation system works—or not—and how it integrates into a neighborhood—or not—than I do from behind a windshield. The same goes for public transit. You've got to ride it to understand it. And it's not just about riding it. You need to test how easy—or hard—it is to access transit in the first place and reach your destination after. You need to see how hard it is to pay fares … and travel in bad weather … and how it feels when your bus doesn't show up … and how it feels when your bus doesn't show up in bad weather. You also need to watch and talk to other people about their experiences.

All these experiences relate to understanding people and understanding their behaviors and choices in different contexts. And it all relates to road safety.

So does empathy. Most traffic engineers would make our streets a little safer if only they had a little more empathy. In a 2020 *Freakonomics* podcast episode, the director of the Empathy and Relational Science Program at Massachusetts General Hospital, Helen Riess, mentions "a company that made this wrist device that helped you experience what it was like to have Parkinsonism. And when I tried it and I couldn't even hold a pen, I realized I had no idea how hard it would even be to write anything or zip up your jacket. And it instantly gave me more empathy for people who can't control their movements."[18]

---

[h] These three different quotes collected from various parts of his book sure seem like they belong together.

I occasionally assign class projects along these lines. Instead of a Parkinson's wrist device, students cross a street using crutches, using a wheelchair, or blindfolded. It quickly transforms their perspective as to what a safe design looks and feels like.

We need to do that for road safety because, to paraphrase Charlton Heston in the 1973 movie *Soylent Green*,

"It's people. Transportation is made out of people.

Listen to me…. You've got to tell them.

Transportation is people!"

And to also paraphrase Phil Hartman in the lesser acclaimed, but still awesome, "Soylent Green II" sketch from *Saturday Night Live*, "Transportation is still made out of people.

They didn't change the recipe like they said they were going to.

It's still people!"

Part 11

Spark Joy

---

81 I Declare Vision Zero!

The Japanese home organizing guru Marie Kondo helps people tidy up. She encourages putting our homes in order by focusing on what sparks joy. The big idea is to end up surrounded by things we love.

As is, we have too many streets that spark the opposite of joy. Like that sweater you wore to your second-grade picture day, our streets and cities are hand-me-downs from a bygone era. I'm not exactly Marie Kondo, but I do want to tidy up our streets. If we are surrounded by streets we love, wouldn't we be safer?

But how? How should we go about making that happen?

Many really good transportation safety people would say that we need to declare **Vision Zero**. How did we get to this point in a book about road safety without talking about Vision Zero? It's because of what I see as a massive gap between Vision Zero in theory and Vision Zero in practice.

In theory, Vision Zero isn't a specific safety intervention as much as it is an ethical shift toward the idea that fatalities and severe injuries are not acceptable on our streets. To accomplish this vision of zero deaths, the thinking is that we need to approach road safety—and street design—in a fundamentally different way.

I like the sound of that. So do the dozens of municipalities that have declared Vision Zero—and a goal of zero traffic-related deaths and serious injuries by some future year—since it debuted in Sweden in 1997. In practice, though, Vision Zero is like when Michael Scott declares bankruptcy on *The Office*. After initially confusing bankruptcy with the witness protection program,[a] Michael walks over to the reception desk and shouts, "I DECLARE BANKRUPTCY!!!"

Once Michael goes back to his desk, Oscar peeks in and has this subdued exchange:

---

[a] "I've always wanted to be in the witness protection program. Fresh start. No debts, no baggage. I've already got my name picked out. Lord Rupert Everton. I'm a shipping merchant who raises fancy dogs. That's the life."—Michael Scott (Steve Carell)

OSCAR MARTINEZ: Hey, I just wanted you to know that you can't just say the word bankruptcy and expect anything to happen.

MICHAEL SCOTT: I didn't say it; I declared it.

OSCAR MARTINEZ: Still, that's not anything.

Dig into what Vision Zero should be, and I'm on board. Dig into how most agencies approach Vision Zero, and it's traffic engineering business as usual under a new slogan. The US Department of Transportation and the American Association of State Highway and Transportation Officials even adopted their more pragmatically titled *Towards Zero Deaths* slogan as a national strategy in 2011. During the ensuing so-called *Decade of Action*, many state Departments of Transportation submitted target safety goals for walking/biking fatalities that were *higher* than their actual number of pedestrian/bicyclist deaths from the previous year.

Yup.

I once saw a Vision Zero presentation about the efforts of Boulder, Colorado, and the speaker talked about how they had "joined leading-edge cities across the country and the world that are working toward this goal" of Vision Zero. Their big change was adding *evaluation* as the fourth E to the classic three E's of *engineering*, *education*, and *enforcement*. Most Vision Zero cities do something similar and maybe release a report or some pretty infographics. Some cities even put together interactive web-based dashboards of crash data.

The evaluation E told Boulder it had a human error problem. Boulder's presenter went on to highlight its new shared responsibility tagline of "it's up to us to make our streets safe" and gave several examples focused on the education E.

Severe injuries and fatalities have *not* improved in Boulder since it declared Vision Zero. Other cities have gotten worse. In Portland, Oregon, "the region failed to meet almost all of its goals…. In fact, pedestrian and motor vehicle deaths and injuries have steadily increased since Vision Zero was implemented."[1]

The same thing can be said for the United States as a whole.

I'm not against Vision Zero, but, as Sidney Dekker says in his field guide to human error, "that vision is exactly that: a vision. It is an ethical commitment to do no harm. As such, it is laudable. But it is not a research result, it is not a scientific prediction. There is no accident theory underpinning it or supporting it."[2]

Vision Zero also seems to encourage people to focus more on the outputs—fatalities and severe injuries—rather than inputs that traffic engineers have more direct control over. Even if safety improves, injury reductions "might feed the illusion that risk is under control"[3] and divert our attention away from what controls the underlying source of risk. Dekker also says that "the risk of having an accident is not dependent on anybody's commitment to zero. The risk of having an accident is a fixed, structural property of the complexity of the systems we choose to build and operate. If we produce really complex things … then we should expect those to produce interactions and couplings that are really hard for us to understand."[4]

If we truly saw Vision Zero as the moral underpinning of transportation design, basing our approach on scientific evidence and physics, we'd design spongey cars that can't go more than 20 mph. Or streets that make doing so exceedingly difficult, particularly in urban areas. Or maybe ban cars from near downtowns or schools altogether. Such solutions aren't on the horizon. And since we don't seem willing to make such changes, it's hard to know whether Vision Zero works or not. Given the lack of progress, I'm suggesting we try something different.

The *Safe Systems* approach gets intertwined with Vision Zero and would be a big step in the right direction. But like Vision Zero, it's too easily co-opted by generic, uncontroversial messaging that lets us continue business as usual. Saying we need safer road users, vehicles, speeds, and streets is all well and good … but how?

It's clear we need to revamp the traffic engineering discipline along with particulars like *design speed* and *level of service*. First, though, we need to start with a mental reset. Instead of focusing on human error and what every road user did wrong or could do better, we need to understand their behaviors as part of the overall system. I'm 100 percent not talking about letting people get away with murder on the streets. What I'm saying is that traffic engineers need to concentrate more on the role of the transportation system as the underlying context for such behaviors. If we flip from our human error-based model to a systems model, we won't end up with interventions focused on *education.*

This change is harder than it looks. Boulder sifted through all its crash data and found human errors all over the place. And when human error looks like the obvious problem, we obviously want to remind people to be safe. But, as Dekker makes clear, this approach rarely works: "Telling people not to have accidents, or getting them to commit to not having an accident, is not very promising."[5]

That's not news. A 1964 *Traffic Quarterly* road safety paper complains about such "sporadic campaigns" focused on "short-cut panaceas" because "experience has shown that attempts at overnight cures of the traffic accident problem produce no lasting results and often do more harm than good."[6]

If we need to constantly remind people in the transportation system to be safe, is it really a safe transportation system? And how can we uncover the systemic problems foiling safety if we blame the humans involved and take few actions beyond education and enforcement?

We talked about asking *why* and then asking *why* again. And then doing so a third, fourth, and fifth time. In the words of James Reason, "When an adverse event occurs, the important issue is not who blundered, but how and why the defenses failed."[7]

I'd take it a step further than focusing on our failed defenses. We also need to untangle the role that the transportation system played in the blundering in the first place. In most cases, the blunder is a symptom. The blunder isn't the root cause. In fact, blunder isn't even the right word. Most blunders epitomize how our system typically functions. They represent a completely normal by-product of that system. Take speeding, for instance. Does it make sense to label speeding a blunder if nearly everyone is doing it, on nearly every street and highway, at nearly every chance they get? Wouldn't it make more sense to design streets and highways that help make speeding the exception as opposed to the norm?

Instead, most Vision Zero cities that step beyond education and enforcement do so by searching for crash hot spots and easy-to-do countermeasures such as leading pedestrian intervals.[b] The result is like a never-ending game of Whac-A-Mole. You rarely get much real

---

[b] A leading pedestrian interval—or LPI—gives pedestrians a short head start when crossing an intersection. We can't be bothered to give pedestrians the time needed to get all the way across the street, so we instead wait until they get to the middle of the street before telling drivers they can proceed. In other words, an LPI is a half-assed way to prioritize pedestrian safety. Let's use our whole ass.

progress, but you do get an illusion of progress because you feel like you are doing something.

Here's where Marie Kondo comes in. She used to tidy one room at a time but said that was a "fatal mistake" because "when we tidy each place separately, we fail to see that we're repeating the same work in many locations and become locked in a vicious cycle." Think about that in terms of the traffic engineer who plays the crash hotspot game. That Whac-A-Mole approach never lets us grasp the sheer scale of the safety problem.

How does Marie Kondo do it? Instead of cleaning one room at a time, she focuses on one category at a time. Instead of starting in the bedroom and moving on to the living room and then the kitchen, she first puts all her energy into a single category: clothes. Congregating *all* the clothes together is the only way to understand the scale of one's clothing problem and begin to tackle it.

What if we tried something similar with road safety? Instead of going from hot spot to hot spot, let's focus on a single category. Instead of clothes, maybe start with kids. In other words, what if we started by putting all our energy into making our city safe for kids? Where do kids live? Where do kids want or need to go? How can we connect these places with a network of infrastructure that is safe for kids to use, maybe even independently? Or maybe your city is short on kids. If so, start with older folks, persons with disabilities, or those living in disadvantaged neighborhoods. Or maybe your city has a lot of kids. If so, break them into subcategories and focus on elementary school kids first before moving on to middle and high school kids. Whatever the case, wouldn't making your city safe for kids, older folks, persons with disabilities, or those living in disadvantaged neighborhoods help give us a more definitive place to start toward the goal of helping make the transportation system safer for everyone?

Once Marie Kondo puts the clothes in order, she moves to her second category: books. And then papers. Then come komono (miscellaneous items) and mementos (items with sentimental value). The thinking is simple: hit the easier categories before moving on to the more difficult ones. As Marie Kondo says, "By starting with the easy things first and leaving the hardest for last, you can gradually hone your decision-making skills, so that by the end, it seems simple."[8]

Why not do something similar with transportation safety? Start with kids. Or older folks. Or persons with disabilities. Or those living in disadvantaged neighborhoods. Then you could focus on Active Transportation, which could lead into more difficult categories like transit or parking. At this point, maybe it's time for the sort of difficult category—like land use—that everyone underestimates in terms of its impact on road safety.

My goal here is to change how we think about transportation safety and give us a fresh place to start. Doing so will mean transforming the traffic engineering mentality, traffic engineering practice, and in turn, transportation system design.

Follow current best practice and you'll end up focusing on problems like speeding, drunk driving, and distracted driving. Such categories will likely lead you down a path toward **education** and **enforcement**. That's not what I'm looking for. We've got enough of that already. What we need is a simpler, more fundamental place to start. The rest of this part of the book will dig into some examples of my fundamental categories. Keep in mind that these categories are a tool to help us head in the right direction. They are not the final destination. But, as Marie Kondo says, "before you start, visualize your destination."[9] The final destination is a vision for your city or town.

To paraphrase Marie Kondo, what places spark joy? What streets spark joy?

Listen to what places and streets spark joy for a wide variety of people, particularly those who live, work, and play in and around the area you are working on. Then, try to figure out what goes into building such places. It will mean going there. Use the place. Watch others do the same. Good science is good observation. What do these places have? What don't they have? How wide are they? What are the sidewalks like? Are there street trees? Think carefully about the role these things play in how people use the streets. Look at the crash data. Does it match what you observe?

Then visualize what your place or street could be. Look for existing examples and measure them. It might mean getting rid of the things you don't need—like extra car lanes or parking—and bringing in more of what you do—like bigger and better sidewalks or bike lanes—to achieve the vision. As Marie Kondo says, "The best way to find out what we really need is to get rid of what we don't."[10]

This won't happen overnight, but keep at it. Slowly but surely, your city will transform.

Look around at the safest cities and countries in the world. Remember that few of them started out the way they are now. Otherwise, you will end up saying, "We can't do that. We're not [insert name of safe city or country here]."

Give it a chance, a real chance, and you will end up with a city where driving fast on a residential street doesn't make any sense.

People will be safer, and if all goes well, they'll be happier, too.

You might remember Chef Gusteau's mantra from *Ratatouille*, my favorite Pixar movie,[c] who says, "Anyone can cook."

The same goes for making our streets and communities safer: you don't have to be an engineer or a planner to make a difference. You just have to be willing to think differently. It also helps to have a little insight into how to effectively argue with a traffic engineer.

Or if you'd prefer to walk out into the street and shout "I DECLARE VISION ZERO!!!"—good luck with that.

---

[c] Much to the annoyance of my kids, for some reason.

82 Department of (Child) Transportation Services

___

Old Enough! is a Japanese reality show on Netflix that follows the adventures of a little kid out in the world running their first errand.[a] When I say little kid, I mean little. We're talking ages 2 through 5 or 6. When I say errand, we're often talking about toddlers crossing major streets by themselves.[b]

To be honest, it's disconcerting to watch these tiny kids take on traffic with a kooky narrator and a laugh track.

Take the fourth Netflix episode of *Old Enough!*, for example. Yuka walks to the Uonotana Fish Market in Akashi, Japan. She has to cross a five-lane arterial to get there. On the way out the door, the mom asks, "How do you use a crosswalk, Yuka? Can you go all the way to Uonotana? Without getting hit by any cars?"

Yuka is 3 years old.

Hiroki is only 2 years, 9 months in the premiere Netflix episode. He also has to cross a busy arterial. Lucky for him, his mom made him a little yellow sign that says "STOP" in Japanese. He holds up his sign, and the narrator says, "It sure does the trick! Just like magic, the cars all stopped. That's the power of Mom's love."

Mom's love? That's nice, but that's not why Hiroki made it home safely. Nor is a lack of mom's love the reason an American parent might be arrested for letting a much older kid make a similar trip.

Take Nicole Gainey. The police charged her with felony child neglect for letting her 7-year-old son Dominic walk a half mile from their house to a playground in Port St. Lucie, Florida. According to CNN, "The arresting officer said that Gainey had failed to provide

___

[a] The original show, *Hajimete No Otsukai*, is based on a 1977 picture book of the same name. Apparently, approximately 20 percent of Japanese viewers watch every 2- or 3-hour episode and have been for years. Recent episodes feature children of alumni from old episodes. The title roughly translates to *First Time Errand* or *My First Errand*, but Netflix went with *Old Enough!* and also cut the multihour episodes down to 10 or 20 minutes.

[b] Albeit with a camera crew and some producers nearby.

her son with care and supervision by allowing him to cross the street."[1] The officer was worried about all the "recent criminal activity in and around the park." But instead of focusing on those criminals, Gainey was "interrogated, arrested, and handcuffed in front of her son" before being "transported to the local jail where she was physically searched, fingerprinted, photographed and held for seven hours and then forced to pay almost $4,000 in bond."[2]

Lori Pierce from Columbus, Mississippi, let her 10-year-old son William walk a half mile to soccer practice at their local elementary school.[3] Pierce wasn't arrested, but the police reprimanded her and told her she would've been charged with child endangerment had anything bad happened.

In Saratoga Springs, New York, Janette Kaddo Marino had to deal with the police for letting her son Adam bike to middle school. At the time, his middle school had a policy forbidding kids from walking or biking to school. In fact, the same policy applied to all Saratoga Springs elementary school students.

Tennessee police threatened to arrest Teresa Tryon for letting her 10-year-old daughter Cynthia bike a mile to school.[4] According to the police report, the officer "observed that vehicles had to slow and negotiate around the bicyclist." Heaven forbid. The officer then goes on to say that "in my opinion, this section of roadway is not a safe place for a child of her age to be riding." Even though Cynthia was on a bike, the sergeant was "especially" concerned about "the lack of sidewalks in the area." The local police chief then admitted that there is "no sensible alternative route or even a safe way to cross that intersection."[5] The police reported Gainey to Child Protective Services.

Cases like these also play out in the court of public opinion. Conversations devolve into an argument about who's right and who's wrong. Some blame the police and schools for overstepping. Others blame the parents for putting their kids in seemingly dangerous situations.

On one hand, I agree with these parents for wanting their kids to gain some confidence and independence while getting some exercise. On the other hand, it's hard to look at these streets and intersections and think anyone—let alone kids—should be walking or biking.

So who do I blame? I blame traffic engineers for putting us in this situation in the first place.

It's no big surprise that only 10 percent of American kids walk to school today.[c] In Japan, it's more like 80 percent.[6] Japanese kids also tend to do so without a parent or guardian along for the trip.[7] Data suggests that only 15 percent of weekday trips for Japanese kids aged 10 and 11 include a parent. For American kids that age, 65 percent would include a parent.[8]

With all these Japanese kids walking around without an adult, they must be dying on the streets in droves, right? Each year, about 20 elementary/early middle school–aged kids die on Japanese streets. That is 20 too many, but in the United States, we see about 600 such deaths each year.

The US population is about 2.6 times that of Japan, yet we kill 30 times more kids on our streets than Japan does, and we do so while chauffeuring most of them around everywhere they go.

The 2- and 3-year-olds we see walking the streets in *Old Enough*? Even in Japan, that's bordering on ridiculous. But elementary and middle school kids getting around by themselves? Japan wouldn't think twice.

Why so different?

Part of it is culture. Japan expects kids to walk and bike on their own at a young age. Culture, expectations—and mom's love—unfortunately aren't enough. But combined, they embolden Japan's traffic engineers and planners to build slow, narrow streets where kids walking and biking is a viable option. Better yet, they do so proactively.

US traffic engineers, however, tend to be reactive. Even near schools or parks, we wait for a kid to get hurt before thinking something might be a safety problem. Truth be told, we usually need more than one kid to get hurt. Two kids may not even be enough. For a location to become a hot spot—such as by triggering a MUTCD[d] *warrant*—we typically need to see at least three injury crashes. And if those injuries—or deaths—are spread out across four or five years instead of two or three at a single location, your friendly neighborhood traffic engineer may want to wait for more kids to get hurt before considering it a safety problem.

---

[c] In 1970, more than 40 percent of US kids walked to school. Noreen McDonald, "Active Transportation to School: Trends among US Schoolchildren, 1969–2001," *American Journal of Preventive Medicine* 32, no. 6 (2007).

[d] As a reminder, the MUTCD is the document that sets standards for traffic signs, signals, and markings. It also presents "warrants" for how many crashes, injuries, or deaths we need before taking action.

We don't need to be reactive. The other way to trigger some action via a MUTCD warrant—and end up with a supposedly safer crossing—is with a high number of pedestrians crossing the street. Our guidelines say we need a certain number of pedestrians crossing at a particular intersection in a single hour to warrant a crosswalk or something like a HAWK beacon.[e] But how do you reach such high pedestrian numbers if our streets are so seemingly dangerous that letting a kid cross by themselves equates to negligence? And how can we respond to pedestrian crashes that never occur because nobody is willing to cross?

It doesn't have to be this way. As we know, the MUTCD gives traffic engineers leeway to use **engineering judgment**. If we want an all-pedestrian phase, for example, "an unacceptable number of conflicting pedestrian movements with right-turn-on-red maneuvers, especially involving children, older pedestrians, or persons with disabilities" would do the trick.[9] The MUTCD doesn't give us much to go on, so how we define an "an unacceptable number" is entirely up to us.

Take the case of sixth-grader Eliza Almuina. Eliza was walking to her school carnival in Albuquerque, New Mexico, the day before school break. Doing so meant crossing Louisiana Boulevard, a high-speed multilane arterial. Eliza tried to cross at the unsignalized crosswalk adjacent to the school. Not that it should matter with a crosswalk, but it was 5 p.m., outside of normal school hours and beyond when the school zone would've been active. One lane of traffic stopped, and Eliza started crossing with her best friend, Hailie. The driver in the second lane didn't stop.[f] Hailie made it across. Eliza didn't.

The city reacted to this catastrophe by saying they'd **study the problem**. They considered a HAWK beacon but didn't find enough pedestrians crossing Louisiana Boulevard to meet the MUTCD **warrant**. The city installed it anyway.[g] Once it did, the number of

---

[e] A HAWK (for *h*igh-intensity *a*ctivated cross*w*al*k*) beacon is the more common name for a pedestrian-activated hybrid beacon. The flashing lights come on, as needed, when a pedestrian tries to cross. They eventually turn red, and we hope drivers stop.

[f] The driver, 76-year-old Revi Pexa, pleaded no contest to careless driving and was sentenced to 45 days in jail. But an unsignalized crosswalk on a multilane arterial designed like a highway? I would argue that traffic engineers put him in a difficult situation as well.

[g] In addition to the HAWK beacon, Albuquerque increased the size of the median

pedestrians crossing there increased to where it would've been warranted anyway.

So yes, these traffic engineers did something and did so using their own engineering judgment rather than a warrant. The previous unsignalized crosswalk on Louisiana Boulevard was heartbreak waiting to happen, especially with a school next door. But it still took the death of Eliza, a 12-year-old girl, for them to take action. One death was enough for them.

For me, one death—especially when we are talking about a kid—is more than enough. But if your city traffic engineer is sitting around, waiting for more people to get hurt or die on our streets before taking action, we can do better.

If we want to make our cities safer—and healthier and happier—we need to stop focusing on making sure streets are wide enough to accommodate some extrapolated future traffic demand. Instead, let's figure out where kids want to go—like schools and parks—and figure out where they are coming from. Then, let's make sure it's safe for them to walk, bike, scooter, hop, roller-skate, or however else they might want to get there.

What does this approach mean in practice?

We know that kids are more vulnerable than what traffic engineers consider to be our most vulnerable road users. We also know that kids don't have the ability to detect the relative speed of fast-moving cars as well as adults.[10] Kids can be more difficult to see, especially near large parked cars, and might have difficulty with complex situations where danger could come from multiple different directions.

Start with these facts and the keys to making a transportation system safe for kids begin to emerge. We need slower streets.[h] Fewer lanes of traffic. Simpler intersections with protected pedestrian crossings. Stop bars painted farther back from the crosswalk. Parked cars set way back from the corner. And not just during the fleeting minutes before and after school. We need it all the time.

I came across a school zone speed limit sign from Fort Lauderdale, Florida, that lists four short bursts of time when the school zone limit applies. And another from Detroit, Michigan, that lists

---

and added fencing to help make sure pedestrians crossed where the traffic engineers want them to.

[h] Not just lower speed limits, but streets specifically designed to keep speeds down.

six separate time periods, including school zone times like 7:52 a.m. through 8:22 a.m. If you happen to arrive at school between 8:23 and 8:36 a.m., good luck, because the school zone speed limit doesn't apply.

In Denver, many school zones don't extend to the arterial abutting to the school. Time and time again, we choose to allow the allure of speed and efficiency to override safety. This prioritization isn't conjecture; it's what we do. Here is a 1952 *Traffic Quarterly* paper telling us how a good engineer deals with the "real and imaginary fears in the hearts of parents": "While they sympathize with parents' fear, they also know their responsibility to taxpayer and motorist to provide an efficient, economically controlled street and highway system, free from unwarranted stop signs, traffic signals and other installations that may cause delay and congestion; or cause unnecessary expenditure of critically insufficient municipal funds."[11]

A 1956 *Traffic Quarterly* paper advises us to dismiss such fears, saying, "An influential and intelligent but difficult citizen whose 7-year-old daughter was to enter a new school ... argued the Parent-Teacher Association into petitioning the council of the city for a fixed time signal to protect the children crossing the highway."[12]

Unfortunately, this parent was only "an average road user" and not experienced with traffic problems like the paper's author, Fred Herring. Yet, for reasons unbeknown to Fred, this parent "persisted in his demand for the fixed time signal." Fortunately, "in the end, reason prevailed." Peak-hour traffic was just too damn high[i] to help these kids cross the street, and from Fred's perspective, "the meeting ended in complete harmony."[13]

Maybe, but I'd bet this kid didn't walk to school the next day.

In a 1952 *Traffic Quarterly* paper, the mayor of Portland, Oregon, brags about quelling fears in a similar situation.[14] Portland had a "high-speed arterial state highway" located adjacent to its school for blind children. The blind kids nicknamed this arterial "Death Valley." Portland's solution? They made the pedestrian beg button bigger. Does the bigger button bring about the walk signal any sooner? Nope, but it did add an audible buzzer to accompany the walk signal.

After a kid got hit in the crosswalk at my own kids' elementary

---

[i] Like rent.

[j] Fortunately, he wasn't badly hurt.

school,[j] the school convened a safety committee to study the problem. At the first of those meetings, I learned that the committee's idea of safety was making it easier for parents to drive and drop their kids off at school. If we want to make it as safe as possible for kids to walk and bike, making it easier for adults to drive won't help the cause.

Unfortunately, this sort of thinking is common to most Safe Routes to School programs. Maybe everyone read the 1958 AAA report that argued against making our streets too safe for kids? How can we teach kids "self-reliance and personal responsibility" if we give them a "golden path to and from school"?[k] But given today's transportation system, how can we teach them "self-reliance and personal responsibility" if we're forced to chauffer kids everywhere they go?[15]

There is an interesting 1993 *Transportation Quarterly* paper by Sigurd Grava, a professor at Columbia University. Sigurd grew up in Germany. As a kid, he says that "there were not many automobiles around, and I climbed trees, rode my bike, and roller skated on the street in front of my house without any worries. I was king of my block."[16]

Sigurd assumed that the world where he grew up was long gone. But when he returned to Germany in the early 1990s, his hometown had undergone an extensive program that can best be described as a combination of traffic calming and living streets, or woonerfs. The idea is to design streets that force drivers to adjust their behaviors, specifically in terms of speed, and give other road users priority over cars. According to Sigurd, the result was that "to drive faster than 30 kph [kilometers per hour] is illegal, unthinkable, and not really possible."[17]

That speed, 30 kph, is 18.6 mph. Imagine a world with streets where 20 mph is "not really possible"? Wouldn't that be a safer world for kids to walk and bike? Wouldn't that be a safer world for everyone?

We don't have to imagine because it's been proven true. Dating back to the 1980s, research into similar programs found that road safety injuries and deaths decrease significantly at lower speeds.[18]

---

[k] Here's the full quote: "Everyone wants to protect school children, and sometimes there is a tendency to overprotect them, virtually giving them a 'golden path' to and from school. Where protection is excessive, there can be failure to equip children with the increasing degree of self-reliance and personal responsibility which they need to acquire as a part of growing up."

We can also look to the countries that systematically employ similar interventions. The Netherlands, Sweden, Norway, the United Kingdom, Germany, Spain, Switzerland, Australia, and Japan all have road fatality rates between 2 and 4.5 fatalities per 100,000 people. The overall US road death rate is nearly 13 fatalities per 100,000 people.[1]

Once we make our cities safe for kids, similar fundamentals—albeit with different destinations—apply to older folks who may move more slowly or be limited in other ways. Once your city tackles kids and older folks, move on to persons with disabilities and then to those living in disadvantaged neighborhoods.

Whatever the category, here's one thing I know I'm not willing to do: sit around and wait for more people to get hurt before trying to make our streets safer.

---

[1] In the early 1970s, the road fatality rates for most of these countries—including the United States—were very similar. They've all improved safety much more than the United States has.

83 Where the Sidewalk Begins

—

Many of my favorite movies and television shows don't work if the kids can't safely walk or bike around town on their own. *The Goonies*. *Home Alone*. *The Sandlot*. *Stranger Things*. Even *E.T.* falls into this category despite the kids living in the midst of suburban California construction sites. These movies and shows all work because some producer or director chose a filming location with streets good for walking or biking. Our problem? Most streets aren't good for walking or biking—especially if the main characters are kids.

How do traffic engineers accommodate walking and biking? Well, the first problem is that word *accommodate*. We don't design *for* these modes; we **accommodate** them by giving up whatever's left over after we design for those in cars. If we don't have any leftovers, we'll throw up a "SHARE THE ROAD" sign or paint a sharrow in the street.[a]

When we do have leftovers, the basic idea is to separate pedestrians and bicyclists from the drivers and their vehicles. We can do so either temporally or spatially.

Temporal separation is as simple as a walk phase on a traffic signal. The idea is for pedestrians to use part of the street at one time and drivers to use it at a different time. This approach rarely ensures safety because our de facto design allows drivers to, for example, turn right on red or take a permissive left at the same time we tell pedestrians it's their turn to cross.

Spatial separation used to mean building something like an expensive pedestrian overpass over a terrible street. Whenever I see one or hear somebody propose one, I take it as an admission of failure. In other words, we can't possibly make this street safe enough for humans to cross on foot. Instead, we need to spend millions building

---

[a] As mentioned earlier in the book, a sharrow is the common name for a shared-lane bike marking that we paint in the vehicle lane to remind drivers that bicyclists have the right to exist in the street.

a bridge that is extremely annoying for people to use. It's like lipstick on a pug.[b]

These days, spatial separation might mean protected bike facilities. In the words of our old pal Paul Hoffman on the first page of his 1939 book, protected bike facilities offer "a degree of safety that the white lines alone could never guarantee."[1]

Paul wasn't talking about protected bike lanes, though. He was trying to bolster his argument for turning our streets into highways. Yet the same thinking applies here.

All protected bike lanes aren't created equal, either. A flimsy plastic post isn't going to protect anyone like a concrete barrier or steel bollard could. But the reason a plastic post isn't good enough is because our streets aren't good enough. It's reasonably feasible to design a street that calms drivers down to a speed where we don't need bike lanes or even sidewalks to safely walk and bike.

It's just not reasonably likely.

It wouldn't be hard to argue that the humble sidewalk is the most fundamental of all transportation infrastructure. Yet when you ask most city traffic engineers about their sidewalks, they put their hands up as if they are trying to get away with an obvious foul in an NBA game.

My student Peyton Gibson interviewed traffic engineers from 16 different cities about their roadway asset management processes.[2] Every city tracked its roadways and did so on a regular basis with hired consultants who drive the entire network using a vehicle mounted with high-tech sensors that collect high-quality, geo-located data.

We then asked the same questions about sidewalks. The most common response was a frustrated sigh and a warning that "it's complicated."

Five cities had no data whatsoever on their sidewalks. Ten cities knew where sidewalks existed but had no information on basics such as width. Only one city was collecting data in a similar ballpark to their roadway data.

Got a pothole? One city fixes 90 percent of them within 24 hours. Most cities do so within 48 hours. The longest was less than five days.

But if your sidewalk is falling apart, good luck getting the city to help. In fact, most cities will turn around and tell you to fix it yourself.

---

[b] I meant to write "pig," but I liked my typo and kept it.

For whatever reason, "the fundamental value of sidewalks and the corresponding need to provide a safe walking path" are "often forgotten or deferred" and "viewed more as a luxury." This 1978 *Traffic Quarterly* paper goes on to say, "While it is apparent that sidewalks serve an essential function, it is equally apparent that in many areas sidewalks are missing, discontinuous, or unsuitable for pedestrian use."[3]

That wasn't news. Even AAA, in 1958, said that "adequate sidewalks are another responsibility of engineers."[4]

You could argue that AAA promoting sidewalks was more about getting the pedestrians off the street and out of the way. Whatever their motivation, it's better than what the Urban Land Institute said at the time. Their 1954 *Community Builders Handbook* worries that when it comes to the safety of kids, sidewalks may do more harm than good: "It has been hinted that sidewalks actually tend to encourage playing in the street rather than in off-street areas such as rear yards or a playground."[5]

When asked for evidence, the director of the Traffic Operations Division of the National Safety Council admits that we just don't know: "Unfortunately there is no good accident information available which would demonstrate the relationship between child safety in areas with sidewalks and without sidewalks."[6]

Why don't we know? Well, we don't collect the sidewalk or exposure data needed to answer such basic questions.[c] That was true when said in 1957, and it's true today.

The bigger issue is that this sort of thinking misses the point. Even if we had the data and found that not having sidewalks or bike facilities helped confine kids to their backyards and led to better road safety, is that really what we are looking for?

Of course not. In his book *Bicycle Diaries*, David Byrne of the Talking Heads says, "Being in a car may feel safer, but when everyone drives it actually makes a less safe city."[7]

David Byrne is talking about crime—along the lines of Jane Jacobs's eyes-on-the-street concept—but the same is true about road safety. Put everyone in a car, and your city isn't going to be safe. But

---

[c] In the same report, the American Society of Planning Officials (precursor to the American Planning Association) said the research doesn't matter because "sidewalks, like babies and cars, are here to stay."

build the sort of infrastructure that gets people out walking and biking, and your city is going to be safer.

This result goes against the grain of conventional traffic engineering thinking. On average, you are about 20 times more likely to die when walking a mile than driving a mile.[8] Bike a mile? That's maybe 10 times more dangerous than a mile in a car. But guess what happens in cities with lots of people walking or biking? They end up as our safest cities, far safer than conventional cities where everyone is conventionally driving.[9]

Safety isn't the only benefit.

My cholesterol level was at 222 when my doctor said it was time for medication. When I asked what I could do exercise-wise instead, the answer was getting my heart rate up with cardio for at least 20 straight minutes every day. I was active before, but not quite like that. I figured that biking to work—about a half hour each way—would do the trick. Six months later, my cholesterol level was at 149.

It's hard to say what exercise does for mental health, but the research suggests good things.[10]

How much money does biking to work save me as compared to driving and paying to park? Look at one day, and it may not seem like much. But it adds up over time.

It saves me time as well. The last time I checked, the average commuter in the Denver region spends about 40 hours each year stuck in rush-hour traffic. I spend zero.[d]

If I drove to work, I'd have to spend time looking for a parking space, paying to park, and then walking from one of the campus parking garages to my office. Instead, I park my bike in my office.

Plus, whatever space my car would've consumed sitting around doing nothing all day in downtown Denver can be used for something else and something better.

And whenever I couldn't figure out how the puzzle pieces of this book fit together, they came together on those bike rides.

I've also heard that walking and biking are good for the planet. But guess what? Most people aren't going to risk their lives on dangerous streets out of the goodness of their environmental heart. It

---

[d] The only exception was the Denver Broncos' Superbowl parade. As a New England Patriots fan, I wanted no part of that nonsense and biking around it was a nightmare.

doesn't matter if walking and biking save money, time, our health, and the planet if you feel like you are going to die the whole time.

So my second Kondo category is simple: focus on making your streets and intersections safe for walking and biking. We've already done so in and around where kids go; now do so for the rest of your city.

But how?

I could go through all the tools in the traffic engineering toolboxes.

Some are simple, like the all-pedestrian phase Barnes Dance. Others get more complicated, like some of the protected bike intersection designs. My favorite interventions tend to focus on slowing drivers down while also prioritizing those walking and biking. Australia's wombat crossing, for instance, places a sometimes-colorful zebra crossing on top of a wide, raised speed hump. Even when there are no pedestrians around, the wombat humps slow drivers down to 20 or 25 mph.

Suffice it to say there are a lot of good—and bad—designs out there.

But before focusing on those, let's change our fundamental design approach.

Many safe street groups use an upside-down triangle diagram with a pedestrian icon at the top and a car icon at the bottom. In between, you see bikes, scooters, buses, and trains closer to the top with trucks and freight closer to the bottom. The idea is that safe street efforts, including *Vision Zero*, should prioritize the needs of the pedestrian first and those of cars last.

This doesn't stop traffic engineers from doing what we always do: thoroughly designing for cars, leaving those walking and biking with the leftovers.

But what if we followed that mantra of the upside-down prioritization triangle and actually designed for pedestrians first? Instead of starting at the centerline of the street—as traffic engineers are taught—we design from the outside in. Start at the edges and figure out how much space the pedestrians need. Then do the same for bicyclists. And then transit. Whatever's left over goes to drivers.

With this reprioritization, the de facto intersection designs could change as well. Making a right turn on red would be prohibited unless a traffic engineer signs off on a variance. The same goes for protected pedestrian signals; allowing permissive left turns becomes the exception, not the norm.

Make the intersections safe to cross and guess what? Pedestrians opt to use them instead of opting for jaywalking and giving us so-called human error to blame.

In 2017, the American Society of Civil Engineers asked me to give a presentation about the future of transportation. Going into that event, it was clear that the other two guys in my session were going to talk all about our amazing, electric, autonomous, high-tech future.

I rained on their parade. Why? Well, the future they laid out isn't as simple as they were saying it was. Even if we get technology to work. Even if it works in all kinds of weather. Even if it does so without having to reboot at inopportune times. Even if we get all that right, we still need everyone on board. We don't get all the benefits we've been promised—including many of the safety ones—without every car on the road being autonomous.

But guess what? People love driving. There's a reason why *The Fast and the Furious* movie franchise made billions of dollars. While driving is not exactly a constitutional right, trying to force people to use autonomous vehicles—or even electric vehicles—might end up mirroring the US gun debate.

The real future of transportation isn't that different than it was 20 or 200 years ago. That's why some of our simplest streets and cities have worked for hundreds of years and continue to do so. The real future of transportation will always come back to the fundamentals of walking and biking.

So let's start there and build in. Make your city safe for walking and rolling—and for kids to do so—and it will be safer for everyone.[e] Let drivers and autonomous vehicles adapt to that future, not the other way around.

Until then, I have a hard time getting excited about this amazing, electric, autonomous, high-tech transportation future of ours while we live in a world where we can't even get the sidewalks right. If a kid

---

[e] Jeff Speck's *Walkable City* books are a great resource for making that happen. It is interesting to note (at least to me) that my stepfather, Richard Bartley, was Jeff's high school track coach. Once we figured out this connection, Jeff spoke so highly of him and told me that my stepfather instilled a confidence in him that he never knew he had. My stepfather recently passed away as I was finishing up this book.

can't safely sell lemonade on the sidewalk in front of where they live, we're doing something wrong.

Let's stop working harder on getting people to Mars and instead work harder on helping them cross the street. I think E.T. would agree.

84   Another One Rides the Bus

—

My next Kondo road safety category to consider is transit. If you want to know why, the answer is simple: cities with good transit have good road safety.

A 1988 paper compares the capital of Illinois, Springfield, to that of Australia, Canberra.[1] Both are about the same size in terms of area and were about the same size in terms of population throughout the 1970s and early 1980s.

Canberra invested heavily in transit,[a] Springfield not so much. By 1985, Canberra had 14 times the transit ridership of Springfield. Recent data shows a transit mode share of 1.2 percent for the Springfield metro area. Canberra's is closer to 10 percent.[b]

In terms of road safety, about 20 people die each year in the Springfield region. In Canberra, that number is fewer than 9.

This difference isn't all that impressive unless you consider that over the last 30 years, Canberra doubled Springfield's population. Take population into account and Springfield kills 9.4 people per 100,000 population, which is safer than the US average. Canberra kills 1.9 people per 100,000 population, which safer than the countries considered the safest in the world.

The overall research suggests that, on average, a 1 percent rise in transit mode share equates to a nearly 3 percent drop in traffic deaths per population.[2] Big cities, small cities, it doesn't matter. Fewer people tend to die on our streets when more people use transit.

When I talk transit, I'm not talking about throwing a train in the middle of a highway. I'm also not talking about adding park-and-ride lots to every transit station, which is why I'd like to suggest parking as the Kondo cocategory to transit.

---

[a] Although not to the extent of most other major Australian cities.

[b] Canberra's walk mode share is about 5 percent and bike mode share about 3 percent. Combine Springfield's walking and biking mode share and you'll get about 2 percent.

A 2001 *Transportation Quarterly* paper talks about how cities want two things: to provide enough parking and to be pedestrian friendly.[3] We can say the same about most **transit-oriented developments**, also known as TODs, but the problem is that these two things rarely go well together.

Denver's former transit head once said that his job was to put the T in TOD.[c] In his mind, the O and the D are for other people to worry about. Walking and biking infrastructure near transit? Housing near transit? Not his job.

What was his job? To increase transit ridership and do so quickly. How does a transit agency boost use quickly? They add parking and make it cheap.

I've studied parking in and around TODs in the Denver area. What I found is tens of thousands of mostly empty parking spaces within what should be the walkable zone near a transit station.[d]

Looking at parking in aggregate reminds me of how Marie Kondo views clutter: "Most people realize that clutter is caused by too much stuff. But why do we have too much stuff? Usually it is because we do not accurately grasp how much we actually own."[4]

Like transit, we don't think about parking as a road safety issue, but it is. Kondo's clutter is our parking, and improving access to transit can't just be about those in cars.

The Englewood station in Denver was a national poster child for good transit-oriented development. As TODs go, it looked the part. But try walking to the apartment complex directly across the tracks from the station platform; it's going to be a long, convoluted route that isn't close to pedestrian friendly. Look at Denver's mode share data, and the results are not surprising. Most people still drive to these train stations even if they live within a mile or two of it.

So how do we make transit better and the world safer?

If you don't have the basics of a transit network, you have some work to do.[e] If you do have a functioning transit network, don't spend

---

[c] The transit agency has changed its tune on this issue.

[d] For example, the busiest day we could find left 60 percent of the 20,000 parking spaces empty within a half mile of the Arapahoe station. Denver's transit agency still built another almost 1,600 spaces next to the station.

[e] I'd suggest that you start by reading Jarrett Walker's *Human Transit*. It is also worth noting that transit can't bring us everywhere, at least not yet. Don't waste your time or money trying to serve the lowest-density suburbs.

tens of millions on parking garages. Instead, focus on low-hanging fruit such as walking and biking connections. Make them really nice and really safe. Then make the transit shelters nicer and safer. Make it easy to find real-time information so that people know when (and if) the bus or train will show up. Make it easy to know where it goes next.

Once the above fundamentals are in place, make sure the traffic engineers are counting people, not cars.

In 1969, northern Virginia opened up a reversible bus-only lane on I-395 that saved commuters to Washington, DC, between 12 and 18 minutes each way.[5] Bus ridership skyrocketed, and more buses were added. Still, it was standing room only on the buses.

The transit planners wanted to extend the dedicated bus lanes farther toward DC to shave off another 15 minutes. The traffic engineers didn't like seeing big, empty headways between buses. Why should these buses—which represent less than 15 percent of the traffic—consume 25 percent of the space? The answer? The buses held 82 percent of the people.[6] Nevertheless, they downgraded the bus-only lane into a combined bus/high-occupancy-vehicle lane. The buses lost their advantage and eventually, their riders.

If only they had been counting people instead of cars.

Denver made this policy change a few years ago. In doing so, it's been able to transform downtown streets from four or five lanes of one-way vehicle traffic into multimodal streets with two red bus-only lanes and a green protected bike lane. Ten years ago, these changes felt like a moonshot. Now, it's like Christmas every day.

Bus-only lanes like these were trendy in the 1950s.[7] Cleveland went in a different direction by inventing parking-focused transit. According to a *Transportation Quarterly* paper, "The Cleveland rapid transit system, built largely in railroad rights-of-way, was the first new system to provide wide station spacing and extensive park-and-ride ... facilities."[8]

As you know, park-and-ride lots won out in most places.[f]

Today, cities are bringing dedicated bus lanes back, and the data shows they are working. The 14th Street busway in New York City

---

[f] I've long been a fan of what I call park-and-pedal lots. The idea is to find an existing, underutilized parking lot on the edge of a city and build an amazing bike facility between there and downtown. People who live in unbikeable locations could get themselves a bike rack and park there before pedaling into the city.

displaced 21,000 cars.[9] Market Street in San Francisco saw similar results, along with a significant reduction in bus travel times and a 25 percent jump in biking.[10] With more than 5,000 daily transit users on Market Street, that shift away from driving exposure starts to add up. So do the road safety benefits.

Transit is about 20 times safer—in terms of fatalities and severe injuries—than driving. Yet, for whatever reason, we never sell transit as a safety intervention.

Instead, we tell you that transit is more efficient. Here is a 1950 quote from the chief of the Urban Road Division for the Public Roads Administration: "The transit vehicle, while it is moving, is a much more efficient user of street space than a private car."[11]

And then we show you various images of *x* number of people in a bus versus that same number of people in cars. The first of these images dates to 1965 England, when Transport London took an overhead photo of 69 people standing in the middle of a street. Above it, you see an image of 69 people in cars. Below it, you see an image of 69 people on a bus. The text reads, "These vehicles are carrying … 69 people who could all … be on this one bus."

This set of images has since been revised and updated dozens of times by cities all around the world. The most famous comes from Münster, Germany. In 1991, its planning department did something similar to London but added bikes. Over the years, I've seen videos, animations, and computer simulations all trying to get across the same point.

But you don't win friends with efficiency.[g]

You do win friends by helping people live a better life by building places that help them do so.

A 1962 *Traffic Quarterly* paper found that during the morning rush into Manhattan, cars represent 85 percent of the total traffic

---

[g] Nor salad, but on a related note, here is a great quote about efficiency—or lack thereof—from a 1956 *Traffic Quarterly* paper by George Anderson: "It is paradoxical that we Americans, priding ourselves on efficiency, persist in using the least efficient means of transportation in congested areas. It is ironic that the private automobile, the most prominent evidence of our mass production genius, is undermining the urban civilization which created it. It is ironic that in pursuit of convenience, we should take a couple tons of steel, glass, and rubber with us wherever we go, thus producing the inconvenience we all suggest in traffic congestion." George W. Anderson, "Urban Mass Transportation," *Traffic Quarterly* 10, no. 2 (1956).

but carry only 9 percent of the total people entering the city.[12] At the other end of the spectrum, my first transportation engineering professor, Lester Hoel, said in a 1968 *Traffic Quarterly* paper that "Atlanta ... by 1970, in the absence of a rapid transit system ... would require 120 expressway lanes radiating from downtown together with a 28-lane central city connector."[13]

A 1956 *Traffic Quarterly* paper then explains what transit efficiency could mean to the bigger picture: "Without the development of elevated and subway lines, New York and Chicago would never have become the cities they are today."[14]

If done right, all the above comes together in a way that helps make transit useful and the city safer. Just make sure it is useful.

When the train station opened a mile from where I live in Denver, a connecting bus route right near my house was also added. The train came every 15 minutes and the bus every 45. Coming home from work would've meant that only one of every three trains was useful. Those 45-minute headways must not have been useful to others because that bus route didn't last long.

We assumed that low ridership meant the route isn't useful. In reality, maybe it was the service that wasn't useful, not the route. It's easy to think that once we make transit nice enough and useful enough to gain ridership, we can then increase headways. But maybe we need to first increase headways to make transit nice and useful.

But be careful because conventional traffic engineering methods will have us provide less useful transit in poor neighborhoods and more useful transit in rich ones. Why? Our calculations include estimating the value of people's time. If wealthy people value their time more and we want to capture their ridership, we need good transit. If lower-income folks have few other viable options, we can provide crappy transit and assume that ridership will hold.

In a 1993 music video, LL Cool J is driving around and gets a flat tire. Instead of waiting for AAA, he tosses his car keys to his buddy in the passenger seat and jumps on the bus. Here he is, 30 years later, talking about that scene in an interview: "Yeah, I'm not sitting around waiting for the flat to get fixed. I got places to go. I ain't too good for public transportation."[15]

If our streets are the bones of our cities, transit is the connective tissue. We just need to make it good enough for LL to use.

There's one more Kondo category to get to, but here's what we have so far:

(Kids) + (Active Transportation) + (Transit) − (Parking) =
Safer Places

85   Won't You Be My Neighbor?

—

In the 1983 movie *Trading Places*, the Duke brothers—Randolph and Mortimer[a]—debate the value of nature versus nurture.

They then witness a misunderstanding gone awry between the guy running their company—the well-educated and well-mannered Louis Winthorpe III (played by Dan Aykroyd)—and a street hustler named Billy Ray Valentine (played by Eddie Murphy).

As Billy Ray gets hauled away by the police,[b] Randolph Duke says, "That man is a product of a poor environment. There's nothing wrong with him. I can prove it."

After some casually racist remarks by Mortimer, they have the following exchange.

> RANDOLPH DUKE: Given the right surroundings and encouragement, I'll bet that that man could run our company as well as your young Winthorpe.
> MORTIMER DUKE: Are we talking about a wager, Randolph?

So these business tycoons make a bet. They are going to flip the lives of Louis Winthorpe III and Billy Ray Valentine to see what happens.

It all goes according to plan until Billy Ray—spoiler alert—sees Mortimer handing over a grand total of one dollar to Randolph to settle their bet.

So Randolph was right. One's current environment can make more of a difference than the environment they came from.

I bring this subject up because I've always wanted to run a similar experiment, albeit without the racism and more focused on transportation and the built environment.

The idea is simple. We find somebody who lives in a car-dependent

---

[a] Played by Ralph Bellamy and Don Ameche.

[b] "Is there a problem, officers?"

neighborhood who drives everywhere they go. We also find some-body who lives in a multimodal neighborhood who drives on occa-sion but usually walks, bikes, or takes transit to get where they need to go.

They then trade places. The driver moves into the multimodal person's house and vice versa. What happens to their transportation behaviors over the short term? What happens over the long term? How do their lives—and their exposure risk in the transportation system—change?[c] What else changes? And what happens when we flip them back?

One of the greatest barriers to road safety is having to drive everywhere, all the time, even for the simplest errand.[d] More spe-cifically, the greatest barriers to road safety are the places we've built that force people to drive everywhere, all the time, even for the simplest errand.

Earlier in this book, we talked about exposure. Conventionally, distance traveled is the denominator in our crash rates. The number of crashes, injuries, or deaths on our roads is the numerator.

You'd think that focusing on reducing the numerator so as to improve crash rates makes the most sense. The problem is that we've instead focused on *increasing* the denominator. In other words, given the way we measure road safety, we all *seem* safer if we all drive a lot more. We built places that force us to do so. Crashes go up a little, but we drive a lot more, so we're convinced we improved safety.

And if the crash rate tells us we're safer, we must be. Right?

So what I want to do with the last—and maybe the most diffi-cult—Kondo category is build the sort of places that help us *reduce* the denominator.

Crashes are a by-product of the quantity and quality of our expo-sure. These things aren't independent. In terms of nature versus nur-ture, how much we drive—and the exposure that comes from this driving—is the rational human response to the transportation and land use situation put in front of us. It's not the only reason for this

---

[c] The best research on something remotely similar was Kevin Krizek's 2003 paper, which found that trading places from a Seattle neighborhood to suburban Bellevue equates to driving approximately five extra miles per day. Kevin J. Krizek, "Resi-dential Relocation and Changes in Urban Travel: Does Neighborhood-Scale Urban Form Matter?," *Journal of the American Planning Association* 69, no. 3 (2003).

[d] It's also one of the greatest barriers to saving planet Earth.

exposure, but it's what we're forced to do to correct the spatial imbalance between where we are and where we go.

Put where we are and where we go closer together and we can decrease the quantity of that denominator. Then make the streets kid-friendly, safe for walking and biking, and add some good transit, and we can increase the quality of that denominator.

It seems simple, but look back at our history. We ripped out the kind of places that would reduce exposure quantity and increase its quality in favor of disconnected networks with segregated land uses and big parking lots surrounded by big arterials and big highways. What we ripped out was the road not taken. The safer road.

The underlying principles aren't new. A 1969 *Traffic Quarterly* paper called "Logic and Road Safety Research" compares the road safety problem to an epidemic: "As in any epidemic, when we have exhausted all means for reducing the virulence of a disease-carrying germ, then the only other way to stop the ravages of the disease is to reduce exposure to the pathogen."[1]

The virus he speaks of is excess driving. If we want to make real headway on road safety, we need to reduce exposure. To do so, "somehow techniques must be found for transferring people from the present types of highway vehicles to other inherently safer, cleaner, and faster means of travel."

I'm not as sure about "faster," but if we stop focusing on getting from A to B as quickly as possible and think more about ways to bring A and B closer together, we might get there faster as well. The author of that 1969 paper goes on to say that "by changing the nature of the game we can change the risk of death and injury. Therefore, we should invest greatly in efforts to reduce risk for the overall system including the driver, pedestrians, road, vehicle, and environment."[2]

All these decades later, the nature of the game remains the same—for most people.

Some cities, states, and countries have road fatality rates—population-based fatality rates—way below average. Others are way higher. These differences aren't flukes. They aren't statistical improbabilities, because they don't change much from year to year. But the gaps also continue to widen.

These examples show we can do better.

You might remember that I grew up in a place where—even as a kid—I could walk or bike all over town. Hop on a bus and I could get to Fenway Park or Harvard Square. From there, the T could take me anywhere I wanted to go.

Today, I live in a neighborhood where I have five grocery stores within a mile and a half of my house. We have schools, parks, a library, a recreation center, a hardware store, and more good restaurants than I can count. Even if I drive to these places, I'm not driving far. But it's often easier not to drive.

On the other hand, I've lived in places where it was a 20-minute drive to the grocery store or a 30-minute drive to the hardware store. But what drove me the craziest were the physically close destinations that still required a car.

Still, we underestimate the road safety detriments of a neighborhood that only serves as the place you live. Flip it and we also underestimate the safety benefits of a neighborhood where you can take care of most of your basic needs.

It's not about turning every place into New York City. It's about connecting people with destinations. The term used for this connection is *access*.[3]

Most traffic engineers focus on *mobility*. Good mobility means we can drive around without sitting in traffic or waiting at red lights. Traffic engineers measure mobility using metrics like speed, delay, and level of service.

For a traffic engineer, thinking about access usually means thinking about how many driveways we should allow on a particular street. But access is really about whether we can reach goods, services, and activities such as schools, jobs, grocery stores, restaurants, doctors, hospitals, shops, concerts, parks, nature trails, and friends.

Good *mobility* leads to an increase in the crash rate denominator. Good *access* decreases the denominator.

When I'm teaching, I bring up the mobility versus access issue by asking my students the seemingly simple question, what's the point of transportation? They all respond with something about transportation being about moving people and goods from A to B.

I press them with follow-up questions until I start seeing the light bulbs go off. At that point, the discussion shifts from questioning how we can improve *mobility* to questioning how we can improve *access*. The answers shift from things that we hope could maybe, possibly fix traffic congestion to issues like active transportation, transit, land use, and community design.

When we revisit that initial question—what's the point of transportation?—the answers the students now give are nothing like when we started. I then show off one of my favorite answers to that question:

"A good transportation system minimizes unnecessary transportation."

My PhD advisor, Norman Garrick, would again and again throw that quote up on the screen when he was presenting. It comes from Lewis Mumford and his 1963 collection of essays *The Highway and the City*. Here is that whole paragraph; it's worth reading:

> What's transportation for? This is a question that highway engineers apparently never ask themselves: probably because they take for granted the belief that transportation exists for the purpose of providing suitable outlets for the motorcar industry. To increase the number of cars, to enable motorists to go longer distances, to more places, at higher speeds has become an end in itself. Does this over-employment of the motorcar not consume ever larger quantities of gas, oil, concrete, rubber, and steel, and so provide the very groundwork for an expanding economy? Certainly, but none of these make up the essential purpose of transportation, which is to bring people or goods to places where they are needed, and to concentrate the greatest variety of goods and people within a limited area, in order to widen the possibility of choice without making it necessary to travel. A good transportation system minimizes unnecessary transportation; and in any event, it offers a change of speed and mode to fit a diversity of human purposes.[4]

That's what I'm trying to do with this last Kondo road safety category. For traffic engineers, it means putting mobility on the back burner in favor of measuring access to destinations. Improved mobility is one way to improve access, but it's not the only way.

Driving 5 minutes to the dentist is better than driving 30 minutes. It helps reduce that denominator, and if done right, it'll be a slower and safer 5-minute drive.

Whatever the specifics, it doesn't work without reenvisioning community design and land use in a way that puts stuff closer together. And we can't bring destinations and amenities closer if we aren't willing to take on the density of neighbors needed to make the destinations and amenities viable and sustainable.[e]

That's what Lewis Mumford was talking about all those years ago with his words "to concentrate the greatest variety of goods and people within a limited area, in order to widen the possibility of choice without making it necessary to travel."[5]

So let's plan for places like this and the traffic safety problems of the current generation instead of traffic congestion problems of a future generation. Doing so might solve both.

And doing it right would mean tying all our Kondo categories together. You've got some good places to go in your neighborhood? Great, let's build some housing. But don't get me wrong, it won't be easy.

In a season 7 episode of *Parks and Recreation*, Ron Swanson wants to build an apartment building next to the park that Leslie Knope built in the early seasons. Leslie is mad: "What the hell, Ron? You're building your stupid building on the block next to Pawnee Commons? The park that I built from scratch out of a pit? This building is gonna ruin the views, you jerk."

After some back and forth on Leslie's emotional connection to one of the houses being torn down, Ron replies, "The world needs apartment buildings. The park you built is nice, and people want to live next to it."

Ron is right. We need more housing near nice things. But the journey is just as important as the destination. So make those places easier to get to. Make the streets safe for kids. Make them safe for people to walk, roll, and jump onto the bus or train. Maybe even design the street from the outside in. As for parking? You don't need as much as you think you do.

Take a good look at the real differences between the safe and unsafe places. You'll start to see that what I'm suggesting works.

What have you got to lose?

---

[e] Detractors will argue that people don't want to live in such places and that they'd prefer a big house in the middle of nowhere. They will also point out that the real places that resemble what I'm suggesting are unaffordable to most people. But that combination doesn't make sense. Why are these places so expensive if nobody wants to live there? Isn't this a supply problem, not a demand problem? Why not build more of the good places and find out?

Part 12

What Matters and What Next?

___

86 Tell the Stories behind the Numbers

—

Given the numbers, our road safety problem seems undeniable. More than 1.35 million people die on the roads every year. An average day ends with 3,700 dead and tens of thousands more injured.[1] But many safety experts think people like me overestimate the road safety problems by pointing out obscure safety statistics, suggesting that "one's chance of being killed while driving a car is only one percent in fifty years of driving."[2]

That may be true, but I still think we underestimate the road safety problem.

By a lot.

As I wrote this book, it was hard not to notice the prevalence of fictional crashes and deaths in movies, television shows, and literature.

For whatever reason, it's easier to overlook the nonfiction ones.

James Dean, Grace Kelly, and Jayne Mansfield all died on our streets. So did Randy "Macho Man" Savage and the Junkyard Dog.

Jackson Pollock and Sam Kinison.

Linda Lovelace and Paul Walker.

Donald Appleyard[a] and Lisa "Left Eye" Lopes.

Steve Prefontaine and Princess Diana.

I looked up the Princess Diana crash and came across the front page of a 1982 New Jersey newspaper. The headline was about Princess Diana going into labor and Prince Charles postponing his polo match. Another front-page article reported on a fatal crash that had nothing to do with Princess Diana, but the combination of both stories is why I found it. The second article showed a picture of Kristin Sickel, a 13-year-old middle school student from Eatontown, New Jersey.

---

[a] Donald Appleyard was a Berkeley professor who wrote the seminal 1981 book *Livable Streets* (also check out the 2.0 version by his son Bruce). He died on the streets of Athens, Greece, a year after *Livable Streets* came out.

I went down this rabbit hole and tried to learn more about Kristin's crash. The next thing you know, I bought a memoir called *Ten Minutes from Home* written by Beth Greenfield decades after the crash happened. Beth was 12 years old and on the way home from her ballet recital. A 25-year-old Navy sailor named Edward Pahuli flipped his car onto the one Beth was in. Her best friend, Kristin, died that day. So did her younger brother, 7-year-old Adam.

Beth describes the crash in detail. She also describes decades of grief and guilt that started that day in 1982. She talks about the enduring struggles experienced by her, her parents, Kristin's parents, friends, family, teachers, and almost anyone close to these families.[3]

So when people tell me we are lucky to live in a world where you have a 1 percent chance of dying on our streets, I think of people like Beth Greenfield. She is one of the lucky 99 percent, but I'd bet it doesn't feel that way to her.

The same can be said for millions of people who don't end up among our fatal crash victims.

Collateral damage isn't the right phrase here. But when traffic engineers consider road safety, we count crashes, injuries, and fatalities. We estimate property damage. We even try to quantify the traffic congestion impacts that crashes burden other drivers. But we don't consider the never-ending anguish suffered by the family and friends of those in our crash statistics. We don't think about the long-term sacrifices incurred by the family and friends of those who are seriously injured. Nor do we put much thought into what the others involved might go through.

Still, the so-called safety experts would tell me that I'm exaggerating the dangers of our streets, especially when compared to actual dangerous activities. Let's say you are a Navy fighter pilot. Flying a jet is about 20 times more dangerous than driving a car. In fact, "Navy statistics indicate that a fighter pilot who spends a 20-year career flying high-performance jets has, incredibly enough, a 23% chance of dying in an aircraft accident."[4]

But guess what? More Navy fighter pilots die on our streets than in jets. In other words, Tom Cruise's *Top Gun* Maverick character is more likely to die on the way to work than at work.[b]

---

[b] Maverick pumping his fist while he speeds on his motorcycle isn't helping his cause.

Maverick may not have much in the way of family or friends. If he did, it's not like our crash statistics would account for their pain and suffering. The social and emotional costs of road safety may not seem real for the family or friends of our fictional characters, but they are very real for our nonfiction ones.

So please don't tell me that I'm overestimating the road safety problem.

Each time I teach my road safety course, I pick a class toward the end of the semester to talk about some of the real people who make up our fatality numbers. It's a hard talk to get through.[c] I tell myself the students have no idea I'm getting choked up. I know that isn't true.

I sometimes end that class with the story of Olivia Schnacker. Olivia was 13 years old when she and her siblings went for a bike ride to a park behind a nearby school. It was Easter Sunday. They tried to cross at a signalized crosswalk, in a school zone, on five-lane Ustick Road in Boise, Idaho. Olivia never made it to the other side.

I didn't know about Olivia or her crash until a friend of mine, Don Kostelec, brought it to my attention. Don was working for the Ada County Highway District when the county decided to expand Ustick Road from two lanes to five. The City of Boise sued the county agency over this widening. So Don went to court to defend the widening, the 20-year time horizon vehicle volume estimates, the need to design for this future level of service problem, and the so-called traffic engineering *standards*. With the help of Don's testimony, the county won the lawsuit and expanded Ustick Road to five lanes.

Disillusioned by the lawsuit and being forced to defend "improvements" he knew weren't improvements, Don moved across the country to start a career focused on the safety of vulnerable road users. Olivia then died on the same widened street whose widening he defended in court.[d]

Olivia's death hit Don hard. To this day, he keeps a copy of his deposition close at hand. He reads it whenever he needs to be reminded about why he does what he does.

---

[c] And it's getting harder as I get older.

[d] The police report placed 100 percent of the blame on Olivia even though she was in the crosswalk, with several witnesses saying the driver had a red light and the kids were visible.

If you want to dig down your own rabbit hole of heartbreaking stories, they aren't hard to find.[e]

Ask somebody on the street how many people die on our roads each year, and the research suggests their answer will be much lower than the real number. Give them the real number and I'm not sure it'll make a difference. It's a different number, but it's still just a number.

What matters more are the real people behind these numbers, and not giving kids like Olivia the opportunity to grow old matters. It isn't about sending out "thoughts and prayers" while going about business as usual and hoping our road safety problems resolve themselves. Rather, it's about inspiring traffic engineers to fix these problems. It's about inspiring us all to fix these problems.

So get out there and tell the stories of the very real people—and the very real kids—who died on our streets while doing completely normal things in their normal, everyday lives. Keep their stories alive. Maybe we can save someone else's life while doing so.

---

[e] Families for Safe Streets is a group that amplifies the voices of those who have lost loved ones due to traffic violence.

87 Reengineer the Traffic Engineers

—

Ned Flanders—arguably the world's most famous neighbor—is the only one to lose his house in the hurricane that hits Springfield.[a] This leads to Ned's nervous breakdown and some insight into his childhood as an out-of-control terror. In a flashback, Ned's beatnik parents take him to a child psychiatrist,[b] and Ned's mom says, "Yeah. You've got to help us Doc. We've tried nothing, and we're all out of ideas."

I kind of feel the same way about road safety.

You might scoff at that, and to some extent, I don't disagree. When it comes to road safety, we've definitely tried some things. Too many crashes? Speeds too high? We'll study the problem and flip through our first aid kit of interventions. If you're lucky, we'll find one we can afford and stick a Band-Aid on the problem.

But what we haven't done is try to fix systematic traffic engineering issues that led to the underlying conditions. So when it comes to that, I'd say Ned's mom is right. Up until now, we've tried nothing, and we're all out of ideas.

Fixing our road safety problems will mean fundamental change. And fundamentally changing our transportation system means fundamentally changing the traffic engineering discipline that created it. Traffic engineers need to reassess:

- Our so-called standards, which have been built on a rickety foundation of pseudoscientific theories.
- How our design guidelines and processes encourage more speed, more capacity, and more driving.

---

[a] "Hurricane Neddy" is the eighth episode of the eighth season of *The Simpsons*. Toward the end of the episode, Ned says, "And if you really tick me off, I'm going to run you down with my car."

[b] The doctor tests an experimental therapy called the University of Minnesota Spankalogical Protocol (that is, eight straight months of spanking).

- How our safety metrics reward more driving, to the detriment of safer humans.
- How our oversimplified crash data reassures us that the safety problems are not the fault of traffic engineers.
- How traffic engineers blame everyone but ourselves and expect *enforcement*, *education*, or *technology* to save the day.
- How traffic engineers let a fear of liability become our built-in excuse not to do better.
- How traffic engineers fail to teach the next generation of traffic engineers to be better.

In other words, traffic engineers need to reassess the fundamentals of … everything we do.

Education and *enforcement*—if done right—can be part of the story, but they shouldn't be the whole story or even the lead story.

But "done right" isn't easy. We follow the oversimplified crash data and create education campaigns that cross into victim blaming. We follow the moving violation data and ramp up enforcement in lower-income and minority neighborhoods. We then create more data points in these neighborhoods—confirming our biases—while the same behaviors go undetected in other neighborhoods. Or worse, we end up with unnecessary and dangerous confrontations that fail to solve the underlying problems we allegedly want to solve.

We also can't sit around and wait for *technology* to save the day.

Even if autonomous vehicles were to take over the streets soon, they will be overhyped, they will be glitchy, and half the population will treat them like some sort of conspiracy theory.

For certain crash types, technology can help the cause. My cell phone, for instance, has far better security than my car.

That's why a 41-year-old woman from New Jersey with no driver's license, 39 license suspensions on her record, and "a history of regular substance abuse, including 13 years of smoking crack cocaine on a daily basis," was able to drink a half-pint of vodka and run down and kill 16-year-old Quason Turner.[1]

Neither education nor enforcement saved Quason, who was trying to walk home from his job at the Cherry Hill Mall in New Jersey. And it wouldn't take much in the way of technology to have kept this woman from driving. Or our cell phones from distracting us. Or our vehicles from being allowed to exceed the speed limit.

But that doesn't mean traffic engineers shouldn't first fix the six-lane, 45 mph design with a concrete median running right through the middle of town. Even worse, this stretch of road where Quason Turner died was already signed as the "'We Shall Never Forget' Memorial ¼-Mile."

What do we need to fix streets like this? ***Money*** is always our answer.

But if we gave today's traffic engineers all the money in the world, I'm not convinced we'd make the world all that much safer. Traffic engineers with unlimited budgets dream about diverging diamond interchanges and continuous flow intersections. And when I say pedestrians and bicyclists need help, they come up with inane five-stage[c] pedestrian crossings.[2]

Traffic engineers think this way because our processes value saving you time more than saving your life.

Last I checked, the US Department of Transportation (USDOT) values a "statistical life" at around $11.8 million, which is less than the number we get when we multiply Denver's population (713,252 in 2022) by the generic USDOT hourly value of time savings number ($17 per hour).

Do the math and you'll see it's better to kill one person every day than it would be for all of Denver to sit in traffic for an hour each day.

In the late 1970s, we valued a life at $300,000[d] and our value of time at right around $4 an hour.[3] Using these numbers, every resident of a city as big as Denver would only have to sit in daily traffic for six or seven minutes before somebody dying each day was a worthwhile trade-off. We built most of our existing transportation system with cost-benefit analyses based on numbers like these.

---

[c] A two-stage crossing refers to a street with a median that provides "refuge" for the pedestrian. The pedestrian crosses the first half of the street in stage 1 and then hangs out in the middle of the street for a chance to cross the second half in stage 2. A five-stage crossing is that system on steroids in the middle of an intersection on steroids.

[d] That is, unless you're over 60 or disabled in some way. If so, your life wasn't valued as highly back then.

We can argue how much a life is worth, or how much people value their time, from now until the end of time. But the real problem is that the traffic engineering discipline still doesn't put our money where our mouth is. Our mouths say safety is our first priority, but we put our money into solving congestion.

Here's the kicker: we've also done an awful job of fixing congestion. And we've squandered most of the money you've given us along the way in trying to do so.

What will it take to shift our basic priorities?

It's easy to think it's just a matter of political will. That is, if our leaders want to put money toward designing safer streets instead of less congested streets, they can earmark the money and put us on a path toward fixing our safety problem.

But it's not that easy. Traffic engineers can sell anything as a safety intervention. We can even sell almost anything as a pedestrian or bicyclist safety intervention. Earmark that safety money all you want. Traffic engineers will do what we do under a different umbrella.

We all need to take a stand if we want to see fundamental change.

Hey traffic engineers—that means you too.

88   Keep Asking Why

—

Despite everything you've read in this book, don't be too hard on your friendly, neighborhood traffic engineer. At least not yet.

I know a lot of traffic engineers. Most are honestly trying to do the best they can with the limited knowledge they've been given.

We don't condemn doctors when a clinical trial isn't effective. Instead, we hope for the best but understand that medicine is an experimental science. When done right, the medical field will use that empirical data and try to get better. The problematic doctors hold on to to discredited approaches and still give kids heroin for a sore throat. We might've solved the sore throat, but we've created a bigger safety issue.

The problematic traffic engineers see the science as settled. They continue to blindly do what we've always done despite empirical data telling us that something isn't right. We might've reduced vehicle delay during the peak 15-minute period, but we've created some bigger safety issues.

We now know that the traffic engineering discipline is not as steeped in experimental science as we'd like to believe. So are we going to follow the evidence, learn from our mistakes—and successes—and try to get better? Or are we going to continue delivering babies with unwashed hands because that's what we've always done?

If you are a traffic engineer, this is your fork in the road.

If you want to ignore the ever-expanding mountains of evidence shouting that things aren't working out, you may eventually do so at your own peril. There is precedent—at least in Europe—for charging civil engineers with negligence, manslaughter, and homicide.[a] While I've only seen cases related to structural or geotechnical civil engineers, usually of dams, why not traffic engineers?

—

[a] France's Malpasset Dam failed in 1959. The head engineer was charged with negligence and homicide. In 1963, Italy's Vaiont Dam failed, and 14 engineers were prosecuted for manslaughter.

Structural and hydrology engineers want you to go to bed each night feeling that all dam things will be good come morning. But traffic engineering isn't the same as living downstream from a dam.

If we tell you we've built a transportation system or a car where it's impossible to die, we've done something wrong. The designers of the *Titanic*—and thus the captain—thought the ship was unsinkable. A minefield of icebergs dead ahead? Barely an inconvenience. Let's continue rearranging the deck chairs.[b]

Traffic engineers don't need to build an unsinkable ship, but we do need to learn from our mistakes. And we've made a lot of damn mistakes over the last 100 years. Some are as obvious as a downtown slip lane built to make the ***level of service*** at a downtown intersection marginally better … and the safety of pedestrians a whole lot worse. Some remain embedded in the ways we measure safety, assign blame, or determine the futuristic demand of a street.

We've done some things right, but traffic engineering remains a discipline stuck in its dark age. We've got work to do to reach our age of enlightenment, and we need your help.

The first time I saw my PhD advisor, Norman Garrick, give a presentation was at the Transportation Research Board conference in Washington, DC, on a warm January morning in 2005. His speech was nominally about functional classification and context-sensitive street design.[1] His speech was really a traffic engineering call to action. I looked around at audience faces not understanding what they were watching, but that was the moment I knew I had come to the right place. This book is my answer to his call. This final chapter is my own traffic engineering call to action.

When Steve Jobs was trying to hire John Sculley away from Pepsi, Jobs said, "Do you want to spend the rest of your life selling sugared water, or do you want a chance to change the world?"[2]

Traffic engineers have a lot of work to do if we want to stop selling sugared water and start changing the world. It needs to start with fundamental change. Keep asking ***why*** we do what we do. Keep asking ***why*** those using our streets do what they do. Once we start answering those ***why*** questions—and connect the dots between what we do as traffic engineers and what they do as road users—I know we can do better. We can be better.

---

[b] 1,500-plus people would disagree.

But traffic engineers aren't the only ones with a chance to change the world.

I was 10 years old when I happened to stay home sick from school and watch the *Challenger* launch live on television. Claus Jensen later wrote about what happened and why it's so easy for engineers to miss the bigger picture:

> We should not expect the experts to intervene, nor should we believe that they always know what they are doing. Often they have no idea, having been blinded to the situation in which they are involved. These days, it is not unusual for engineers and scientists working within systems to be so specialized that they have long given up trying to understand the system as a whole, with all its technical, political, financial and social aspects.[3]

Traffic engineers aren't immune from specialization. It's easy to get stuck in the weeds of our methods and processes and miss the bigger picture. So even if you aren't a traffic engineer, this call to action means you too. Show us our blind spots. Call us out on our *engineer speak*. Make sure we see the bigger picture.

In his 1939 book *Seven Roads to Safety*, our old friend Paul Hoffman daydreamed about what the world would be like in 30 years. His thinking was that "traffic engineering will probably be the most important single contributing element to safety."[4]

He wasn't wrong. But like doctors for their first 1,000 years, we spent our first 100 years killing more people than we saved. It's about time we flip the script and start contributing to better safety.

So what are you gonna do about it?

About the Author

—

Wes Marshall, PhD, PE, is a professor of civil engineering at the University of Colorado Denver, where he holds a joint appointment in urban planning. He plays a pivotal role as director of the CU Denver Human-Centered Transportation program and the Transportation Research Center at CU Denver. Wes is a licensed Professional Engineer and focuses on transportation teaching and research dedicated to creating safer and more sustainable transportation systems.

Photograph by Paul Wedlake

Wes has more than 80 peer-reviewed journal publications and book chapters to his name and has received millions of dollars in research funding. He was the winner of the campus-wide CU Denver Outstanding Faculty in Research Award. He also has a passion for teaching and mentoring students and is the only three-time winner of the CU Denver College of Engineering Outstanding Faculty in Teaching Award.

A native of Watertown, Massachusetts, Wes is a graduate of the University of Virginia (BS) and the University of Connecticut (MS and PhD). He is a recipient of the Eisenhower Transportation Fellowship, Australia's Endeavour Fellowship, and the Transportation Research Board's Wootan Award for the outstanding paper in policy and organization.

He has no plans to stop working on any of this anytime soon.

Acknowledgments

—

Whenever my graduate students can't come up with a research topic, I ask them, what drives you crazy? This strategy may not work as well for nontransportation research, but when it comes to what I do, the answer can lead to some of the best research questions.

This endeavor started off with something very big that drove me very crazy. As I look back at the last few years, I admit that I put an insane amount of time and energy into this book. But other than this acknowledgments section, not one minute ever felt like work.

So thank you to the early traffic engineers. Even though a lot of your work wasn't easy to track down, it was worth the effort. Digging in helped me understand what was going through your minds and see that, in many instances, your hearts were in the right place.

Thank you to the midcentury traffic engineers who made a mountain of guidebooks out of a molehill of empirical evidence. Without you, this book would not have been possible.

And thank you to the enlightened traffic engineers of today as well as anyone else fighting to right the wrongs I harp on in this book. We may be few and far between, but if we keep learning from one another and pushing in the right direction, our numbers will grow and grow quickly.

I also want to thank both the teachers I've had through the years and my students. My PhD advisor hates it when my PhD students call him their grand advisor, but that is the legacy we are building. And it will be a good one.

Thank you to my colleagues and collaborators, especially for your patience during the times my mental energy kept shifting back to transportation books from the 1930s and 1940s. Steve Jobs once said that "focusing is about saying no," but I am looking forward to more yeses (it's also hard to believe that "yeses" is really a word).

Thank you to my long-suffering friends for indulging my nonsense, offering to voice the audiobook, and always responding with gusto whenever I asked for their two cents or whether one of my many pop culture or sports references was too obscure.

On that note, thank you to Heather Boyer of Island Press for challenging the obscurity of some of those references, as well as for being a fantastic editor and champion of this book. Likewise, thank you to Sharis Simonian for her efficient project supervision and Kathleen Lafferty for her careful copyediting, as well as everyone else at Island Press for believing in my vision and the need to tell these stories.

I wish this part wasn't our reality, but I need to acknowledge and extend my deepest sympathy to the millions and millions of people whose stories were cut short by traffic violence. It sucks. You didn't deserve it, and your family and friends don't either. I just hope traffic engineers can start doing better so that fewer and fewer people end up joining this unfortunate club.

The rest of my love and appreciation all goes to my family—who pretend not to mind when I stop to take a dozen pictures of a street or intersection when we are out and about. Thank you to my parents and grandparents, who helped pave the way and put me in this position, as well as to my kids for walking and biking to school even when it's really cold outside. And thank you most of all to my wife, Debbie, for her love, patience, and understanding as she took on the real world while I put together this big, dumb puzzle of a book.

Now, I am going to bike home....

Wes Marshall
September 1, 2023
Denver, Colorado

Endnotes

Chapter 1

1. Michael Lewis, "The Data Coach," *Against the Rules*, podcast, 2020.
2. Charles Wright, "Urban Transport, Health, and Synergy," *Transportation Quarterly* 45, no. 3 (1991) (italics added).
3. Laurence Morehouse and Leonard Gross, *Total Fitness in 30 Minutes a Week* (New York: Simon and Schuster, 1975).
4. Ralph Paffenbarger et al., "Physical Activity, All-Cause Mortality, and Longevity of College Alumni," *New England Journal of Medicine* 314, no. 10 (1986).
5. Rose George, *The Big Necessity* (New York: Macmillan, 2008).
6. Road Research Laboratory, *Research on Road Traffic* (London: H.M. Stationery Office, 1965).
7. Seymour Taylor, "Freeways Alone Are Not Enough," *Traffic Quarterly* 13, no. 3 (1959).
8. Road Research Laboratory, *Research on Road Traffic*.

Chapter 3

1. Alvin Goldstein, "Traffic Safety, the Courts, and the Schools," *Traffic Quarterly* 22, no. 1 (1968).
2. Centers for Disease Control and Prevention, "WISQUARS Injury Data," database, last reviewed November 8, 2023.
3. Patricia Waller, "Speed Limits: How Should They Be Determined?," *Transportation Quarterly* 56, no. 3 (2002).
4. Howard Matthias, "A Different Look at America's Highway Accident Record," *Transportation Quarterly* 37, no. 4 (1983).
5. Earl Campbell, "Highway Traffic Safety—Is It Possible?," *Traffic Quarterly* 19, no. 3 (1965).
6. Richard Retting, "Urban Traffic Crashes: New York City Responds to the Challenge," *Transportation Quarterly* 45, no. 4 (1991).
7. Maria Segui-Gomez et al., "Exposure to Traffic and Risk of Hospitalization Due to Injuries," *Risk Analysis: An International Journal* 31, no. 3 (2011).
8. Organisation for Economic Co-operation and Development, *OECD Factbook—Quality of Life: Transport* (Paris: OECD, 2006).

Chapter 4

1. Edward Fisher and Robert Reeder, *Vehicle Traffic Law* (Evanston, IL: Traffic Institute, Northwestern University, 1974).
2. William Eno, "Suggested Rules and Reforms for the Management of Street Traffic—Designed to Minimize Accidents and Prevent Confusion, Delay and Blockades," *Rider and Driver* 12, no. 6 (1902).
3. Robert Goetz, "An Editorial," *Traffic Quarterly* 1, no. 1 (1947).

4. John Carter, "The Urban Traffic Problem," *Traffic Quarterly* 16, no. 2 (1962).
5. Scheffer Lang, "A New Transportation Paradigm," *Transportation Quarterly* 53 (1999).

Chapter 5

1. Sidney Dekker, *The Field Guide to Understanding "Human Error"* (Boca Raton, FL: CRC Press, 2017).
2. Ibid.
3. James Reason, *Human Error* (Cambridge, UK: Cambridge University Press, 1990).
4. Wes Marshall, Daniel Piatkowski, and Aaron Johnson, "Scofflaw Bicycling: Illegal but Rational," *Journal of Transport and Land Use* 10, no. 1 (2017).

Chapter 6

1. National Highway Traffic Safety Administration, "Fatality Analysis Reporting System (FARS)" US Department of Transportation database, 1975–present.
2. Richard Retting, "Pedestrian Traffic Fatalities by State" (2020).
3. David Schwebel, Aaron Davis, and Elizabeth O'Neal, "Child Pedestrian Injury," *American Journal of Lifestyle Medicine* 6, no. 4 (2012).
4. Robert Goetz, "Accidents," *Traffic Quarterly* 14, no. 1 (1960).
5. Robert Goetz, "Safety: An Editorial," *Traffic Quarterly* 15, no. 4 (1961).
6. Goetz, "Accidents."

Chapter 7

1. J. Edgar Hoover, "The First Consideration," *Traffic Quarterly* 25, no. 4 (1971).
2. Paul G. Hoffman and Neil M. Clark, *Seven Roads to Safety* (New York: Harper and Brothers, 1939).
3. Kenneth Todd, "Modern Rotaries," *Transportation Research Record* 737 (1979).
4. Institute of Traffic Engineers, "About ITE: Jack E. Leisch," 2023, www.ite.org/about-ite/history/honorary-members/jack-e-leisch/.
5. William Sacks, "Testimony to Committee on Public Works," US Senate (1973).
6. Mike Isaac, *Super Pumped* (New York: Norton, 2019).
7. Charles Rotter, "Self-Driving Uber Running Red Light," December 14, 2016, YouTube video, www.youtube.com/watch?v=_CdJ4oae8f4.
8. Mike Isaac and Daisuke Wakabayashi, "A Lawsuit Against Uber," *New York Times* (2017).
9. Ibid.
10. Sidney Dekker, *The Field Guide to Understanding "Human Error"* (Boca Raton, FL: CRC Press, 2017).
11. Hallie Myers, "Promoting Traffic Safety in Indiana," *Traffic Quarterly* 9, no. 4 (1955).
12. Delbert Taebel, "Citizen Groups, Public Policy and Urban Transportation," *Traffic Quarterly* 27, no. 4 (1973).
13. Ibid.
14. Seymour S. Taylor, "Freeways Alone Are Not Enough," *Traffic Quarterly* 13, no. 3 (1959).
15. Barry Wellman, "Public Participation in Transportation Planning," *Traffic Quarterly* 31, no. 4 (1977).

Chapter 8

1. Dan Brown, *Origin* (New York: Anchor, 2017).
2. Ibid.
3. Kenneth Todd, "A History of Roundabouts," *Transportation Quarterly* 42, no. 4 (1988).
4. David Klein, "The Teen-Age Driver," *Traffic Quarterly* 22, no. 1 (1968).
5. John Grimaldi, "The Road Ahead for Traffic Safety?," *Traffic Quarterly* 25, no. 4 (1971).
6. Richard Jones, "Federal-State Partnership in Highway Safety," *Traffic Quarterly* 24, no. 4 (1970).
7. Ibid.
8. William Haddon and David Klein, "Assessing the Efficacy of Accident Measurements," *Traffic Quarterly* 19, no. 3 (1965).
9. John Garrett, "Evaluation of Effectiveness of Door Locks," *Public Health Reports* 77, no. 5 (1962).
10. John Moore and Boris Tourin, "A Study of Automobile Doors Opening under Crash Conditions" (1954).
11. Jane Jacobs, *Dark Age Ahead* (New York: Random House, 2004).
12. Jane Jacobs, *The Death and Life of Great American Cities* (New York: Random House, 1961).
13. William Haddon and David Klein, "Assessing the Efficacy of Accident Measurements," *Traffic Quarterly* 19, no. 3 (1965).
14. Leonard Evans, "Future Predictions and Traffic Safety Research," *Transportation Quarterly* 47, no. 1 (1993).
15. Committee on Public Works, US House of Representatives, "Highway Safety, Design and Operations" (1973).

Chapter 9

1. Ezra Hauer, "Safety in Geometric Design Standards I: Three Anecdotes," International Symposium on Highway Geometric Design (2000).
2. Ibid.
3. Ibid.
4. American Society of Civil Engineers (ASCE), "Code of Ethics" (2017).
5. National Society of Professional Engineers (NSPE), "Code of Ethics" (2019).
6. Institute of Transportation Engineers (ITE), "Code of Ethics" (2003).

Chapter 10

1. Robert Goetz, "An Editorial," *Traffic Quarterly* 1, no. 1 (1947).
2. American Association of State Highway Officials, *A Policy on Arterial Highways in Urban Areas* (Washington, DC: AASHO, 1957).
3. Bernard Fox, "The Problem of Countermeasures in Drinking and Driving," *Traffic Quarterly* 19, no. 3 (1965).
4. William Greene, "Getting Support for Traffic Safety Education in the Community," *Traffic Quarterly* 1, no. 2 (1947).
5. Ibid.
6. Sidney Dekker, *The Field Guide to Understanding "Human Error"* (Boca Raton, FL: CRC Press, 2017).
7. Ibid.

8. Hallie Myers, "Promoting Traffic Safety in Indiana," *Traffic Quarterly* 9, no. 4 (1955).
9. Arthur Henderson, "European and North American Traffic, Engineering and Design," *Traffic Quarterly* 16, no. 4 (1962).
10. Phil Ellis, "Traffic Safety," *Traffic Quarterly* 17, no. 1 (1963).
11. AAA, "Planned Pedestrian Program," AAA Foundation for Traffic Safety (1958).
12. Ezra Hauer, "The Reign of Ignorance in Road Safety," in *Transportation Safety in an Age of Deregulation*, ed. Leon M. Moses and Ian Savage (New York: Oxford University Press, 1989).
13. Dekker, *Field Guide*.

Chapter 11

1. James Malfetti, "Scare Technique and Traffic Safety," *Traffic Quarterly* 15, no. 2 (1961).
2. Ibid.
3. Neo Film Reviews, "MCL Cinema Hong Kong Mobile Phone Car Crash Advertising Effective," June 29, 2014, YouTube video, www.youtube.com /watch?v=5Gtio4V1L3o.
4. Malfetti, "Scare Technique."
5. T. J. Carmichael, "This Business of Safe Driving," *Traffic Quarterly* 2, no. 4 (1948).

Chapter 12

1. David Baldwin, "Can We Afford More Accidents?," *Traffic Quarterly* 3, no. 4 (1949).
2. Frank McGlade, "Traffic Accident Research," *Traffic Quarterly* 16, no. 4 (1962).
3. Kenneth Cook and Wayne Ferguson, "What Do Teen-Agers Really Think about Traffic Safety?," *Traffic Quarterly* 22, no. 2 (1968).
4. National Highway Traffic Safety Administration, "Teen Driving," accessed December 4, 2023, www.nhtsa.gov/road-safety/teen-driving.
5. Paul G. Hoffman and Neil M. Clark, *Seven Roads to Safety: A Program to Reduce Automobile Accidents* (New York: Harper and Brothers, 1939).
6. Joseph Grillo, "Highway Safety Goals of Motor Vehicle Registration," *Traffic Quarterly* 27, no. 3 (1973).
7. Joseph Little, "Highway Safety Programs and the Public Trust," *Traffic Quarterly* 22, no. 4 (1968).
8. Chris McCahill, "Driver's Licenses Are Too Often a Tool of Oppression," State Smart Transportation Initiative (February 18, 2021).
9. "Tag the Reckless Motorist," editorial, *Saturday Evening Post* (September 3, 1921).

Chapter 13

1. J. Edgar Hoover, "The First Consideration," *Traffic Quarterly* 25, no. 4 (1971).
2. Richard Raub, "Projecting Police Traffic Enforcement Workload," *Transportation Quarterly* 42, no. 2 (1988); Anuja Sarode et al., "Traffic Stops Do Not Prevent Traffic Deaths," *Journal of Trauma and Acute Care Surgery* 91, no. 1 (2021).
3. Lawrence Hince, "Seven Steps to Safety," *Traffic Quarterly* 1, no. 1 (1947).

4. W. D. Hoback, "Accidents on Oklahoma Turnpikes," *Traffic Quarterly* 13, no. 4 (1959).
5. AAA, "Planned Pedestrian Program," AAA Foundation for Traffic Safety (1958).
6. James Slavin, "The Role of Traffic Law Enforcement," *Traffic Quarterly* 21, no. 4 (1967).
7. Ibid.
8. Paul G. Hoffman and Neil M. Clark, *Seven Roads to Safety: A Program to Reduce Automobile Accidents* (New York: Harper and Brothers, 1939).
9. Dorothy Lee, "Portland Protects Its Pedestrians," *Traffic Quarterly* 6, no. 3 (1952).
10. John Leeming, "Road Accidents, Other Accidents, and Our Laws," *Traffic Quarterly* 11, no. 3 (1957).
11. Lee, "Portland Protects Its Pedestrians."
12. Ibid.
13. Delaney Hall, "Curb Cuts," *99% Invisible*, episode 308, podcast, April 27, 2021, https://99percentinvisible.org/episode/curb-cuts/.
14. Marcel Honore, "Honolulu Is Slashing Crosswalks," *Honolulu Civil Beat* (2019).
15. Richard Retting, "Traffic Engineering Approaches to Improving Pedestrian Safety," *Transportation Quarterly* 53, no. 2 (1999).

Chapter 14

1. Quinn Tamm, "The Police: Pivot for Highway Safety Efforts," *Traffic Quarterly* 18, no. 2 (1964).
2. American Association of State Highway and Transportation Officials, *A Policy on Geometric Design of Highways and Streets*, 4th ed. (Washington, DC: AASHTO, 2001).
3. American Association of State Highway and Transportation Officials, *A Policy on Geometric Design of Highways and Streets*, 6th ed. (Washington, DC: AASHTO, 2011).
4. Ibid.
5. Solomon Morris, "Freeways and the Urban Traffic Problem," *Traffic Quarterly* 27, no. 4 (1973).
6. Theodore Caplow, "The Theory of Traffic Control," *Traffic Quarterly* 6, no. 4 (1952).
7. James Antoniou, "Planning for Pedestrians," *Traffic Quarterly* 25, no. 1 (1973).

Chapter 15

1. National Transportation Safety Board, "Single-Vehicle Run-Off-Road Crash and Fire, Fort Lauderdale, Florida, May 8, 2018," Highway Accident Brief, 2019.
2. "Automatic Speed Governor," *Scientific American* 97, no. 19 (1907).
3. Peter Norton, *Fighting Traffic: The Dawn of the Motor Age in the American City* (Cambridge, MA: MIT Press, 2011).
4. AAA, "Planned Pedestrian Program," AAA Foundation for Traffic Safety (1958).
5. Bruce Greenshields, "Attitudes, Emotions, Accidents," *Traffic Quarterly* 13, no. 2 (1959).

6. Kenneth Haase, "Characteristics of Persons Injured in Motor Vehicle Accidents," *Traffic Quarterly* 17, no. 4 (1963).

Chapter 16

1. Leonard Evans, *Traffic Safety and the Driver* (Bloomfield Hills, MI: Science Serving Society, 1991).
2. Charles Lave, "Economic Considerations," *Transportation Research Circular* 264 (1983).
3. E. Desapriya et al., "Do Light Truck Vehicles Impose Greater Risk of Pedestrian Injury?," *Traffic Injury Prevention* 11, no. 1 (2010).
4. Marc Fredette et al., "Safety Impacts Due to the Incompatibility of SUVs, Minivans, and Pickup Trucks," *Accident Analysis and Prevention* 40, no. 6 (2008).
5. Desapriya et al., "Do Light Truck Vehicles Impose Greater Risk?"
6. Ronald Burns, "Realities of Automotive Safety," *Transportation Quarterly* 53, no. 1 (1999).
7. William Luneburg, "Responsible Concern for Safety," *Traffic Quarterly* 25, no. 4 (1971).
8. Burns, "Realities of Automotive Safety."
9. Ibid.

Chapter 17

1. "Airbag Crisis Confronts Auto Industry," *Geneva Times* (1972).
2. Leo Wolinsky, "Big Lobbies Clash in Fight on Seat Belts," *Los Angeles Times* (February 19, 1985).
3. Statement of General Motors Corporation to the U.S. Senate Committee on Finance (1981).
4. Ibid.
5. *Motor Vehicles Manufacturers Association v. State Farm Mutual Automobile Insurance Co.* (1983).
6. NHTSA, "Lives Saved in 2017 by Restraint Use and Minimum-Drinking-Age Laws," DOT-HS-812-683 (2019).
7. WTHR.com Staff, "13 Investigates Reveals Hidden Dangers in Your Vehicle's 'Blind Zone,'" WTHR 13, April 25, 2019, https://www.wthr.com/article/wthrs-blind-zone-check.
8. Michael Baltes, "The Cost-Effectiveness of Safety Options for Big Public School Buses," *Transportation Quarterly* 53, no. 3 (1999).

Chapter 18

1. James Taylor, "Moral Dilemmas in Highway Safety Decisions," *Traffic Quarterly* 35, no. 1 (1981).
2. Austin Frakt, "Putting a Dollar Value on Life? Governments Already Do," *New York Times* (2020).
3. Taylor, "Moral Dilemmas."
4. A. E. Johnson, "The Interstate System," *Traffic Quarterly* 10, no. 2 (1956).
5. D. J. Reynolds, "The Cost of Road Accidents," *Journal of the Royal Statistical Society* 119, no. 4 (1956).
6. Ibid.
7. Lawrence Blincoe et al., *The Economic and Societal Impact of Motor Vehicle*

Crashes (DOT-HS-813-403), National Highway Traffic Safety Administration (2023).

8. Robert Anderson, "Economics of Highway Safety," *Traffic Quarterly* 29, no. 1 (1975).

Chapter 19

1. Craig Gilbert, "Johnson Is Telling People to Keep Coronavirus in Perspective," *Milwaukee Journal Sentinel* (2020).
2. Herbert Hoover, "Opening Address," National Conference on Street and Highway Safety (1924).
3. M. Earl Campbell, "Traffic Safety—Is It Possible?," *Traffic Quarterly* 19, no. 3 (1965).
4. Ibid.
5. Philip Fleming, "After the President's Highway Safety Conference," *Traffic Quarterly* 1, no. 1 (1947).
6. Ibid.
7. Harry S. Truman, "Harry S. Truman: Containing the Public Messages, Speeches, and Statements of the President, 1945–53," US Government Printing Office (1966).
8. Robert Goetz, "Our Millionth Victim," *Traffic Quarterly* 5, no. 3 (1951).
9. Charles Perrow, *Normal Accidents: Living with High-Risk Technologies* (Princeton, NJ: Princeton University Press, 1999).
10. Bruce Greenshields, "Traffic Accidents—The Uncommon Common Events Problem," *Traffic Quarterly* 27, no. 2 (1973).
11. Robert Baker, *Highway Risk Problem: Policy Issues in Highway Safety* (Hoboken, NJ: Wiley, 1971).

Chapter 20

1. Robert Barkley, "The Civil Engineering Function in Urban Renewal," *Traffic Quarterly* 18, no. 3 (1964).
2. William McGrath, "Transportation and Urban Development," *Traffic Quarterly* 18, no. 4 (1964).
3. Ibid.
4. Robert Holmes, "Traffic Engineering as a Profession," *Traffic Quarterly* 3, no. 4 (1949).
5. Herbert Thriscutt, "The Training of Traffic Engineers," *Traffic Quarterly* 16, no. 3 (1962).
6. AreaVibes, "Top 10 Worst Cities—Worst Places to Live 2013" (2013).
7. Nick Aresco, "Springfield among Worst Cities to Drive in Across the U.S., Study Says," WWLP News (2019).
8. MassDOT, "Top Crash Locations, 2018–2020"; MassDOT, "Top Crash Location Reports by Year" (2006–2017).
9. John Forester, "The Bicycle Transportation Controversy," *Transportation Quarterly* 55, no. 2 (2001).
10. Krys Ochia, "Bicycle Programs and Provision of Bikeway Facilities," *Transportation Quarterly* 47, no. 3 (1993).
11. Forester, "Bicycle Transportation Controversy."
12. David Byrne, *Bicycle Diaries* (New York: Penguin, 2010).
13. Jane Jacobs, "Downtown Is for People," *Fortune* (April 1958).

14. Alina Tugend, *Better by Mistake: The Unexpected Benefits of Being Wrong* (New York: Penguin, 2011).
15. Wes Marshall, Norman Garrick, and Gilbert Hansen, "Reassessing On-Street Parking," *Transportation Research Record* 2046, no. 1 (2008).
16. Harold Marks, "Subdividing for Traffic Safety," *Traffic Quarterly* 11, no. 3 (1957).

Chapter 21

1. Bahar Barami and Clark Merrefield, "Beyond Traffic 2045," DOT-VNTSC-16-01 (2016).
2. Lawrence Hince, "Seven Steps to Safety," *Traffic Quarterly* 1, no. 1 (1947).
3. National Highway Traffic Safety Administration, "Fatality Analysis Reporting System (FARS)" (Washington, DC: US Department of Transportation); Federal Highway Administration, "Traffic Volume Trends," database, https://www.fhwa.dot.gov/policyinformation/travel_monitoring/tvt.cfm.
4. David Baldwin, "Highway Engineering and Traffic Safety," *Traffic Quarterly* 31, no. 3 (1977).
5. Ibid.
6. Ezra Hauer, "Engineering Judgment and Road Safety," *Accident Analysis and Prevention* 129 (2019).
7. Bureau of the Census, "American Time Use Survey (ATUS)," 2018.

Chapter 22

1. Laurence Aurbach, *A History of Street Networks: From Grids to Sprawl and Beyond* (Hyattsville, MD: PedShed Press, 2020).
2. Anthony Peranio, "Conceptualization and Use of Road Safety and Traffic Engineering Formulas," *Traffic Quarterly* 25, no. 3 (1971).

Chapter 23

1. Laurence Aurbach, *A History of Street Networks: From Grids to Sprawl and Beyond* (Hyattsville, MD: PedShed Press 2020).
2. Joseph Furnas, "—And Sudden Death," *Reader's Digest* (1935).
3. James Malfetti, "Scare Technique and Traffic Safety," *Traffic Quarterly* 15, no. 2 (1961).
4. Aurbach, *A History of Street Networks.*
5. Ibid.
6. Ibid.
7. Ibid.
8. Paul G. Hoffman and Neil M. Clark, "The White Line Isn't Enough," *Saturday Evening Post* (1938).
9. Paul G. Hoffman and Neil M. Clark, *Seven Roads to Safety: A Program to Reduce Automobile Accidents* (Harper and Brothers, 1939).
10. Ibid.
11. Arthur Stevens, *Highway Safety and Automobile Styling* (Boston: Christopher Publishing, 1941).
12. S. M. Breuning and A. J. Bone, "Interchange Accident Exposure," *Highway Research Board Bulletin* 240 (1960).
13. David Klein, "The Teen-Age Driver," *Traffic Quarterly* 22, no. 1 (1968).

Chapter 24

1. Wes Marshall and Nicholas Ferenchak, "Assessing Equity and Urban/Rural Road Safety Disparities in the US," *Journal of Urbanism* 10, no. 4 (2017).

Chapter 25

1. Herbert Heinrich, *Industrial Accident Prevention: A Scientific Approach* (New York: McGraw-Hill, 1931).
2. Ibid.
3. Norman Damon, "How Much Further Can We Reduce Traffic Accidents?," *Traffic Quarterly* 4, no. 4 (1950).
4. American Association of State Highway Officials, *A Policy on Arterial Highways in Urban Areas* (Washington, DC: AASHO, 1957).
5. Sidney Dekker, *The Field Guide to Understanding "Human Error"* (Boca Raton, FL: CRC Press, 2017).
6. David Schoppert, *Traffic Control and Roadway Elements: Their Relationship to Highway Safety* (Automotive Safety Foundation, 1963).
7. Arnold Vey, "Effect of Signalization on Motor Vehicle Accident Experience," New Jersey Department of Motor Vehicles (1933).
8. John Grimaldi, "The Road Ahead for Traffic Safety?," *Traffic Quarterly* 25, no. 4 (1971).
9. Ibid.
10. Dekker, *The Field Guide*.

Chapter 26

1. Herbert Heinrich, *Industrial Accident Prevention* (New York: McGraw-Hill, 1931).
2. Frank E. Bird and George L. Germain, *Damage Control* (American Management Association, 1966).
3. John Phillips, "The Story of Hans Monderman and the Safety of Insecurity," *Car and Driver* (2016).
4. As quoted in Project for Public Spaces, "Hans Monderman," obituary, December 31, 2008.

Chapter 27

1. James I. Taylor, "Moral Dilemmas in Highway Safety Decisions," *Traffic Quarterly* 35, no. 1 (1981).
2. American Association of State Highway Officials, *A Policy on Arterial Highways in Urban Areas* (Washington, DC: AASHO, 1957).
3. Ibid.
4. Ibid.
5. Ibid.
6. Arnold Vey, "Relationship between Daily Traffic and Accident Rates," *American City* 52, no. 9 (1937).
7. Neil Arason, *No Accident: Eliminating Injury and Death on Canadian Roads* (Waterloo, Canada: Wilfrid Laurier University Press, 2014).
8. Jack Hart, "Right Turns at Urban Intersections," *Traffic Quarterly* 3, no. 1 (1949).
9. Wilbur Smith, "Influence of Parking on Accidents," *Traffic Quarterly* 1, no. 2 (1947).

10. John Rae, "The Mythology of Urban Transportation," *Traffic Quarterly* 26, no. 1 (1972).
11. "Automobiles May Solve the Traffic Problem," *The Automobile* 20, no. 4 (1909).
12. Robert Goetz, "City Traffic," *Traffic Quarterly* 14, no. 2 (1960).
13. Robert Goetz, "Our Road Problem," *Traffic Quarterly* 10, no. 2 (1956).
14. Wilbur Smith and Jack Hart, "A Case Study of One-Way Streets," *Traffic Quarterly* 3, no. 4 (1949).
15. Laurence Aurbach, *A History of Street Networks: From Grids to Sprawl and Beyond,* (Hyattsville, MD: PedShed Press, 2020).
16. Ibid.
17. William Adkins, "Economic Impacts of Expressways," *Traffic Quarterly* 13, no. 3 (1959).
18. Tim Lomax, David Schrank, and Bill Eisele, "2021 Urban Mobility Report," Texas Transportation Institute (2021).
19. Farhad Atash, "Mitigating Traffic Congestion in Suburbs. An Evaluation of Land-Use Strategies," *Transportation Quarterly* 47, no. 4 (1993).
20. Natalie McConnell-Fay, "Tackling Traffic Congestion," *Transportation Quarterly* 40, no. 2 (1986).
21. Ibid.
22. Atash, "Mitigating Traffic."
23. Leonard Evans, "Future Predictions and Traffic Safety Research," *Transportation Quarterly* 47, no. 1 (1993).
24. Wes Marshall and Eric Dumbaugh, "Revisiting the Relationship between Traffic Congestion and the Economy," *Transportation* 47, no. 1 (2020).
25. Ibid.
26. Evans, "Future Predictions."
27. Ibid.
28. Ibid.

Chapter 28

1. AAA, *Planned Pedestrian Program*, AAA Foundation for Traffic Safety (1958).
2. Paul Carlson, Eun Sug Park, and Carl Andersen, "Benefits of Pavement Markings," *Transportation Research Record* 2107, no. 1 (2009).
3. Geni Bahar et al., "Pavement Marking Materials and Markers," NCHRP-17-28 (2006).
4. Ezra Hauer, "On the Relationship between Road Safety Research and the Practice of Road Design and Operation," *Accident Analysis and Prevention* 128 (2019).
5. James Musick, "Effect of Pavement Edge Marking on Two-Lane Rural State Highways in Ohio," *Highway Research Board Bulletin* 266 (1960).
6. A. J. Basile, "Effect of Pavement Edge Markings on Traffic Accidents in Kansas," *Highway Research Board Bulletin* 308 (1962).
7. Margie Peden et al., eds., *World Report on Road Traffic Injury Prevention* (Geneva: World Health Organization, 2004).

Chapter 29

1. Santokh Singh, "Critical Reasons for Crashes," (DOT-HS-812). NHTSA (2018).

2. Sidney Dekker, *The Field Guide to Understanding "Human Error"* (Boca Raton, FL: CRC Press, 2017).
3. Norman Damon, "How Much Further Can We Reduce Traffic Accidents?," *Traffic Quarterly* 4, no. 4 (1950).
4. Ibid.
5. Anthony Peranio, "Logic and Road Safety Research," *Traffic Quarterly* 23, no. 1 (1969).
6. John Treat and Kent Joscelyn, "A Study to Determine the Relationship between Vehicle Defects and Crashes," NHTSA (1971).
7. David Shinar and Amos Drory, "Sign Registration in Daytime and Nighttime Driving," *Human Factors* 25, no. 1 (1983).
8. John Treat et al., "Tri-Level Study of the Causes of Traffic Accidents," NHTSA (1979).
9. Ibid.
10. David Shinar, "Crash Causes, Countermeasures, and Safety Policy Implications," *Accident Analysis and Prevention* 125 (2019).
11. Ibid.
12. Amos Drory and David Shinar, "The Effects of Roadway Environment and Fatigue on Sign Perception," *Journal of Safety Research* 13, no. 1 (1982); David Shinar and Amos Drory, "Sign Registration in Daytime and Nighttime Driving," *Human Factors* 25, no. 1 (1983).
13. Shinar and Drory, "Sign Registration."
14. George Taoka, "Asleep at the Wheel," *Transportation Quarterly* 52, no. 1 (1998).
15. John Lauber and Phyllis Kayten, "Sleepiness, Circadian Dysrhythmia, and Fatigue," *Sleep: Journal of Sleep Research and Sleep Medicine* (1988).
16. Charlie Klauer et al., "The Impact of Driver Inattention on Near-Crash/Crash Risk: An Analysis Using the 100-Car Naturalistic Driving Study Data," DOT HS 810 594 (2006).
17. Thomas Dingus et al., "Driver Crash Risk Factors and Prevalence Evaluation Using Naturalistic Driving Data," *National Academy of Sciences Proceedings* 113, no. 10 (2016).

Chapter 30

1. "Kobe Bryant: Breaking Down Ben Simmons," *Detail: Kobe Bryant*, ESPN, March 24, 2020.
2. David Shinar, *Psychology on the Road* (Hoboken, NJ: Wiley, 1978); Dominique Lord, Alison Smiley, and Antoine Haroun, "Pedestrian Accidents with Left-Turning Traffic," TRB Annual Meeting (1998).
3. Zusha Elinson, "Bay Area Drivers Who Kill Pedestrians Rarely Face Punishment, Analysis Finds," Center for Investigative Reporting (2013).
4. Helena Stigson, Maria Krafft, and Claes Tingvall, "Use of Fatal Real-Life Crashes to Analyze a Safe Road Transport System Model," *Traffic Injury Prevention* 9, no. 5 (2008).
5. Bruce D. Greenshields, "Attitudes, Emotions, Accidents," *Traffic Quarterly* 13, no. 2 (1959).

Chapter 31

1. James Reason, *A Life in Error: From Little Slips to Big Disasters* (Boca Raton, FL: CRC Press, 2013).

2. James Reason, *Human Error* (Cambridge, UK: Cambridge University Press, 1990); James Reason, "Human Error: Models and Management," *BMJ* 320, no. 7237 (2000).

3. Alina Tugend, *Better by Mistake: The Unexpected Benefits of Being Wrong* (New York: Penguin, 2011).

Chapter 32

1. James Reason, *Human Error* (Cambridge, UK: Cambridge University Press, 1990).

2. Ibid.

3. National Transportation Safety Board, "Highway Accident Report: Collision between Vehicle Controlled by Developmental Automated Driving System and Pedestrian, Tempe, Arizona, March 18, 2018" (2019).

4. Ibid.

5. Neil Arason, *No Accident: Eliminating Injury and Death on Canadian Roads* (Waterloo, Ontario: Wilfrid Laurier University Press, 2014).

6. Victoria Gitelman, Etti Doveh, and David Zaidel, "An Examination of Billboard Impacts on Crashes on a Suburban Highway," *Traffic Injury Prevention* 20, no. 2 (2019).

7. Paul Staffeld, "Accidents Related to Access Points and Advertising Signs," *Traffic Quarterly* 7, no. 1 (1953).

Chapter 33

1. Arnold H. Vey, "Effect of Signalization on Motor Vehicle Accident Experience," New Jersey Department of Motor Vehicles (1933).

2. Anthony Peranio, "Logic and Road Safety Research," *Traffic Quarterly* 23, no. 1 (1969).

3. Hallie Myers, "Promoting Traffic Safety in Indiana," *Traffic Quarterly* 9, no. 4 (1955).

4. Ibid.

5. Bruce D. Greenshields, "Traffic Accidents—The Uncommon Common Events Problem," *Traffic Quarterly* 27, no. 2 (1973).

6. John Blatnik, "The Need for Highway Safety Consciousness," *Traffic Quarterly* 22, no. 2 (1968).

7. Stanley Polanis, "Reducing Traffic Accidents through Traffic Engineering," *Transportation Quarterly* 46, no. 2 (1992).

8. Jeff Hecht, "Managing Expectations of Artificial Intelligence," *Nature* 563, no. 7733 (2018).

9. James Reason, *A Life in Error: From Little Slips to Big Disasters* (Boca Raton, FL: CRC Press, 2013).

10. Sidney Dekker, *The Field Guide to Understanding "Human Error"* (Boca Raton, FL: CRC Press, 2017).

Chapter 34

1. Stephen Dubner, "Is it Too Late for General Motors to Go Electric?," *Freakonomics*, podcast (2020).

2. Gabe Klein with David Vega-Barachowitz, *Start-Up City: Inspiring Private and Public Entrepreneurship, Getting Projects Done, and Having Fun* (Washington, DC: Island Press, 2015).

3. National Transportation Safety Board, "The Use of Forward Collision Avoidance Systems to Prevent and Mitigate Rear-End Crashes" (2015).
4. David Zipper, "Life-Saving Car Technology No One Wants," *Bloomberg News* (2021).
5. AAA, "Automatic Emergency Braking with Pedestrian Detection," AAA NewsRoom, October 2019.
6. Wilson, Hoffman, and Morgenstern, "Predictive Inequity in Object Detection," arXiv:1902.11097 (2019).
7. AAA, "Automatic Emergency Braking."
8. Vikas Bajaj, "The Bright, Shiny Distraction of Self-Driving Cars," *New York Times* (2018).
9. Peter Keating, "The Danger of Safer Equipment," *ESPN The Magazine* (2001).
10. Oliver Connolly, "Helmets Don't Eliminate Concussions," *The Guardian* (2019).
11. Sidney Dekker, *The Field Guide to Understanding "Human Error"* (Boca Raton, FL: CRC Press, 2017).
12. Lucian Leape and Donald Berwick, "Five Years after To Err Is Human: What Have We Learned?," *JAMA* 293, no. 19 (2005).
13. James Reason, *Human Error* (New York: Cambridge University Press, 1990).

Chapter 35

1. Malcolm Gladwell, "Blame Game," *Revisionist History*, podcast (2016).
2. Brian Ross et al., "Owners of Toyota Cars in Rebellion," ABC News (2009).
3. Gladwell, "Blame Game."
4. Bruce D. Greenshields, "Attitudes, Emotions, Accidents," *Traffic Quarterly* 13, no. 2 (1959).
5. Phil Koopman, "A Case Study of Toyota Unintended Acceleration" (2014).
6. Michael Perel, "Analyzing the Role of Driver/Vehicle Incompatibilities," NHTSA (1976).
7. AAA, "Event Data Recorder," 2022, https://exchange.aaa.com/automotive /automotive-trends/event-data-recorder.
8. Gustavo Ruffo, "Chinese Tesla Owners Are Installing Brake Cameras," *InsideEVs* (2021).
9. James Reason, *Human Error* (Cambridge, UK: Cambridge University Press, 1990).
10. Mark Matousek, "Elon Musk Doubled Down," *Business Insider* (2019).
11. Johnny Diaz, "Man Riding in Driverless Tesla Is Arrested," *New York Times* (2021).
12. Bryan Pietsch, "2 Killed in Driverless Tesla," *New York Times* (2021).
13. Tesla, "Autopilot and Full Self-Driving Capability," 2023, www.tesla.com /support/autopilot.
14. Isobel Hamilton, "Hackers Stuck Tape on a 35-mph Speed Sign," *Business Insider* (2020).
15. Reason, *Human Error*.
16. Ibid.
17. Charles Perrow, *Normal Accidents: Living with High-Risk Technologies* (New York: Basic Books, 1984).

Chapter 36

1. Sidney Dekker, *The Field Guide to Understanding "Human Error"* (Boca Raton, FL: CRC Press, 2017).
2. Ibid.
3. Anthony Hidden, "Investigation into the Clapham Junction Railway Accident," Her Majesty's Stationery Office (1989).
4. Dekker, *Field Guide.*
5. Paul G. Hoffman and Neil M. Clark, *Seven Roads to Safety: A Program to Reduce Automobile Accidents* (New York: Harper and Brothers, 1939).

Chapter 38

1. Paul G. Hoffman and Neil M. Clark, *Seven Roads to Safety: A Program to Reduce Automobile Accidents* (New York: Harper and Brothers, 1939).
2. Ibid.
3. Ibid.
4. Sidney Dekker, *The Field Guide to Understanding "'Human Error"* (Boca Raton, FL: CRC Press, 2017).
5. Ibid.
6. Howard Anderson, "Let's Try to Dispel Some Highway Safety Myths," *Traffic Engineering* 46, no. 12 (1976).

Chapter 39

1. David Solomon, "Accidents on Main Rural Highways Related to Speed, Driver, and Vehicle," Federal Highway Administration (1964).
2. Laurence Aurbach, *A History of Street Networks: From Grids to Sprawl and Beyond* (Hyattsville, MD: PedShed Press, 2020).
3. Paul G. Hoffman and Neil M. Clark, *Seven Roads to Safety: A Program to Reduce Automobile Accidents* (New York: Harper and Brothers, 1939).
4. B. A. Lefeve, "Relation of Accidents to Speed Habits and Other Driver Characteristics," *Highway Research Board Bulletin* 120 (1956).
5. Roger Stewart, "Are We Over-Emphasizing Speed as an Accident Cause?," *Traffic Quarterly* 11, no. 4 (1957).
6. Ibid.
7. Libert Ehrman, "Causes of Highway Accidents," *Traffic Quarterly* 12, no. 1 (1958).
8. "Federal Role in Highway Safety," 86th Congress, 1st session, House Document 93 (Washington, DC: US Government Printing Office, 1959).
9. Craig Kloeden, Giulio Ponte, and Jack McLean, "Travelling Speed and Risk of Crash Involvement on Rural Roads," Australian Transport Safety Bureau (2001); Letty Aarts and Ingrid Van Schagen, "Driving Speed and the Risk of Road Crashes," *Accident Analysis and Prevention* 38, no. 2 (2006).

Chapter 40

1. Ginna Roe, "Wrong-Way Crashes Are Too Common," 2KUTV, October 21, 2019.
2. Carter Williams, "Speed Limit Jump Will 'Eliminate the Safety Risk,'" KSL, Salt Lake City, Utah, October 18, 2019.
3. Ibid.

4. David W. Schoppert, *Traffic Control and Roadway Elements: Their Relationship to Highway Safety* (Washington, DC: Automotive Safety Foundation, 1963).
5. Ibid.
6. Federal Highway Administration, *Manual on Uniform Traffic Control Devices* (Washington, DC: US Department of Transportation, 2009).

Chapter 41

1. National Transportation Safety Board, "Reducing Speeding-Related Crashes Involving Passenger Vehicles" (2017).
2. John J. Leeming, "Road Accidents, Other Accidents, and Our Laws About Them," *Traffic Quarterly* 11, no. 3 (1957).
3. Seymour S. Taylor, "Freeways Alone Are Not Enough," *Traffic Quarterly* 13, no. 3 (1959).
4. David W. Schoppert, *Traffic Control and Roadway Elements: Their Relationship to Highway Safety* (Washington, DC: Automotive Safety Foundation, 1963).
5. Wilbur Smith and Charles Le Craw Jr., "Travel Speeds and Posted Speeds in Three States," *Traffic Quarterly* 2, no. 1 (1948).
6. Ibid.
7. National Transportation Safety Board, "Reducing Speeding-Related Crashes."

Chapter 42

1. National Transportation Safety Board, "Reducing Speeding-Related Crashes Involving Passenger Vehicles" (2017).
2. National Transportation Safety Board, "Reducing Speeding-Related Crashes Involving Passenger Vehicles" (2017).
3. California Vehicle Code, "Division 17. Offenses and Prosecution; Chapter 3. Illegal Evidence" (1959).
4. Laura Nelson, "As L.A. Struggles to Reduce Traffic Deaths, Speed Limits Keep Going Up," *Los Angeles Times*, July 22, 2018.
5. Ibid.
6. Ibid.
7. California Assembly Bill 43, "Traffic Safety" (2021).

Chapter 43

1. Fred Herring, "Human Relationships Are the Key to Effective Traffic Engineering," *Traffic Quarterly* 10, no. 3 (1956).
2. Ibid.
3. Ibid.
4. American Association of State Highway and Transportation Officials, *A Policy on Geometric Design of Highways and Streets*, 7th ed. (Washington, DC: AASHTO, 2018).
5. Ibid.
6. Ibid.
7. American Association of State Highway and Transportation Officials, *A Policy on Geometric Design of Highways and Streets*, 1st, 2nd, and 3rd eds. (Washington, DC: AASHTO, 1984, 1990, and 1994).
8. American Association of State Highway Officials, *A Policy on Design of Urban Highways and Arterial Streets* (Washington, DC: AASHO, 1973).

9. Joseph Barnett, E. R. Haile, and R. Moyer, "Safe Side Friction Factors and Superelevation Design," Highway Research Board Proceedings (1936).
10. Frederick Cron, "Highway Design for Motor Vehicles—A Historical Review," *Public Roads* (1975–1976).
11. Kay Fitzpatrick et al., "Design Speed, Operating Speed, and Posted Speed Practices," NCHRP-504 (2003).
12. American Association of State Highway and Transportation Officials, *A Policy on Geometric Design of Highways and Streets*, 4th and 5th eds. (Washington, DC: AASHTO, 2001 and 2004).

Chapter 44

1. R. J. Porter, Eric Donnell, and John Mason, "Geometric Design, Speed, and Safety," *Transportation Research Record* 2309, no. 1 (2012).
2. Ezra Hauer, "Engineering Judgment and Road Safety," *Accident Analysis* and *Prevention* 129 (2019).

Chapter 45

1. Ronald Smothers, "Officer's Blood Alcohol Level Was Over Limit in Fatal Crash," *New York Times*, June 14, 2003.
2. National Transportation Safety Board, "Passenger Vehicle Median Crossover and Head-On Collision with Another Passenger Vehicle, Linden, New Jersey, May 1, 2003," (2003).
3. American Association of State Highway and Transportation Officials, *A Policy on Geometric Design of Highways and Streets.* (Washington, DC: AASHTO, 2001).
4. Kay Fitzpatrick et al., "Compatibility of Design Speed, Operating Speed, and Posted Speed," Texas Transportation Institute, FHWA/TX-95/1465-2F (1995).

Chapter 46

1. Henry Seiff, "Status Report on Large-Truck Safety," *Transportation Quarterly* 44, no. 1 (1990).
2. Thomas Deen and Stephen Godwin, "Safety Benefits of the 55 mph Speed Limit," *Transportation Quarterly* 39, no. 3 (1985).
3. James O'Day and Jairus Flora, "Alternative Measures of Restraint System Effectiveness," SAE 820798, National Transportation Safety Board (1982).
4. Frank Palmer, "Physics and Math for Drivers," *Safety Reviews* 15 (1958).
5. National Transportation Safety Board, "Reducing Speeding-Related Crashes Involving Passenger Vehicles" (2017).
6. Linda Dultz et al., "Vulnerable Roadway Users Struck by Motor Vehicles," *Journal of Trauma and Acute Care Surgery* 74, no. 4 (2013).
7. Shankuan Zhu et al., "Obesity and Risk for Death due to Motor Vehicle Crashes," *American Journal of Public Health* 96, no. 4 (2006).
8. Andrew McFarlane, "How the UK's First Fatal Car Accident Unfolded," *BBC News*, August 17, 2010.
9. Rune Elvik, Peter Christensen, and Astrid Amundsen, "Speed and Road Accidents," Transportøkonomisk Institutt (2004).
10. M. Earl Campbell, "Highway Traffic Safety—Is It Possible?," *Traffic Quarterly* 19, no. 3 (1965).
11. Elvik, Christensen, and Amundsen, "Speed and Road Accidents."

Chapter 47

1. Charles Norcross, "Owners of America: The Vanderbilts," *Cosmopolitan* 45, no. 4 (1908).
2. Neil Arason, *No Accident: Eliminating Injury and Death on Canadian Roads* (Waterloo, Ontario: Wilfrid Laurier University Press, 2014).
3. Spencer Miller, "History of the Modern Highway in the U.S.," in *Highways in Our National Life: A Symposium*, ed. Jean Labatut and Wheaton J. Lane (Princeton, NJ: Princeton University Press, 1950).
4. City of Boston Bylaws, City Document-137 (1701).
5. Eric Ravenscraft, "Does Speeding Really Get You There Any Faster?," *Lifehacker* (2014).
6. Jacques Nouvier, "The OECD-ECMT Joint Working Group on Speed Management," ITS World Congress (2006).
7. Jeff Speck, *Walkable City Rules* (Washington, DC: Island Press, 2018).
8. William Greene, "Getting Support for Traffic Safety Education in the Community," *Traffic Quarterly* 1, no. 2 (1947).
9. Walter Gropius and Martin Wagner, "A Program for City Reconstruction," *Architectural Forum* 79, no. 1 (1943).
10. Vincent Tofany, "Life Is Best at 55," *Traffic Quarterly* 35, no. 1 (1981).
11. Thomas B. Deen and Stephen Godwin, "Safety Benefits of the 55 mph Speed Limit," *Transportation Quarterly* 39, no. 3 (1985).
12. Lee Friedman, Donald Hedeker, and Elihu Richter, "Long-Term Effects of Repealing the National Maximum Speed Limit," *American Journal of Public Health* 99, no. 9 (2009).
13. Patricia F. Waller, "Speed Limits: How Should They Be Determined?," *Transportation Quarterly* 56, no. 3 (2002).
14. National Transportation Safety Board, "Reducing Speeding-Related Crashes Involving Passenger Vehicles" (2017).
15. Ibid.
16. Hallie Myers, "Promoting Traffic Safety in Indiana," *Traffic Quarterly* 9, no. 4 (1955).
17. Robert Herman, Ronald Rule, and Marvin Jackson, "Fuel Economy and Exhaust Emissions under Two Conditions of Traffic Smoothness," SAE 780614 (1978).

Chapter 48

1. American Association of State Highway Officials, *A Policy on Arterial Highways in Urban Areas* (Washington, DC: AASHO, 1957).
2. L. E. Peabody and O. K. Normann, "Applications of Automatic Traffic Recorder Data," *Public Roads* 20, no. 11 (1941).
3. Frederick Cron, "Highway Design for Motor Vehicles—A Historical Review," *Public Roads* (1975–1976).
4. Ibid.
5. AASHO, *A Policy on Arterial Highways in Urban Areas*; American Association of State Highway Officials, *A Policy on Design of Urban Highways and Arterial Streets* (Washington, DC: AASHO, 1973).
6. AASHO, *A Policy on Arterial Highways* (1957).
7. AASHO, *A Policy on Design of Urban Highways* (1973).
8. American Association of State Highway and Transportation Officials, *A*

Policy on Geometric Design of Highways and Streets, 7th ed. (Washington, DC: AASHTO, 2018).

9. Louis Shallal and Ata Khan, "Predicting Peak-Hour Traffic," *Traffic Quarterly* 34, no. 1 (1980).

10. Ibid.

11. AASHTO, *A Policy on Geometric Design*.

12. C. D. Curtiss, "Urban Highway Planning," *Traffic Quarterly* 11, no. 4 (1957).

13. Maxwell Halsey, "Old Roads," *Traffic Quarterly* 11, no. 4 (1957).

14. George Burpee and J. L. Ray, "Urban Highway Construction and Its Regional Effects," *Traffic Quarterly* 12, no. 1 (1958).

15. Rask Overgaard, "Urban Transportation Planning: Traffic Estimation," *Traffic Quarterly* 21, no. 2 (1967).

16. John Hassell, "How Effective Has Urban Transportation Planning Been?," *Traffic Quarterly* 34, no. 1 (1980).

Chapter 49

1. American Association of State Highway Officials, *A Policy on Arterial Highways in Urban Areas* (Washington, DC: AASHO, 1957).

2. AAA, *Planned Pedestrian Program* (Washington, DC: AAA Foundation for Traffic Safety, 1958).

3. Robert Schneider et al., "U.S. Fatal Pedestrian Crash Hot Spot Locations and Characteristics," *Journal of Transport and Land Use* 14, no. 1 (2021).

4. Ibid.

5. Gabe Klein with David Vega-Barachowitz, *Start-Up City: Inspiring Private and Public Entrepreneurship, Getting Projects Done, and Having Fun* (Washington, DC: Island Press, 2015).

Chapter 50

1. American Association of State Highway Officials, *A Policy on Arterial Highways in Urban Areas* (Washington, DC: AASHO, 1957).

2. Ezra Hauer, "Safety in Geometric Design Standards I: Three Anecdotes," in 2nd International Symposium on Highway Geometric Design, no. FGSV 002/67 (2000).

3. American Association of State Highway Officials, *A Policy on Geometric Design of Rural Highways* (Washington, DC: AASHO, 1954).

4. AASHO, *A Policy on Arterial Highways*.

5. Asriel Taragin, "Effect of Roadway Width on Traffic Operations," Highway Research Board (1945).

6. AASHO, *A Policy on Geometric Design*.

7. Taragin, "Effect of Roadway Width."

8. Hauer, "Safety in Geometric Design Standards I."

9. Robert Noland, "Traffic Fatalities and Injuries," *Accident Analysis and Prevention* 35, no. 4 (2003).

10. Kay Fitzpatrick et al., "Design Factors that Affect Driver Speed," *Transportation Research Record* 1751, no. 1 (2001).

11. Morton Raff, "Interstate Highway-Accident Study," *Highway Research Board Bulletin* 74 (1953).

12. D. M. Belmont, "Effect of Shoulder Width on Accidents on Two-Lane Tangents," *Highway Research Board Bulletin* 91 (1954).

13. Roy Jorgensen Associates, "NCHRP-197: Cost and Safety Effectiveness of Highway Design Elements" (1978).

14. Hauer, "Safety in Geometric Design Standards I."

15. Ingrid Potts, Douglas Harwood, and Karen Richard, "Relationship of Lane Width to Safety," *Transportation Research Record* 2023, no. 1 (2007).

16. Nicholas J. Garber and Lester A. Hoel, *Traffic and Highway Engineering*, 2nd ed. (CL Engineering, 1997).

17. Ibid.

18. Highway Research Board, *Highway Capacity Manual*, Special Report 87 (1965).

19. Ibid.

Chapter 51

1. Wilbur Smith and Jack Hart, "A Case Study of One-Way Streets," *Traffic Quarterly* 3, no. 4 (1949).

2. Ibid.

3. Ibid.

4. American Association of State Highway Officials, *A Policy on Arterial Highways in Urban Areas* (Washington, DC: AASHO, 1957).

5. Ibid.

6. Institute of Transportation Engineers, *Toolbox on Intersection Safety and Design* (Washington, DC: ITE, 2004).

7. Wade Walker, Walter Kulash, and Brian McHugh, "Downtown Streets: Are We Strangling Ourselves on One-Way Networks?," *Transportation Research Circular*, no. E-C019 (2000).

8. AAA, *Planned Pedestrian Program* (Washington, DC: AAA Foundation for Traffic Safety, 1958).

9. Ibid.

10. Macklin Hancock and Werner Billing, "Suburban Town Centers," *Traffic Quarterly* 18, no. 2 (1964).

11. Ibid.

Chapter 52

1. Anthony Downs, "The Law of Peak-Hour Expressway Congestion," *Traffic Quarterly* 16, no. 3 (1962).

2. A. E. Johnson, "The Interstate System," *Traffic Quarterly* 10, no. 2 (1956).

3. William Mortimer, "The Influence of Expressways," *Traffic Quarterly* 10, no. 3 (1956).

4. Russell Singer, "Future Role of the Automobile in Urban Transportation" *Traffic Quarterly* 18, no. 2 (1964).

5. Martin Wohl, "Must Something Be Done About Traffic Congestion?," *Traffic Quarterly* 25, no. 3 (1971).

6. Lew Pratsch, "Reducing Commuter Traffic Congestion," *Transportation Quarterly* 40, no. 4 (1986).

7. Hauer Ezra, "On the Relationship between Road Safety Research and the Practice of Road Design and Operation," *Accident Analysis and Prevention* 128 (2019).

8. American Association of State Highway Officials, *Highway Design and Operational Practices Related to Highway Safety* (Washington, DC: AASHO, 1967).

9. David W. Schoppert, *Traffic Control and Roadway Elements: Their Relationship to Highway Safety* (Washington, DC: Automotive Safety Foundation, 1963).

10. Ibid.

11. American Association of State Highway and Transportation Officials, *A Policy on Geometric Design of Highways and Streets,* 7th ed. (Washington, DC: AASHTO, 2018).

12. Brent Ogden and Chelsey Cooper, *Highway-Rail Crossing Handbook*, 3rd ed., FHWA-SA-18-040 (Washington, DC: Institute of Transportation Engineers, 2019).

13. Illinois Administrative Code (1535.205): Right-of-Way to Be Kept Clear.

14. Robert Taggart et al., "Evaluating Grade-Separated Rail and Highway Crossing Alternatives," NCHRP Report 288 (1987).

15. Ibid.

16. David Schoppert and D. W. Hoyt, "Factors Influencing Safety at Highway-Rail Grade Crossings," NCHRP Report 50 (1968).

17. Ibid.

18. Nicholas Ward and Gerald Wilde, "Driver Approach Behaviour at an Unprotected Railway Crossing," *Safety Science* 22, nos. 1–3 (1996).

Chapter 53

1. David W. Schoppert, *Traffic Control and Roadway Elements: Their Relationship to Highway Safety* (Washington, DC: Automotive Safety Foundation, 1963).

2. Ibid.

3. Ibid.

4. Ibid.

5. American Association of State Highway Officials, *Highway Design and Operational Practices Related to Highway Safety* (Washington, DC: AASHO, 1967).

6. Ibid.

7. David Baldwin, "Highway Engineering and Traffic Safety," *Traffic Quarterly* 31, no. 3 (1977).

8. Bruce Appleyard, *Livable Streets 2.0* (New York: Elsevier, 2021).

9. AASHO, *Highway Design and Operational Practices related to Highway Safety.*

10. Ibid.

11. American Association of State Highway and Transportation Officials, *Roadside Design Guide* (Washington, DC: AASHTO, 2011).

12. Nicholas J. Garber and Lester A. Hoel, *Traffic and Highway Engineering,* 2nd ed. (Boston: PWS Publishing, 1997).

13. Neil Arason, *No Accident: Eliminating Injury and Death on Canadian Roads* (Waterloo, Ontario: Wilfrid Laurier University Press, 2014).

14. AASHTO, *Roadside Design Guide.*

15. AASHTO, Errata to *Roadside Design Guide.*

16. Eric Dumbaugh et al., "The Influence of the Built Environment on Crash Risk in Lower-Income and Higher-Income Communities," Collaborative Sciences Center for Road Safety (2020).

17. Wes Marshall, Nicholas Coppola, and Yaneev Golombek, "Urban Clear Zones, Street Trees, and Road Safety," *Research in Transportation Business and Management* 29 (2018).

18. AASHO, *Highway Design and Operational Practices related to Highway Safety.*

19. David Baldwin, "Highway Engineering and Traffic Safety," *Traffic Quarterly* 31, no. 3 (1977).

20. Kenneth Haase, "Characteristics of Persons Injured in Motor Vehicle Accidents," *Traffic Quarterly* 17, no. 4 (1963).
21. American Association of State Highway Officials, *A Policy on Arterial Highways in Urban Areas* (Washington, DC: AASHO, 1957).
22. James Reason, *A Life in Error: From Little Slips to Big Disasters* (Boca Raton, FL: CRC Press, 2013).
23. William Marston, "Traffic Planning for Chicago," *Traffic Quarterly* 14, no. 3 (1960).
24. Doyle Arthur Conan, "A Scandal in Bohemia," *Strand* (1891).

Chapter 54

1. J. O. Mattson, "The Traffic Accident Outlook," *Traffic Quarterly* 18, no. 3 (1964).
2. Rex M. Whitton, "The Role of Highway and Traffic Engineering and Highway Safety," *Traffic Quarterly* 19, no. 1 (1965).
3. Ibid.
4. Ibid.
5. Richard F. Weingroff, "President Dwight D. Eisenhower and the Federal Role in Highway Safety—Epilogue: The Changing Federal Role," Federal Highway Administration (2003).
6. American Association of State Highway Officials, *A Policy on Arterial Highways in Urban Areas* (Washington, DC: AASHO, 1957).

Chapter 55

1. David W. Schoppert, *Traffic Control and Roadway Elements: Their Relationship to Highway Safety* (Washington, DC: Automotive Safety Foundation, 1963).
2. Ibid.
3. Ibid.
4. Paul R. Staffeld, "Accidents Related to Access Points and Advertising Signs in Study," *Traffic Quarterly* 7, no. 1 (1953).
5. Harold Marks, "Subdividing for Traffic Safety," *Traffic Quarterly* 11, no. 3 (1957).
6. Ibid.
7. Public Works Administration, *Manual on Uniform Traffic Control Devices for Streets and Highways*, Federal Works Agency (1948).
8. Marks, "Subdividing for Traffic Safety."
9. Ibid.
10. Ibid.
11. Wes Marshall and Norman Garrick, "Does Street Network Design Affect Traffic Safety?," *Accident Analysis and Prevention* 43, no. 3 (2011).
12. Wes Marshall and Norman Garrick, "Community Design and How Much We Drive," *Journal of Transport and Land Use* 5, no. 2 (2012).
13. Wes Marshall and Norman Garrick, "Effect of Street Network Design on Walking and Biking," *Transportation Research Record* 2198, no. 1 (2010).
14. Wes Marshall, Daniel Piatkowski, and Norman Garrick, "Community Design, Street Networks, and Public Health," *Journal of Transport and Health* 1, no. 4 (2014).

Chapter 56

1. Institute of Traffic Engineers, "Recommended Practice for Subdivision Streets" (1965).

2. Laurence Aurbach, *A History of Street Networks: From Grids to Sprawl and Beyond* (Hyattsville, MD: PedShed Press, 2020).
3. Michael Southworth and Eran Ben-Joseph, *Streets and the Shaping of Towns and Cities* (Washington, DC: Island Press, 1996).
4. Aurbach, *A History of Street Networks*.
5. Southworth and Ben-Joseph, *Streets and the Shaping of Towns and Cities*.
6. Richard Rothstein, *The Color of Law: A Forgotten History of How Our Government Segregated America* (New York: Liveright, 2017).
7. Aurbach, *A History of Street Networks*.
8. Edward Hall, "Streets for the Urban Traveler," *Traffic Quarterly* 15, no. 4 (1961).
9. Irving Hand and C. D. Hixon, "Planning, Traffic and Transportation in Metropolitan Areas," *Traffic Quarterly* 17, no. 2 (1963).
10. Reid Ewing, "Impediments to Context-Sensitive Main Street Design," *Transportation Quarterly* 56, no. 4 (2002).

Chapter 57

1. American Association of State Highway Officials, *Policies on Geometric Highway Design* (Washington, DC: AASHO, 1950).
2. Laurence Aurbach, *A History of Street Networks: From Grids to Sprawl and Beyond* (Hyattsville, MD: PedShed Press, 2020).
3. USDOT, "National Highway Needs Report" (1968).
4. USDOT, "Supplement to the 1968 National Highway Needs Report" (1968).
5. Aurbach, *A History of Street Networks*.
6. M. W. Eastburn et al., "Report on Joint Conference," *Traffic Quarterly* 29, no. 2 (1975).
7. Ibid.
8. Ibid.
9. Aurbach, *A History of Street Networks*.
10. American Association of State Highway Officials, *A Policy on Arterial Highways in Urban Areas* (Washington, DC: AASHO, 1957).
11. American Association of State Highway Officials, *A Policy on Design of Urban Highways and Arterial Streets* (Washington, DC: AASHO, 1973).
12. Ibid.

Chapter 58

1. Paul G. Hoffman and Neil M. Clark, "America Goes to Town," *Saturday Evening Post*, 211 (1939).
2. Ibid.
3. Ibid.
4. Paul G. Hoffman and Neil M. Clark, *Seven Roads to Safety* (New York: Harper and Brothers, 1939).
5. Edward Logue, "Can Cities Survive Automobile Age?," *Traffic Quarterly* 13, no. 2 (1959).
6. Brian Wolshon and Anurag Pande, *Traffic Engineering Handbook* (Hoboken, NJ: Wiley, 2016) (italics added).

Chapter 59

1. Bureau of Public Roads, *Toll Roads and Free Roads* (Washington, DC: US Government Printing Office, 1939).

2. Laurence Aurbach, *A History of Street Networks: From Grids to Sprawl and Beyond* (Hyattsville, MD: PedShed Press, 2020).

3. Ibid.

4. Ibid.

5. Robert Goetz, "Action for Safety," *Traffic Quarterly* 12, no. 1 (1958).

6. Robert Goetz, "City Traffic," *Traffic Quarterly* 15, no. 1 (1961).

7. M. Earl Campbell, "Highway Traffic Safety—Is It Possible?," *Traffic Quarterly* 19, no. 3 (1965).

8. A. E. Johnson, "The Interstate System," *Traffic Quarterly* 10, no. 2 (1956).

9. Ibid.

10. Val Peterson, "Plans for Evacuating a Large City in Case of Atomic Attack," *Traffic Quarterly* 10, no. 1 (1956).

11. Eugene Maier, "Urban Transportation Planning Can Succeed," *Traffic Quarterly* 16, no. 3 (1962).

12. Harold Hansen, "Guides for Transportation Planning," *Traffic Quarterly* 16, no. 2 (1962).

13. Edward J. Logue, "Can Cities Survive Automobile Age? New Haven Used as a Test Case," *Traffic Quarterly* 13, no. 2 (1959).

14. Irving Hand and C. D. Hixon, "Planning, Traffic and Transportation in Metropolitan Areas," *Traffic Quarterly* 17, no. 2 (1963).

15. "Eisenhower's Meeting with General Bragdon, 4/6/60," Memorandum for the Record, Federal Highway Administration (1960).

16. Russell E. Singer, "Future Role of the Automobile in Urban Transportation," *Traffic Quarterly* 18, no. 2 (1964).

17. Ibid.

18. Ibid.

19. Weizhou Lu, "Thoroughfare Planning and Goal Definition," *Traffic Quarterly* 17, no. 2 (1963).

20. Angie Schmitt, "Rochester Wins Parking Madness Title," *Streetsblog* (2014).

Chapter 60

1. Lawrence Hott and Tom Lewis, *Divided Highways: The Interstates and the Transformation of American Life*, PBS (1997).

2. Laurence Aurbach, *A History of Street Networks: From Grids to Sprawl and Beyond,* (Hyattsville, MD: PedShed Press, 2020).

3. American Association of State Highway Officials, *A Policy on Arterial Highways in Urban Areas* (Washington, DC: AASHO, 1957).

4. Ibid.

5. John Clarkeson, "Urban Expressway Location," *Traffic Quarterly* 7, no. 2 (1953).

6. Ibid.

7. Wilbur Smith, "Traffic and Rebuilding Cities," *Traffic Quarterly* 13, no. 1 (1959).

8. William R. McGrath, "Transportation and Urban Development," *Traffic Quarterly* 18, no. 4 (1964).

9. Robert E. Barkley, "The Civil Engineering Function in Urban Renewal," *Traffic Quarterly* 18, no. 3 (1964).

10. Ibid.

11. John B. Rae, "The Mythology of Urban Transportation," *Traffic Quarterly* 26, no. 1 (1972).

12. Bureau of Public Roads, *Toll Roads and Free Roads* (Washington, DC: US Government Printing Office, 1939).
13. Nathaniel Owings, "Urban Transportation Planning Concepts," *Traffic Quarterly* 21, no. 2 (1967).
14. Ibid.
15. Richard Steiner, "Traffic Improvement, Urban Renewal," *Traffic Quarterly* 13, no. 1 (1959).
16. Ibid.
17. William Claire, "The Contribution of the Engineer to City Planning," *Traffic Quarterly* 17, no. 3 (1963).
18. Ibid.
19. John Sheehy, "Highway Displacement Relocation Experience," *Traffic Quarterly* 28, no. 2 (1974).
20. Thomas Deen, "Search for Sustainable Transportation," *Transportation Quarterly* 50, no. 4 (1996).
21. Ibid.
22. Ibid.
23. Ibid.
24. Ibid.

Chapter 61

1. Wes Marshall and Nicholas Ferenchak, "Assessing Equity and Urban/Rural Road Safety Disparities in the US," *Journal of Urbanism* 10, no. 4 (2017); Eric Dumbaugh et al., "The Influence of the Built Environment on Crash Risk in Lower-Income and Higher-Income Communities," Collaborative Sciences Center for Road Safety (2020); Carolyn McAndrews et al., "Revisiting Exposure," *Accident Analysis and Prevention* 60 (2013); Rebecca Naumann and Laurie Beck, "Motor Vehicle Traffic-Related Pedestrian Deaths—United States, 2001–2010," MMWR.62.15 (2013).
2. Nathan Smith, "Blighted Areas and Traffic," *Traffic Quarterly* 1, no. 4 (1947).
3. James Cornehls, "The Automobile Society," *Traffic Quarterly* 31, no. 4 (1977).
4. Robert Caro, *The Power Broker: Robert Moses and the Fall of New York* (New York: Random House, 1974).
5. Ibid.
6. Ibid.
7. Ibid.
8. Laurence Aurbach, *A History of Street Networks: From Grids to Sprawl and Beyond* (Hyattsville, MD: PedShed Press, 2020).

Chapter 62

1. Masatoshi Abe, "A New Perspective on Urban Transportation," *Traffic Quarterly* 29, no. 4 (1975).
2. David Witheford, "Engineers, Urban Freeways, and the Public," *Traffic Quarterly* 27, no. 1 (1973).
3. Ibid.
4. Daniel Moynihan, "New Roads and Urban Chaos," *The Reporter* 22, no. 14 (1960).
5. Seymour Taylor, "Freeways Alone Are Not Enough," *Traffic Quarterly* 13, no. 3 (1959).
6. Moynihan, "New Roads."

7. Richard F. Weingroff, "Original Intent: Purpose of the Interstate System 1954–1956," Federal Highway Administration (2011).
8. Edmund Mantell, "Economic Biases in Urban Transportation Planning and Implementation," *Traffic Quarterly* 25, no. 1 (1971).
9. Ibid.
10. S. S. Morris, "Freeways and the Urban Traffic Problem," *Traffic Quarterly* 27, no. 4 (1973).
11. Moynihan, "New Roads."

Chapter 63

1. Michael Bloomberg, "You Can't Manage What You Can't Measure," LinkedIn, January 20, 2020.
2. Bruce Greenshields, "Attitudes, Emotions, Accidents," *Traffic Quarterly* 13, no. 2 (1959).
3. Ezra Hauer, "Crash Causation and Prevention," *Accident Analysis and Prevention* 143 (2020).
4. Kevin Chang et al., "Streamlining the Crash Reporting," *PacTrans* (2018).
5. Ibid.
6. David W. Schoppert, *Traffic Control and Roadway Elements: Their Relationship to Highway Safety* (Washington, DC: Automotive Safety Foundation, 1963).
7. National Transportation Safety Board, "Reducing Speeding-Related Crashes Involving Passenger Vehicles" (2017).
8. Ibid.
9. Heidi Coleman and Krista Mizenko, "Pedestrian and Bicyclist Data Analysis," *Traffic Safety Facts: Research Note*, DOT-HS-812-205, National Highway Traffic Safety Administration (2018).
10. Paul Sorlie, Eugene Rogot, and Norman Johnson, "Validity of Demographic Characteristics on the Death Certificate," *Epidemiology* 3, no. 2 (1992).
11. Wes Marshall and Nicholas Ferenchak, "Assessing Equity and Urban/Rural Road Safety Disparities in the US," *Journal of Urbanism* 10, no. 4 (2017).
12. National Transportation Safety Board, "Pedestrian Safety," NTSB/SIR-18/03 (2018).
13. Neil Arason, *No Accident: Eliminating Injury and Death on Canadian Roads* (Waterloo, Ontario: Wilfrid Laurier University Press, 2014).
14. Nicholas Ferenchak and Robin Osofsky, "Police-Reported Pedestrian Crash Matching and Injury Severity Misclassification," *Accident Analysis and Prevention* 167 (2022).

Chapter 64

1. Ben Markus and Veronica Penney, "How Humans Start Most of Colorado's Wildfire—and Get Away with It," *CPR News* (2021).
2. NHTSA, "Traffic Safety Facts 2019," https://crashstats.nhtsa.dot.gov/Api/Public/ViewPublication/813141.
3. Heidi Coleman and Krista Mizenko, "Pedestrian and Bicyclist Data Analysis," *Traffic Safety Facts: Research Note*, DOT HS 812 205, National Highway Traffic Safety Administration (2018).
4. Colorado Department of Transportation (@Colorado DOT), "Do you know," Twitter, August 24, 2015.
5. A. Soica, "Casuistic Analysis of the Passenger's Throw-Off Distance at Car Collision," *Materials Science and Engineering* 444, no. 7 (2018).

6. Paul G. Hoffman and Neil M. Clark, *Seven Roads to Safety: A Program to Reduce Automobile Accidents* (New York: Harper and Brothers, 1939).
7. AAA, *Planned Pedestrian Program* (Washington, DC: AAA Foundation for Traffic Safety, 1958).
8. James Reason, *Human Error* (Cambridge, UK: Cambridge University Press, 1990).
9. Geoff Gibson et al., "Motorist-Cyclist Crash Data Needs," Transportation Research Board Annual Meeting (2017).
10. Gersh Kuntzman, "NYPD Admits: There Are More Injured Cyclists Than You Realize," *NYC Streetsblog* (2019).
11. Markus and Penney, "How Humans Start Most of Colorado's Wildfires."
12. AAA, *Planned Pedestrian Program*.
13. Ibid.

Chapter 65

1. *Jeff Garlin: Our Man in Chicago*, Netflix (2019).
2. Anne Harris et al., "Comparing the Effects of Infrastructure on Bicycling Injury at Intersections and Non-Intersections," *Injury Prevention* 19, no. 5 (2013).
3. Nicholas Ferenchak and Wes Marshall, "An Examination of Shared Lane Markings," *International Journal of Transportation Science and Technology* 8, no. 2 (2019).
4. Josh Cohen, "A Brief History of the American Sharrow," episode 4, *The Bicycle Story*, podcast, March 9, 2016.
5. Jane Stutts and William Hunter, "Injuries to Pedestrians and Bicyclists Based on Hospital Data," FHWA-RD-99-078 (1999).

Chapter 66

1. Ezra Hauer, "Crash Causation and Prevention," *Accident Analysis and Prevention* 143 (2020).
2. Peter Valdes-Dapena, "Volvo Promises Deathproof Cars," *CNN Business* (2016).

Chapter 67

1. Reid Ewing, "Impediments to Context-Sensitive Main Street Design," *Transportation Quarterly* 56, no. 4 (2002).
2. Ibid.
3. Hauer Ezra, "On the Relationship between Road Safety Research and the Practice of Road Design and Operation," *Accident Analysis and Prevention* 128 (2019).
4. Stephen Dubner, "The Perfect Crime," episode 165, *Freakonomics*, podcast, May 1, 2014.
5. Ibid.
6. Charles Komanoff and members of Right of Way, "Killed by Automobile" (1999).
7. Dubner, "The Perfect Crime."
8. A. H. Goldstein Jr., "Traffic Safety, the Courts, and the Schools," *Traffic Quarterly* 22, no. 1 (1968).
9. Ibid.

10. Kenneth Todd, "Pedestrian Regulations in the U.S.," *Transportation Quarterly* 46, no. 4 (1992).
11. Richard Kuhlman, *Killer Roads: From Crash to Verdict* (Charlottesville, VA: Michie Company, 1986).
12. Ibid.
13. Ewing, "Impediments."

Chapter 68

1. *Turturro v. City of New York and Pascarella*, no. 196, 2016, https://www .nycourts.gov/ctapps/Decisions/2016/Dec16/196opn16-Decision.pdf.
2. Ibid.
3. Ibid.
4. Ibid.
5. Ibid.
6. Ibid.
7. Reid Ewing, "Legal Status of Traffic Calming," *Transportation Quarterly* 57, no. 2 (2003).
8. Nikiforos Stamatiadis, "A European Approach to Context Sensitive Design," *Transportation Quarterly* 55, no. 4 (2001).
9. Ibid.
10. *Manna v. State of New Jersey*, 129 NJ 341, 1992, https://law.justia.com/cases /new-jersey/supreme-court/1992/129-n-j-341-1.html.
11. Reid Ewing, "Impediments to Context-Sensitive Main Street Design," *Transportation Quarterly* 56, no. 4 (2002).

Chapter 69

1. Princess Fortin et al., "NYC Child Fatality Report," New York City Department of Health and Mental Hygiene (2010).
2. Richard Retting, "Urban Traffic Crashes," *Transportation Quarterly* 45, no. 4 (1991).
3. Reid Ewing, "Legal Status of Traffic Calming," *Transportation Quarterly* 57, no. 2 (2003).
4. Ibid.
5. Milton Lunch, "Engineering and the Law," *Transportation Quarterly* 41, no. 1 (1987).
6. Alina Tugend, *Better by Mistake: The Unexpected Benefits of Being Wrong* (New York: Penguin, 2011).
7. Nancy Lamo, "Disclosure of Medical Errors," *Clinical Risk* 763 (2011).
8. Steve Kraman and Ginny Hamm, "Risk Management: Extreme Honesty May Be the Best Policy," *Annals of Internal Medicine* 131, no. 12 (1999).
9. Ibid.
10. Ibid.
11. James Reason, *A Life in Error: From Little Slips to Big Disasters* (Boca Raton, FL: CRC Press, 2013).

Chapter 70

1. Gillian Kleiman, "Woman Suffers Grisly Leg Injury after Car Jumps Curb," *New York Post* (2015).
2. Ben Feuerherd, "Woman Dies from Injuries after Car Jumps Curb," *New York Post* (2015).

3. Tara Goddard et al., "Does News Coverage of Traffic Crashes Affect Perceived Blame and Preferred Solutions?," *Transportation Research Interdisciplinary Perspectives* 3 (2019).

4. James Reason, *Human Error* (Cambridge, UK: Cambridge University Press, 1990).

5. Sidney Dekker, *The Field Guide to Understanding "Human Error"* (Boca Raton, FL: CRC Press, 2017).

6. Sharon O'Brien, Richard Tay, and Barry Watson, "An Exploration of Australian Driving Anger," Road Safety Research, Policing and Education Conference Proceedings (2002).

7. Richard Retting et al., "Classifying Urban Crashes for Countermeasure Development," *Accident Analysis and Prevention* 27, no. 3 (1995).

8. Kenneth Todd, "Pedestrian Regulations in the United States: A Critical Review," *Transportation Quarterly* 46, no. 4 (1992).

9. Snehamay Khasnabis, Charles Zegeer, and Michael Cynecki, "Effects of Pedestrian Signals on Safety, Operations, and Pedestrian Behavior," *Transportation Research Record* 847 (1982).

10. Todd, "Pedestrian Regulations."

11. AAA, *Planned Pedestrian Program* (Washington, DC: AAA Foundation for Traffic Safety, 1958).

12. Paul Bartholomew, "The Legal Aspects of Traffic Safety," *Traffic Quarterly* 20, no. 2 (1966).

13. AAA, *Planned Pedestrian Program.*

14. Todd, "Pedestrian Regulations."

15. P. A. Koushki, R. L. Smith, and A. M. Al-Ghadeer, "Urban Stop Sign Violations," *Transportation Quarterly* 46, no. 3 (1992).

16. Seymour S. Taylor, "Freeways Alone Are Not Enough," *Traffic Quarterly* 13, no. 3 (1959).

17. John Joseph Leeming, "Road Accidents, Other Accidents, and Our Laws About Them," *Traffic Quarterly* 11, no. 3 (1957).

18. Mike Isaac, *Super Pumped: The Battle for Uber* (New York: Norton, 2019).

19. Reason, *Human Error.*

Chapter 71

1. John Pucher and Lewis Dijkstra, "Making Walking and Cycling Safer: Lessons from Europe," *Transportation Quarterly* 54, no. 3 (2000).

2. Ibid.

3. Ibid.

4. Ibid.

5. Geoffrey Sant, "Driven to Kill: Why Drivers in China Intentionally Kill the Pedestrians They Hit," *Slate* (2015).

6. Ibid.

7. David Zipper, "Life-Saving Car Technology No One Wants," *Bloomberg News* (2021), www.youtube.com/watch?v=EMmWoU4RhXE.

8. AAA, "Automatic Emergency Braking with Pedestrian Detection" (Washington, DC: American Automobile Association, 2019).

9. Benjamin Wilson, Judy Hoffman, and Jamie Morgenstern, "Predictive Inequity in Object Detection," arXiv preprint arXiv:1902.11097 (2019).

10. Reid Ewing, "Legal Status of Traffic Calming," *Transportation Quarterly* 57, no. 2 (2003).

11. Ibid.
12. Tom McGhee, "Classmates Remember Girl Hit, Killed by Car," *Denver Post,* October 23, 2008.
13. Ibid.

Chapter 72

1. Ezra Hauer, "The Reign of Ignorance in Road Safety: A Case for Separating Evaluation from Implementation," in *Transportation Safety in an Age of Deregulation*, ed. Leon Moses and Ian Savage (Oxford, UK: Oxford University Press, 1989), chapter 6.
2. Richard F. Weingroff, "Celebrating a Century of Cooperation," *Public Roads* 78, no. 2 (2014).
3. Jerome Hall and Daniel Turner, "Development and Adoption of Early AASHO Design Criteria," *Transportation Research Record* 1612, no. 1 (1998).
4. Ibid.
5. Joseph Barnett, "Safe Side Friction Factors and Superelevation Design," *Proceedings of the 17th Annual Meeting*, vol. 16 (Washington, DC: Highway Research Board, 1936).
6. A. L. Luedke and J. L. Harrison, "Superelevation and Easement as Applied to Highway Curves," *Public Roads* 3, no. 31 (1920).
7. Frederick Cron, "Highway Design for Motor Vehicles—A Historical Review," *Public Roads* (1975–1976).
8. Barnett, "Safe Side Friction."
9. American Association of State Highway and Transportation Officials, *A Policy on Geometric Design of Highways and Streets*, 7th edition (Washington, DC: AASHTO, 2018).
10. F. A. Moss and H. H. Allen, "The Personal Equation in Automobile Driving," *SAE Transactions* (1925).
11. MIT, Report of Massachusetts Highway Accident Survey (1934).
12. Paul Olson and Michael Sivak, "Perception-Response Time to Unexpected Roadway Hazards," *Human Factors* 28, no. 1 (1986).
13. Ibid.
14. Ibid.
15. Hugh McGee et al., *Highway Design and Operational Standards Affected by Driver Characteristics* (Washington, DC: Federal Highway Administration, 1983).
16. Olson and Sivak, "Perception-Response Time."
17. Cron, "Highway Design."
18. Ibid.
19. Ibid.
20. Ibid.
21. Hauer, "The Reign of Ignorance."
22. Cron, "Highway Design."

Chapter 73

1. Reid Ewing, "Impediments to Context-Sensitive Main Street Design, *Transportation Quarterly* 56, no. 4 (2002).
2. Frederick Cron, "Highway Design for Motor Vehicles—A Historical Review," *Public Roads* (1975–1976).
3. Ibid.

4. Asriel Taragin, "Effect of Roadway Width on Traffic Operations—Two-Lane Concrete Roads," *Highway Research Board Proceedings* (1945); Roger Morrison, "The Effect of Pavement Widths upon Accidents," *Highway Research Board Proceedings* (1934); A. N. Johnson and L. A. Lyon, "Notes on Traffic Speeds," *Highway Research Board Proceedings* (1934).
5. Cron, "Highway Design."
6. Morrison, "The Effect of Pavement Widths."
7. Johnson and Lyon, "Notes on Traffic Speeds."
8. Taragin, "Effect of Roadway Width."
9. Cron, "Highway Design."
10. Sara Bronin, "The Rules That Made U.S. Roads So Deadly," *Bloomberg News*, March 30, 2021.

Chapter 74

1. Edward Mueller, "A New Look at Highway Capacity," *Traffic Quarterly* 20, no. 3 (1966).
2. Eugene Maier, "Urban Transportation Planning Can Succeed," *Traffic Quarterly* 16, no. 3 (1962).
3. Albert Forde and Janice Daniel, "Pedestrian Walking Speed," *Journal of Traffic and Transportation Engineering* 8, no. 1 (2021).
4. Ibid.
5. AAA, *Planned Pedestrian Program*, AAA Foundation for Traffic Safety (Washington, DC: AAA Foundation for Traffic Safety, 1958).
6. Federal Highway Administration, *Manual on Uniform Traffic Control Devices* (Washington, DC: US Department of Transportation, 2009).
7. Henry Barnes, *The Man with Red and Green Eyes* (New York: Dutton, 1965).
8. Ibid.
9. AAA, *Planned Pedestrian Program*.
10. Ibid.
11. Charles V. Zegeer, Kenneth S. Opiela, and Michael J. Cynecki, "Effect of Pedestrian Signals and Signal Timing on Pedestrian Accidents," *Transportation Research Record* 847 (1982).
12. Ibid.
13. Rune Elvik et al., eds., *The Handbook of Road Safety Measures* (West Yorkshire, UK: Emerald Group, 2009).
14. AAA, *Planned Pedestrian Program*.

Chapter 75

1. Charles Keese, Charles Pinnell, and Donald Drew, "Highway Capacity: The Level of Service Concept," Texas Transportation Institute (1964).
2. Wayne Kittelson, "Historical Overview of the Committee on Highway Capacity and Quality of Service," *Transportation Research Circular* E-C018, 4th International Symposium on Highway Capacity (2000).
3. Keese, "Highway Capacity."
4. Kittelson, "Historical Overview."
5. Ibid.
6. Ibid.
7. Highway Research Board, *Highway Capacity Manual* (1965).
8. Kittelson, "Historical Overview."
9. Keese, "Highway Capacity."

10. Kittelson, "Historical Overview."
11. Aimee Flannery, Kathryn Wochinger, and Angela Martin, "Driver Assessment of Service Quality on Urban Streets," *Transportation Research Record* 1920, no. 1 (2005).
12. Kellen Browning, "Uber Says Sexual Assaults Are Down but Rate of Traffic Deaths Is Up," *New York Times*, June 30, 2022.
13. Highway Research Board, *Highway Capacity Manual.*

Chapter 76

1. Ezra Hauer, "Engineering Judgment and Road Safety," *Accident Analysis and Prevention* 129 (2019).
2. Ibid.
3. Ibid.
4. David Gwynn, "Relationship of Accident Rates and Accident Involvements with Hourly Volumes," *Traffic Quarterly* 21, no. 3 (1967).
5. John Sewell, *How We Changed Toronto* (Toronto, ON: James Lorimer, 2015).
6. Ibid.
7. Sharon McMillan, "Alan," Dinner with Jane, n.d., www.jamii.ca.
8. Ibid.
9. Paul Krizek, "Death on George Washington Memorial Parkway," *Fort Hunt [VA] Herald*, May 9, 2019.
10. Herbert S. Thriscutt, "The Training of Traffic Engineers," *Traffic Quarterly* 16, no. 3 (1962).

Chapter 77

1. James Reason, *A Life in Error: From Little Slips to Big Disasters* (Boca Raton, FL: CRC Press, 2013).
2. Accreditation Board for Engineering and Technology (ABET), "Criteria for Accrediting Engineering Programs," Baltimore, MD (2022–2023).
3. Robert Francis, "Report of the Mid Staffordshire NHS Foundation Trust Public Inquiry" (London: Stationery Office, 2013).
4. John F. Kennedy, Address before the American Newspaper Publishers Association, Waldorf-Astoria Hotel, New York City, April 27, 1961.
5. Frank Gross and Paul Jovanis, "Current State of Highway Safety Education," *Journal of Professional Issues in Engineering Education and Practice* 134, no. 1 (2008).
6. National Academies of Sciences, Engineering, and Medicine, "Core Competencies for Highway Safety Professionals," *Research Digest* 302 (2006).
7. Ibid.
8. Transportation Research Board, "Special Report 289: Building the Road Safety Profession in the Public Sector" (2007).
9. National Academies of Sciences, "Core Competencies."
10. Ibid.
11. Gross and Jovanis, "Current State of Highway Safety Education."
12. Ezra Hauer, "The Road Ahead," *Journal of Transportation Engineering* 131, no. 5 (2005).
13. Ibid.
14. Transportation Research Board, "Special Report 289."
15. Ibid.
16. Ibid.

17. Ibid.

Chapter 78

1. American Association of State Highway and Transportation Officials, *A Policy on Geometric Design of Highways and Streets*, 7th ed. (Washington, DC: AASHTO, 2018).
2. Ezra Hauer, "Engineering Judgment and Road Safety," *Accident Analysis and Prevention* 129 (2019); Ralph Peck, Advice to a Young Engineer," *Geo-Strata*, Geo Institute of ASCE 6, no. 3 (2006).
3. Federal Highway Administration, *Manual on Uniform Traffic Control Devices* (Washington, DC: US Department of Transportation, 2023).
4. Ibid.
5. Ibid.
6. Paul Baybutt, "The Validity of Engineering Judgment and Expert Opinion in Hazard and Risk Analysis," *Process Safety Progress* 37, no. 2 (2018).
7. Alina Tugend, *Better by Mistake: The Unexpected Benefits of Being Wrong* (New York: Penguin, 2011).
8. Ibid.

Chapter 79

1. Raymond Unwin, *Town Planning in Practice* (London: T. F. Unwin, 1909).
2. Andrés Duany, "On the Edge: Latest View from Andrés Duany," Public Lecture at Simon Fraser University, Vancouver, BC (2008).
3. Hans-Georg Retzko, "Interdisciplinary Work in Traffic and Transport," *Transportation Quarterly* 50, no. 1 (1996).
4. Duany, "On the Edge."
5. Ibid.
6. Ibid.
7. Institute of Traffic Engineers, "Report of Special Committee on Purpose and Scope," *Traffic Engineering* (March 1964).
8. John S. Hassell, "How Effective Has Urban Transportation Planning Been?" *Traffic Quarterly* 34, no. 1 (1980).
9. Delbert Taebel and James Cornehls, "Ideological and Policy Perspectives of Urban Transportation," *Traffic Quarterly* 29, no. 4 (1975).
10. Ibid.

Chapter 80

1. Benedikt Horn, Robert Matthews, and Donald Symmes, "Assessing the Effects of Heavy Trucks," *Transportation Quarterly* 37, no. 4 (1983).
2. Highway Research Board, *Highway Capacity Manual,* 7th Edition (2022).
3. Ezra Hauer, "The Road Ahead," *Journal of Transportation Engineering* 131, no. 5 (2005).
4. Transportation Research Board, "Special Report 289: Building the Road Safety Profession in the Public Sector," vol. 289 (2007).
5. Ibid.
6. Sophia Chen, "Physicists Rise to the Challenge," *APS News*, 29, no. 5 (2020).
7. Ben Zigterman, "UI Scientists Modeling COVID-19 Say Campus Can Safely Reopen," *News-Gazette* (Champaign, IL, August 7, 2020).
8. Chen, "Physicists Rise."

9. Kenneth Chang, "Good, but Not Great: Taking Stock of a Big Ten University's COVID Plan," *New York Times*, August 22, 2021.

10. Andy Weir, *Project Hail Mary* (New York: Ballantine, 2021).

11. Hauer, "The Road Ahead."

12. Frank McGlade, "Traffic Accident Research: Review and Prognosis," *Traffic Quarterly* 16, no. 4 (1962).

13. Hauer, "The Road Ahead."

14. Ibid.

15. McGlade, "Traffic Accident Research."

16. Anthony Peranio, "Conceptualization and Use of Road Safety and Traffic Engineering Formulas," *Traffic Quarterly* 25, no. 3 (1971).

17. David Byrne, *Bicycle Diaries* (New York: Penguin, 2010).

18. Stephen Dubner, "How Do You Cure a Compassion Crisis?," episode 444, *Freakonomics*, podcast, December 16, 2020.

Chapter 81

1. Shannon Gormley, "Metro Acknowledges That a Multi-Million Dollar Vision Zero Plan Isn't Working," *Willamette Week Newspaper* (Portland, OR), March 7, 2020.

2. Sidney Dekker, *The Field Guide to Understanding "Human Error"* (Boca Raton, FL: CRC Press, 2017).

3. Ibid.

4. Ibid.

5. Ibid.

6. J. O. Mattson, "The Traffic Accident Outlook," *Traffic Quarterly* 18, no. 3 (1964).

7. James Reason, *Human Error* (Cambridge, UK: Cambridge University Press, 1990).

8. Marie Kondo, *The Life-Changing Magic of Tidying Up: The Japanese Art of Decluttering and Organizing* (Berkeley, CA: Ten Speed Press, 2014).

9. Ibid.

10. Ibid.

Chapter 82

1. Caitlin Schmidt, "Florida Mom Arrested after Letting 7-Year-Old Walk," *CNN*, August 1, 2014.

2. William Patrick, "Florida Mom Could Face 5 Years in Prison," *Daily Signal*, August 7, 2014.

3. Jan Hoffman, "Why Can't She Walk to School?," *New York Times*, September 12, 2009; Dispatch Editorial Board, "The Walk Felt 'Round the World," *The Dispatch*, March 23, 2009.

4. Jonathan Maus, "Read Police Report, Hear More from Mom in Tennessee Child Biking Case," BikePortland, September 2, 2011.

5. Tanya Snyder, "Tennessee Mom Threatened with Arrest," *Streetsblog USA*, September 1, 2011.

6. E. O. D. Waygood and Ayako Taniguchi, "Japan: Maintaining High Levels of Walking," in *Transport and Children's Wellbeing*, ed. E. O. D. Waygood et al. (New York: Elsevier, 2020).

7. E. O. D. Waygood, "What Is the Role of Mothers in Transit-Oriented Development?," *Women's Issues in Transportation* (2010).

8. Ibid.
9. Federal Highway Administration, *Manual on Uniform Traffic Control Devices* (Washington, DC: US Department of Transportation, 2023).
10. John Wann, Damian Poulter, and Catherine Purcell, "Reduced Sensitivity to Visual Looming Inflates the Risk," *Psychological Science* 22, no. 4 (2011).
11. Robert Dier, "A Study of a School Crossing Hazard," *Traffic Quarterly* 6, no. 1 (1952).
12. Fred Herring, "Human Relationships Are the Key to Effective Traffic Engineering," *Traffic Quarterly*, 10, no. 3 (1956).
13. Ibid.
14. Dorothy Lee, "Portland Protects Its Pedestrians," *Traffic Quarterly* 6, no. 3 (1952).
15. AAA, *Planned Pedestrian Program* (Washington, DC: AAA Foundation for Traffic Safety, 1958).
16. Sigurd Grava, "Traffic Calming: Can It Be Done in America?," *Transportation Quarterly* 47, no. 4 (1993).
17. Ibid.
18. S. T. M. C. Janssen, "Effects of Road Safety Measures in Urban Areas in the Netherlands," 85, no. 11 (1985).

Chapter 83

1. Paul G. Hoffman and Neil M. Clark, *Seven Roads to Safety: A Program to Reduce Automobile Accidents* (New York: Harper and Brothers, 1939).
2. Peyton Gibson and Wes Marshall, "Disparate Approaches to Maintaining Roads and Sidewalks: An Interview Study of 16 US Cities," *Transportation Research Record* 2676, no. 9 (2022).
3. David Davis and Lawrence Pavlinski, "Improving Prospects for Pedestrian Safety," *Traffic Quarterly* 32, no. 3 (1978).
4. AAA, *Planned Pedestrian Program* (Washington, DC: AAA Foundation for Traffic Safety, 1958).
5. Urban Land Institute, *The Community Builders Handbook* (Washington, DC: Urban Land Institute, 1954).
6. American Society of Planning Officials, "Sidewalks in the Suburbs," information report 95 (1957).
7. David Byrne, *Bicycle Diaries* (New York: Penguin, 2010).
8. Kay Teschke et al., "Exposure-Based Traffic Crash Injury Rates by Mode of Travel," *Canadian Journal of Public Health* 104 (2013).
9. Wes Marshall and Nicholas Ferenchak, "Why Cities with High Bicycling Rates Are Safer for All Road Users," *Journal of Transport and Health* 13 (2019).
10. Liang Ma and Runing Ye, "Utilitarian Bicycling and Mental Wellbeing," *Journal of the American Planning Association* 88, no. 2 (2022); Liang Ma, Runing Ye, and Hongyu Wang, "Exploring the Causal Effects of Bicycling for Transportation on Mental Health," *Transportation Research Part D* 93 (2021).

Chapter 84

1. William Warren, "Impacts of Land Use on Mass Transit Development," *Transportation Quarterly* 42, no. 2 (1988).
2. Todd Litman, "Safer Than You Think!," VTPI (2016); Jim Stimpson et al., "Share of Mass Transit Miles Traveled and Reduced Motor Vehicle Fatalities," *Journal of Urban Health* 91, no. 6 (2014).

3. Kent Robertson, "Parking and Pedestrians," *Transportation Quarterly* 55, no. 2 (2001).

4. Marie Kondo, *The Life-Changing Magic of Tidying Up: The Japanese Art of Decluttering and Organizing* (Berkeley, CA: Ten Speed Press, 2014).

5. Francis Turner, "Moving People on Urban Highways," *Traffic Quarterly* 24, no. 3 (1970).

6. Ibid.

7. Laurence Aurbach, *A History of Street Networks: From Grids to Sprawl and Beyond* (Hyattsville, MD: PedShed Press, 2020).

8. Herbert Levinson, "Cities, Transportation, and Change," *Transportation Quarterly* 50, no. 4 (1996).

9. Winnie Hu, "Major Traffic Experiment in NYC," *New York Times*, August 8, 2019.

10. Laura Bliss, "What Happened after Market Street Went Car-Free?," *CityLab*, March 10, 2020.

11. Turner, "Moving People."

12. Joseph Leiper, "The Role of the Automobile in Midtown Manhattan," *Traffic Quarterly* 16, no. 2 (1962).

13. Lester A. Hoel, "Transportation Technology and Systems Planning," *Traffic Quarterly* 22, no. 2 (1968).

14. W. T. Rossell and D. Gaul, "Rapid Transit's Value to a City," *Traffic Quarterly* 10, no. 1 (1956).

15. LL Cool J, "Breaking My Silence," TikTok, video, March 10, 2022, www.tiktok.com/@llcoolj/video/7073629644506893610.

Chapter 85

1. Anthony Peranio, "Logic and Road Safety Research," *Traffic Quarterly* 23, no. 1 (1969).

2. Ibid.

3. David Levinson, Wes Marshall, and Kay Axhausen, *Elements of Access: Transport Planning for Engineers, Transport Engineering for Planners* (Sydney: Network Design Lab, 2017).

4. Lewis Mumford, *The Highway and the City* (London: Martin Secker and Warburg, 1963).

5. Ibid.

Chapter 86

1. Centers for Disease Control and Prevention, "Road Traffic Injuries and Deaths—A Global Problem" (2023).

2. Charles Perrow, *Normal Accidents: Living with High Risk Technologies*, updated ed. (Princeton, NJ: Princeton University Press, 1999).

3. Beth Greenfield, *Ten Minutes from Home* (New York: Crown, 2010).

4. Perrow, *Normal Accidents.*

Chapter 87

1. Greg Adomaitis, "Driver Who Lost License 30 Times Gets Probation," NJ.com, February 25, 2017.

2. National Academies of Sciences, Engineering, and Medicine, *Guide for Pedestrian and Bicyclist Safety at Alternative and Other Intersections and Interchange* (Washington, DC: National Academies Press, 2021).

3. Bernard Cohen and I-Sing Lee, "A Catalog of Risks," *Health Physics* 36, no. 6 (1979); David Hensher, "Review of Studies Leading to Existing Values of Travel Time," *Transportation Research Board* 587 (1976).

Chapter 88

1. Norman Garrick and Jianhong Wang, "New Concepts for Context-Based Design," *Transportation Research Record* 1912, no. 1 (2005).
2. Walter Isaacson, *Steve Jobs* (New York: Simon and Schuster, 2021).
3. Claus Jensen, *No Downlink: A Dramatic Narrative about the* Challenger *Accident and Our Time* (New York: Farrar, Straus and Giroux, 1996).
4. Paul G. Hoffman and Neil M. Clark, *Seven Roads to Safety: A Program to Reduce Automobile Accidents* (New York: Harper and Brothers, 1939).